MW01435107

DATE DUE

DEC 1 9 1997	

BRODART, CO. Cat. No. 23-221-003

Analog Circuit Design

ANALOG CIRCUIT DESIGN

MOST RF Circuits,
Sigma-Delta Converters and
Translinear Circuits

Edited by

WILLY SANSEN
K.U. Leuven, Belgium

RUDY J. VAN DE PLASSCHE
Philips Research Laboratories,
T.U. Eindhoven, The Netherlands

and

JOHAN H. HUIJSING
T.U. Delft, The Netherlands

KLUWER ACADEMIC PUBLISHERS
BOSTON / DORDRECHT / LONDON

A C.I.P. Catalogue record for this book is available from the Library of Congress

ISBN 0-7923-9776-2

Published by Kluwer Academic Publishers,
P.O. Box 17, 3300 AA Dordrecht, The Netherlands.

Kluwer Academic Publishers incorporates
the publishing programmes of
D. Reidel, Martinus Nijhoff, Dr W. Junk and MTP Press.

Sold and distributed in the U.S.A. and Canada
by Kluwer Academic Publishers,
101 Philip Drive, Norwell, MA 02061, U.S.A.

In all other countries, sold and distributed
by Kluwer Academic Publishers Group,
P.O. Box 322, 3300 AH Dordrecht, The Netherlands.

Printed on acid-free paper

All Rights Reserved
© 1996 Kluwer Academic Publishers
No part of the material protected by this copyright notice may be reproduced or
utilized in any form or by any means, electronic or mechanical,
including photocopying, recording or by any information storage and
retrieval system, without written permission from the copyright owner.

Printed in the Netherlands

Table of Contents

Preface vii

Part I: MOST RF Circuit Design
Introduction 1

RF modelling of MOSFETs
DBM Klaassen, B. Nauta and RRJ Vanoppen 3

High Integration CMOS RF Transceivers
F. Brianti, G. Chien, T. Cho, S. Lo, S. Mehta, J. Ou, J. Rudell, T. Weigandt, J. Weldon and P. Gray 25

2 GHz RF Circuits in BiCMOS Process
J.M. Fournier and P. Senn 39

RF CMOS Design, Some Untold Pitfalls
Michiel Steyaert, Marc Borremans, Jan Craninckx, Jan Crols, Johan Janssens and Peter Kinget 63

Silicon Integration for Digital Cellular Communication
Jan Sevenhans, Jacques Wenin, Damien Macq, and Jacques Dulongpont 89

A Monolithic 900 Mhz Spread-Spectrum Wireless Transceiver in 1-µm CMOS
Asad Abidi 101

Part II: Bandpass Delta-Sigma and other Data Converters
Introduction 133

Low-Power CMOS ΣΔ modulators for speech coding 135

Passive Sigma-Delta Modulators with Built-in Passive Mixers for Mobile Communications
Bosco Leung and Feng Chen 153

Design of Continuous Time Bandpass ΔΣ modulators in CMOS
Vincenzo Peluso, Michiel Steyaert and Will Sansen 171

Bandpass Delta-Sigma Converters in IF Receivers
Armond Hairapetian 193

Design and Optimization of a Third-Order Switched-capacitor Reconstruction Filters for Sigma-Delta DAC's
Tom Kwan ... 211

Tools For Automated Design of ΣΔ Modulators
F. Medeiro, J.M. de la Rosa, B. Pérez-Verdú and A. Rodríguez-Vázquez ... 231

Part III: Translinear Circuits
Introduction ... 255

Aspects of Translinear Amplifier Design
Barrie Gilbert ... 257

Variable-Gain, Variable-Transconductance, and Multiplication Techniques: A Survey
Max Hauser, Eric Klumperink, Robert Meyer and William Mack ... 291

CMOS Translinear Circuits
Evert Seevinck ... 323

Design of MOS Translinear Circuits Operating in Strong Inversion
Remco Wiegerink ... 337

Translinear Circuits in Low-Voltage Operational Amplifiers
Klaas-Jan de Langen, Ron Hogervorst and Johan Huijsing ... 357

Low-Voltage Continuous-Time Filters
R. Castello ... 387

Preface

This book contains the revised contributions of all the speakers of the fifth AACD Workshop which was held in Lausanne on April 2-4, 1996. It was organized by Dr Vlado Valence of the EPFL University and MEAD of Lausanne. The program consisted of six tutorials per day during three days. The tutorials were presented by experts in the field. They were selected by a program committee consisting of Prof. Willy Sansen of the Katholieke Universiteit Leuven, Prof. Rudy van de Plassche of Philips Research and the University of Technology Eindhoven and Prof. Johan Huijsing of the Delft University of Technology.

The three topics mentioned above have been selected because of their importance in present days analog design. The other topics that have been discussed before are:

in 1992 :	Operational amplifiers
	Analog to digital convereters
	Analog computer aided design
in 1993 :	Mixed A/D cicuit design
	Sensor interface circuits
	Communication circuits
in 1994 :	Low-power low-voltage design
	Integrated filters
	Smart power circuits
in 1995 :	Low-noise, low-power, low-voltage design
	Mixed-mode design with CAD tools
	Voltage, current and time references

Each AACD workhop has given rise to the publication of a book by Kluwer entitled "Analog Circuit Design". This is thus the fifth book. This series of books provides a valuable overview of all analog circuit design techniques and achievements. It is a reference for whoever is engaged in this discipline. The aim of the workshop has been to brainstorm on new possiblities and future developments in the area of analog circuit design. We sincerely hope that this fifth book continues the tradition to make a valuable contribution to the insight in analog circuits in Europe and in the world.

Willy M.C.Sansen
K.U.Leuven

MOST RF CIRCUIT DESIGN

Willy Sansen

Introduction

Radio-frequency design is going through a renaissance because of the explosion of the telecommunication field. They used to be realized with discrete bipolar devices and later on in integrated bipolar technologies. Recently they have become a reality in submicron CMOS technologies as well. This first day of the AACD '96 was therefore entirely devoted to circuit design of telecommunication subsystems in CMOS technologies.

The first presentation is given by Dick Klaassen on aspects of high-frequency modelling of MOST devices. Details are given on Philips' model 9 both at DC as well as at RF frequencies up to 20 GHz. Considerable attention is paid to parasitic effects and to the behavior of elementary circuit combinations such as cascades and cascodes.

The first paper addressing the integration of full transceivers in CMOS technology is given by Paul Gray. It provides an overview of technical challenges in portable battery-powered transceivers for personal communications.

It is followed by a presentation by Patrice Senn on 2 GHz RF circuits in BICMOS technology. Extensive discussion is given on a LNA (low-noise amplifier), a mixer, a low-phase noise VCO, etc.

Michiel Steyaert then gives a presentation on the pitfals of RF CMOS design. He started with a review of the high-frequency limitations of MOST devices. This is applied to the realization of a monochip receiver, a synthesizer and a transmitter.

The fifth paper is given by Jan Sevenhans. He discusses the various implementations currently in use in commercial GSM products. He also sketches the possible merits of GaAs and future battery technologies.

Finally Asad Abidi shows several integrated circuit examples such as a digital frequency synthesizer, an upconversion mixer, an RF power amplifier, etc., all in CMOS to prove that a 900 MHz spread-spectrum wireless transceiver has become reality. Other critical problems such as an integrated inductor are discussed as well.

It can be concluded that these texts amply show that CMOS has become a viable technology for high-frequency communications applications. With the advent of more advanced submicron CMOS, even higher frequency realizations can be expected.

RF modelling of MOSFETs

D.B.M. Klaassen, B. Nauta and R.R.J. Vanoppen

E-mail: klaassen@natlab.research.philips.com / Tel: +31-40-2744251

Philips Research Laboratories, Eindhoven, The Netherlands

ABSTRACT

The accuracy of the Philips compact MOS model, MOS MODEL 9, has been investigated for a number of quantities, that are important for RF circuit design. On-wafer S-parameter measurements have been performed on MOS devices as a function of the frequency up to the GHz-range. From these S-parameters important RF quantities such as input impedance, transconductance, current and voltage gain etc., have been obtained. A comparison between experimental results and model calculations will be presented.

Introduction

RF circuit design in CMOS will be restricted mainly to mainstream CMOS processes. These mainstream CMOS processes are provided by foundries, which commonly supply the compact model parameters used in the circuit design. Most CMOS foundries supply parameters for rather simple compact MOS models, which are suited for digital design only. In fact there are only two public-domain compact MOS models, which are really suited for analog circuit design: the BSIM3v3 model from UC-Berkeley and the Philips compact MOS model, MOS MODEL 9 [1, 2, 3]. These models try to combine a good description of the geometry dependence of the transistor behaviour with continuous derivatives of current with respect to bias voltages.

Literature on high-frequency S-parameter measurements of MOSFETs is scarce [4]. Publications available are focused on characterization of device and process performance (see e.g. [5] to [9]) or on test structures, measurement techniques and special effects (see e.g. [10, 11]). Measurements are compared only with calculations using special small-signal equivalent circuits. In [12] we presented the first comparison of high-frequency measurements on MOSFETs with calculations using an analog compact MOS model, i.e. MOS MODEL 9. Here we extend this comparison to a number of experimental quantities, which are important for RF circuit design.

In the following we will discuss the compact model, MOS MODEL 9, and the experimental method. Next we will present the comparison of measurements and calculations of important quantities such as input impedance, current and voltage gain as a function of frequency and bias conditions for a number of basic transistor configurations.

MOS MODEL 9

This Philips compact MOS model has been introduced within Philips in 1990 and became available in the public domain in 1993. Many of its features and capabilities have been elucidated in publications (see e.g. [12] to [18]), while the derivation of many equations and the underlying physical mechanisms are described in [19]. The complete set of equations of MOS MODEL 9 has been published in [3].

The following physical effects are taken into account by MOS MODEL 9:

- body-effect for implanted substrate;
- mobility reduction due to transversal field;
- velocity saturation;
- subthreshold region;
- drain-induced barrier-lowering;
- static feedback;
- channel length modulation;
- avalanche multiplication and substrate current.

Due to its physical basis MOS MODEL 9 (or MM9) has a minimum number of parameters per phenomenon modelled. This results in a total number of 18 parameters to model a transistor with a specific geometry. For homogeneous substrate dope the body-effect model with one k-factor is used, which implies that this number is reduced to 16. In these numbers the three parameters for the modelling of the avalanche multiplication are included. All these parameters are extracted from the dc characteristics of the transistor. Except for the oxide thickness no additional parameters are needed for the charge model. It should be noted that the charge model of MOS MODEL 9 has a bias-dependent charge partitioning between source and drain. All MOSFET capacitances are derived from this charge model.

Simple scaling rules describe the 18 electrical parameters mentioned above, as a function of channel geometry (i.e. length and width). Due to the physical basis of the model these scaling rules contain only 46 geometry-independent parameters. Additional scaling rules for the temperature dependence have also been established.

		n-channel	p-channel
Current in	$V_{sb} = 0\,V$	1.2%	1.9%
linear region	$V_{sb} \neq 0\,V$	2.8%	3.2%
Current in	$V_{sb} = 0\,V$	6.1%	4.7%
saturation region	$V_{sb} \neq 0\,V$	5.9%	5.4%
Current in	$V_{sb} = 0\,V$	12%	17%
subthreshold region	$V_{sb} \neq 0\,V$	20%	31%
Output	$V_{sb} = 0\,V$	24%	15%
conductance	$V_{sb} \neq 0\,V$	23%	16%
Substrate current		26%	27%

Table 1: Mean absolute deviation (in %) between measured and simulated (using MOS MODEL 9) characteristics averaged over several bias conditions and over 14 geometries of an 0.8 μm process (see [14]).

MOS MODEL 9 describes the transistor characteristics over the whole geometry range of a CMOS process very accurately with only one geometry-independent parameter set. This is demonstrated in Table 1, where for the five operating regions from which the parameters are extracted, the accuracy is given (see also [14]).

As MOS MODEL 9 has been developed especially for analog applications, great care has been taken to obtain continuous derivatives of currents and charges. Consequently, MM9 complies with most benchmark tests for analog models [20, 21].

RF measurements and simulations

Small-signal scattering (S) parameters in two-port configuration have been measured with an HP8510 Network Analyzer using air coplanar high-frequency probes in ground-signal-ground configuration. Special test structures have been designed with the MOSFETs in common source-bulk configuration. For each structure also dummy structures have been designed to correct for parallel and series parasitics. The measured S-parameters have been converted to admittance (Y) parameters. The measurements on the dummy structures have been used to correct the transistor measurements for the interconnect parasitics.

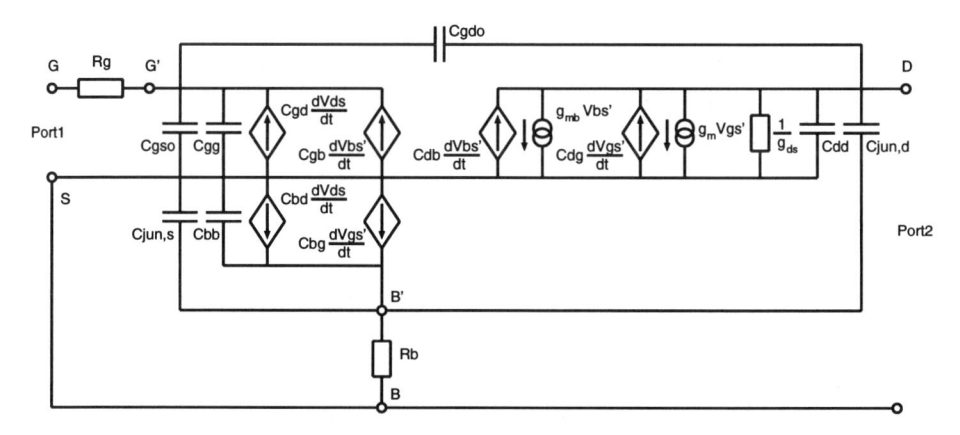

Figure 1: Equivalent circuit of MOS MODEL 9 in two-port common source-bulk configuration. The overlap capacitances, indicated with C_{gso} and C_{gdo}, are included in this model. The junction capacitances, indicated with $C_{jun,s}$ and $C_{jun,d}$, are described by a separate model, JUNCAP (see text). The gate resistance, indicated with R_g, and the bulk resistance, indicated with R_b, are not included in these models.

S-parameter measurements have been performed at fixed bias conditions as function of frequency as well as at fixed frequency as function of bias voltage. The overlap and depletion capacitances have been measured at low frequencies with an HP4284 LCR-meter.

For the simulations we used MOS MODEL 9 with parameters obtained from dc current measurements. The equivalent circuit in two-port common source-bulk configuration is shown in Figure 1. For the depletion capacitances the Philips model, JUNCAP, has been used [22], while the gate and bulk resistances have been added in all simulations.

Basic RF circuits

In this section we will discuss four basic RF circuits: i) common source-bulk configuration; ii) common gate configuration; iii) cascade configuration; and iv) cascode configuration. For these basic configurations a comparison will be presented of measurements and calculations of important quantities such as input impedance, current and voltage gain as a function of frequency and bias conditions.

Common source-bulk configuration

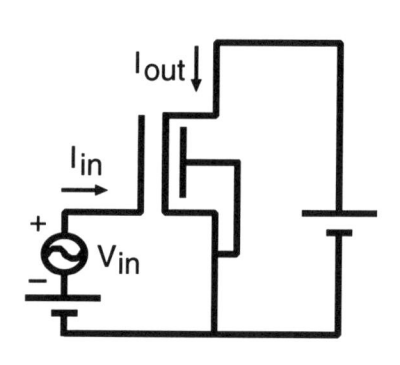

In Figure 2 a MOSFET in common source-bulk configuration is shown. The gate is voltage driven, while the drain is ac short-circuited. The input impedance, $Z_{in} = v_{in}/i_{in}$, of a MOSFET in common source-bulk configuration is shown in Figure 3 as a function of frequency. The drain bias has been chosen at 2 V, being 60% of the supply voltage of the process, while for the gate bias three values equally spaced between the threshold voltage, $V_{TO} \approx 0.6\,V$, and about one Volt above V_{TO}, have been used. For the input impedance the influence of the bulk resistance is negligible. The expression for the input impedance, Z_{in}, yields (see Figures 2 and 1)

Figure 2: MOSFET in common source-bulk configuration.

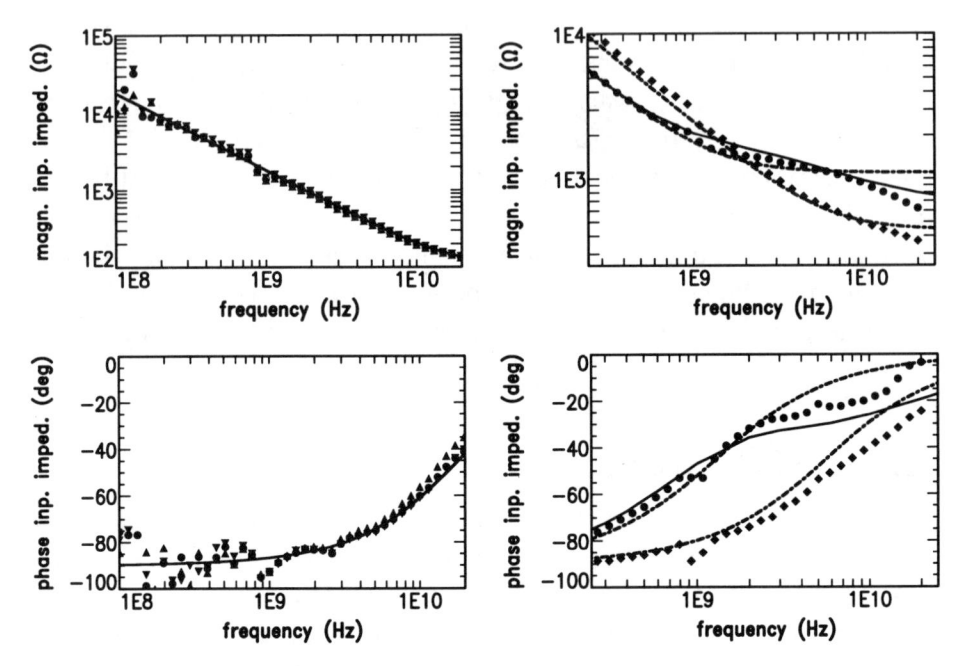

Figure 3: Magnitude (top) and phase (bottom) of the input impedance of a 60/0.5 N-channel transistor with $V_{TO} = 0.6\,V$ in common source-bulk configuration. Symbols represent measurements at $V_{ds} = 2.0\,V$ and $V_{gs} = 0.9\,V$ (downwards-directed triangles), $1.2\,V$ (solid circles) and $1.5\,V$ (upwards-directed triangles). Lines represent MOS MODEL 9 simulations. Note that the transistor has a salicidated poly-Si gate.

Figure 4: Magnitude (top) and phase (bottom) of the input impedance of a 40/1 (diamonds) and 100/1 (solid circles) N-channel transistor in common source-bulk configuration at $V_{ds} = V_{dd} = 5\,V$ and $V_{gs} = 2\,V$. Dashed lines represent MOS MODEL 9 simulations with one segment only, while solid lines represent MOS MODEL 9 simulations with five "distributed" parallel segments of $20\,\mu$m wide. Note that the transistors have non-salicidated poly-Si gates.

$$Z_{in} = \frac{v_{in}}{i_{in}} \approx \frac{1}{j\omega C_{gg}^{eff}} + R_g \quad . \tag{1}$$

Herein C_{gg}^{eff} is the effective capacitance

$$C_{gg}^{eff} = C_{gg} + C_{gso} + C_{gdo} \quad , \tag{2}$$

where C_{gso} and C_{gdo} are the gate-source and gate-drain overlap capacitances, respectively, and C_{gg} is the intrinsic capacitance

$$C_{gg} = \frac{\partial Q_g}{\partial V_g} \quad . \tag{3}$$

Here (and throughout this paper) Q_i is the charge of intrinsic terminal i, which is given by MOS MODEL 9 (see Figure 1).
From Figure 3 and Eq. 1 one sees that up to very high frequencies the input impedance shows a capacitive behaviour. However, for processes with non-salicidated poly-Si as gate material or transistors with a very high W/L ratio, the gate resistance may become very high. In that case the gate resistance can no longer be treated as a lumped element (see Figure 4 and Eq. 8). In Figure 5 the current gain, i_{out}/i_{in}, as a function of frequency is shown for the device and bias conditions from Figure 3. From this figure it is clear that MOS MODEL 9 describes both frequency and bias dependence accurately. Neglecting again the bulk resistance, one finds for the expression for the current gain (see Figures 2 and 1)

$$\frac{i_{out}}{i_{in}} \approx \frac{g_m}{j\omega C_{gg}^{eff}} \left(1 - j\omega \frac{C_{dg}^{eff}}{g_m} \right) \quad . \tag{4}$$

Herein g_m is the dc transconductance and C_{dg}^{eff} is the effective capacitance

$$C_{dg}^{eff} = C_{dg} + C_{gdo} = -\frac{\partial Q_d}{\partial V_g} + C_{gdo} \quad . \tag{5}$$

For the cut-off or unity current-gain frequency, f_T, the familiar expression is found from Eq. 4 (see e.g. [23])

$$f_T \approx \frac{g_m}{2\pi C_{gg}^{eff}} \quad . \tag{6}$$

In Figure 6 this cut-off frequency is shown as function of gate voltage at three different drain voltages. From this figure it can be seen that MOS MODEL 9 predicts the bias dependence of f_T quite accurately.

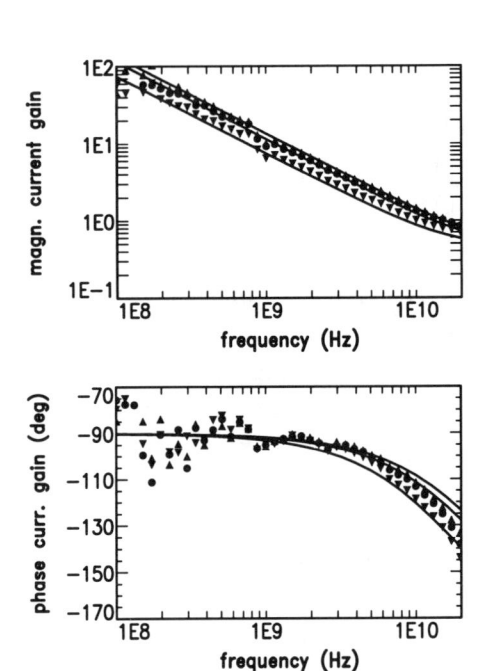

Figure 5: Magnitude (top) and phase (bottom) of the current gain of a 60/0.5 N-channel transistor ($V_{TO} = 0.6\,V$) in common source-bulk configuration. Symbols represent measurements at $V_{ds} = 2.0\,V$ and $V_{gs} = 0.9\,V$ (downwards-directed triangles), $1.2\,V$ (solid circles) and $1.5\,V$ (upwards-directed triangles). Lines represent MOS MODEL 9 simulations.

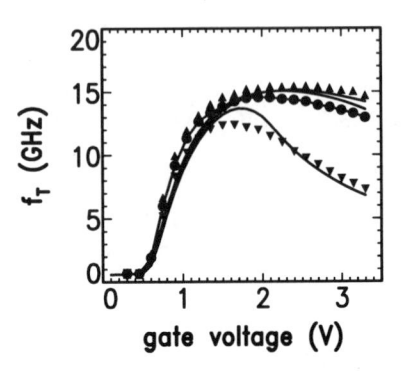

Figure 6: The f_T of a 60/0.5 N-channel transistor ($V_{TO} = 0.6\,V$) in common source-bulk configuration as function of gate voltage for drain voltages of $1\,V$ (downwards-directed triangles), $2\,V$ (solid circles) and $3.3\,V$ (upwards-directed triangles). Lines represent MOS MODEL 9 simulations.

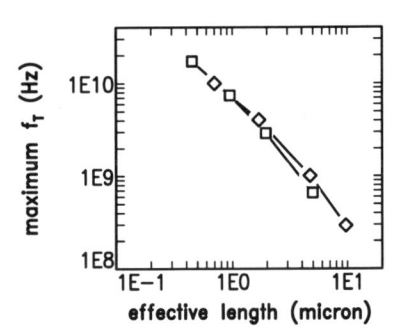

Figure 7: Measured maximum f_T as function of effective channel length. Diamonds represent N-channel devices from a $1\,\mu m$ CMOS process ($V_{dd} = 5\,V$) and squares represent N-channel devices from a $0.5\,\mu m$ CMOS process ($V_{dd} = 3.3\,V$), respectively. The measurements have been performed at $V_{ds} = V_{dd}$.

In Figure 7 the maximum f_T is shown as a function of effective gate length. From this figure it can be seen that the maximum f_T is inversely proportional to the square of the effective channel length. This implies that f_T is quite low for long-channel devices. It should be noted that due to mobility reduction for the smallest channel lengths, the length dependency of the maximum f_T is less steep.

The transconductance, i_{out}/v_{in}, of the 60/0.5 N-channel device is shown in Figure 8 as a function of frequency. From Figures 5 or 6 and Figure 8 it can be seen that below the cut-off frequency of the current gain, again the description of MOS MODEL 9 for frequency and bias dependence is quite accurate. The expression found for the transconductance is (see Figures 2 and 1)

$$\frac{i_{out}}{v_{in}} \approx \frac{g_m - \omega^2 R_g C_{dg}^{eff} C_{gg}^{eff} - j\omega\left(g_m R_g C_{gg}^{eff} + C_{dg}^{eff}\right)}{1 + \left(\omega R_g C_{gg}^{eff}\right)^2} . \quad (7)$$

From Eq. 7 follows that both the real and imaginary parts of the transconductance are affected by the gate resistance. This is demonstrated in Figure 9, where the transconductance for a 40/1 N-channel transistor from a non-salicidated process is shown. From Figure 9 it is also clear that the correct expression for the resistance of a single-sided contacted gate, is (as can be shown from theory)

$$R_g = \frac{W \rho_{\square,poly}}{3L}, \quad (8)$$

where $\rho_{\square,poly}$ is the sheet resistance of the poly-Si gate material. It should be noted that for a double-sided contacted gate the factor 3 in the denominator of Eq. 8 should be replaced by 12.

Once we have established the correct value of the gate resistance, it is interesting to look into non-quasi static effects, which are observed most easily from the transconductance. In Figure 10 the transconductance is shown for a number of P-channel devices with gate length varying from 0.6 µm to 30 µm. The cut-off frequency, f_T, varies between 5 GHz and 6 MHz (indicated by the arrows in Fig. 10). From Figure 10 it can be seen that for the long-channel devices both the magnitude and the phase of the transconductance show a sharp dip at frequencies above f_T. This typical non-quasi static behaviour is not described by the present quasi static implementation of MOS MODEL 9. However,

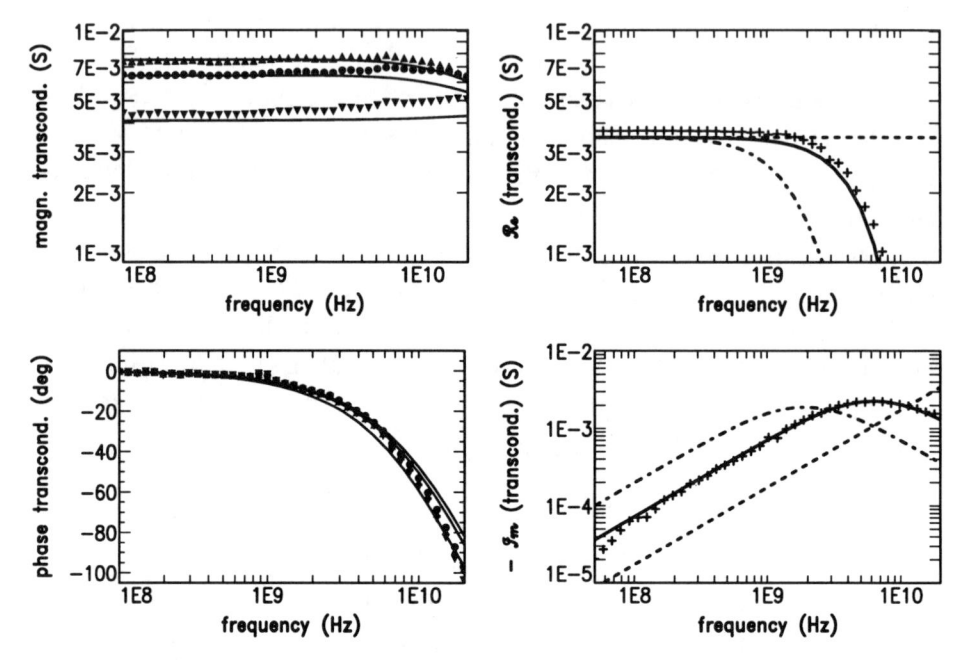

Figure 8: Magnitude (top) and phase (bottom) of the transconductance of a 60/0.5 N-channel transistor ($V_{TO} = 0.6\,V$) in common source-bulk configuration. Symbols represent measurements at $V_{ds} = 2.0\,V$ and $V_{gs} = 0.9\,V$ (downwards-directed triangles), 1.2 V (solid circles) and 1.5 V (upwards-directed triangles). Lines represent MOS MODEL 9 simulations.

Figure 9: Real (top) and imaginary (bottom) parts of the transconductance a 40/1 N-channel transistor ($V_{dd} = 5\,V$) in common source-bulk configuration at $V_{ds} = 4\,V$ and $V_{gs} = 4\,V$. Symbols represent measurements, lines represent MOS MODEL 9 simulations with $R_g = 0$ (dashed line); $R_g = (W\rho_{\Box,\text{poly}})/(3L)$ (solid line); and $R_g = (W\rho_{\Box,\text{poly}})/L$ (dashed-dotted line).

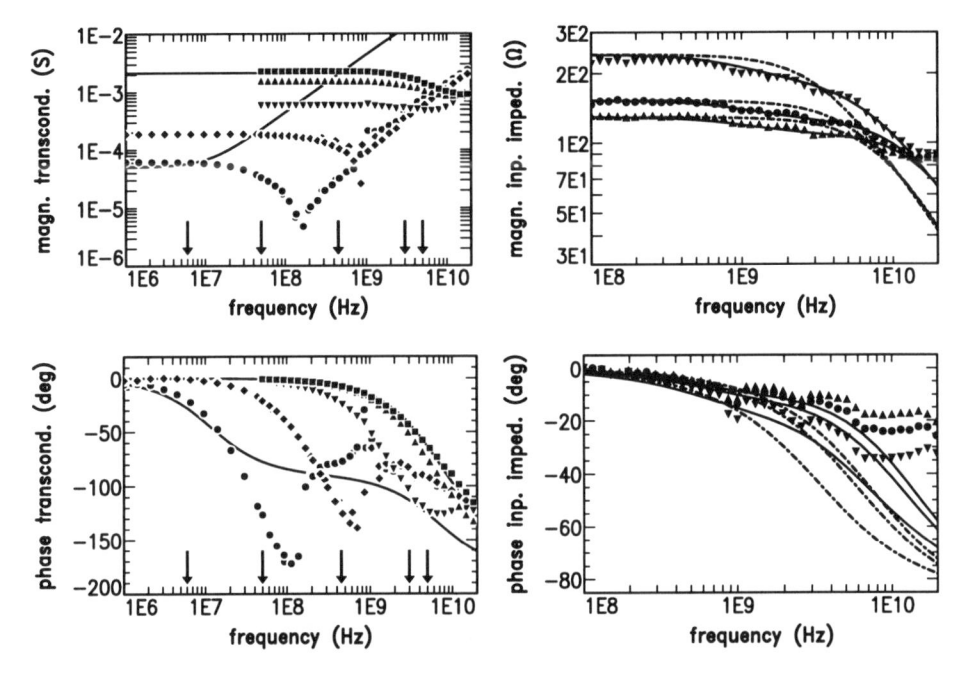

Figure 10: Magnitude (top) and phase (bottom) of the transconductance of PMOSFETs in common source-bulk configuration at $V_{ds} = 4\,V$ and $V_{gs} = 4\,V$ ($V_{dd} = 5\,V$). Gate width is 30 μm, while the gate length is 30 μm (solid circles), 10 μm (diamonds), 3 μm (downwards-directed triangles), 1 μm (upwards-directed triangles) and 0.6 μm (solid squares). Lines represent MM9 simulations for gate lengths of 30 and 0.6 μm, respectively. Arrows indicate f_T for each device.

Figure 11: Magnitude (top) and phase (bottom) of the input impedance of a 60/0.5 N-channel transistor ($V_{TO} = 0.6\,V$) in common gate configuration. Symbols represent measurements at $V_{ds} = 2.0\,V$ and $V_{gs} = 0.9\,V$ (downwards-directed triangles), 1.2 V (solid circles) and 1.5 V (upwards-directed triangles). Dashed lines represent MOS MODEL 9 simulations without bulk resistance and solid lines represent MOS MODEL 9 simulations with bulk resistance.

the formulation of a non-quasi static extension of MOS MODEL 9 is already available [24].

Common gate configuration

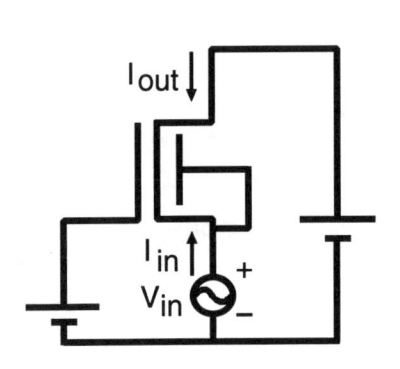

In Figure 12 a MOSFET in common gate configuration is shown. The source and bulk are voltage driven, while the drain is ac short-circuited. This common gate configuration may not be used in practice, but the admittance parameters for this common gate configuration are easily obtained from the admittance parameters for the common source-bulk configuration (see e.g. [25]). This procedure has been followed for both measurements and simulations. The input impedance, $Z_{in} = v_{in}/i_{in}$, of a MOSFET in common gate configuration is shown in Figure 11 as a function of the frequency. The drain bias has been chosen at 2 V, being 60% of the supply voltage of the process, while for the gate bias three values equally spaced between the threshold voltage, V_{TO}, and about one Volt above V_{TO}, have been used. From Figure 11 it can be seen that MOS MODEL 9 gives an accurate description of both frequency and bias dependence of the input impedance. However, for the input impedance in common gate configuration the influence of the bulk resistance is no longer negligible. Nevertheless, neglecting both gate and bulk resistance, one finds for the input impedance

Figure 12: MOSFET in common gate configuration.

$$Z_{in} = \frac{v_{in}}{i_{in}} \approx (g_m + g_{ds})^{-1} \left(1 + j\omega \frac{C_{ss}^{eff}}{g_m + g_{ds}}\right)^{-1}. \quad (9)$$

Herein g_{ds} is the dc output conductance and C_{ss}^{eff} is the effective capacitance

$$C_{ss}^{eff} = \frac{\partial Q_s}{\partial V_s} + \frac{\partial Q_b}{\partial V_s} + \frac{\partial Q_s}{\partial V_b} + \frac{\partial Q_b}{\partial V_b} + C_{gso} + C_{jun,d}, \quad (10)$$

where $C_{jun,d}$ is the drain-bulk junction capacitance. From Eq. 9 we

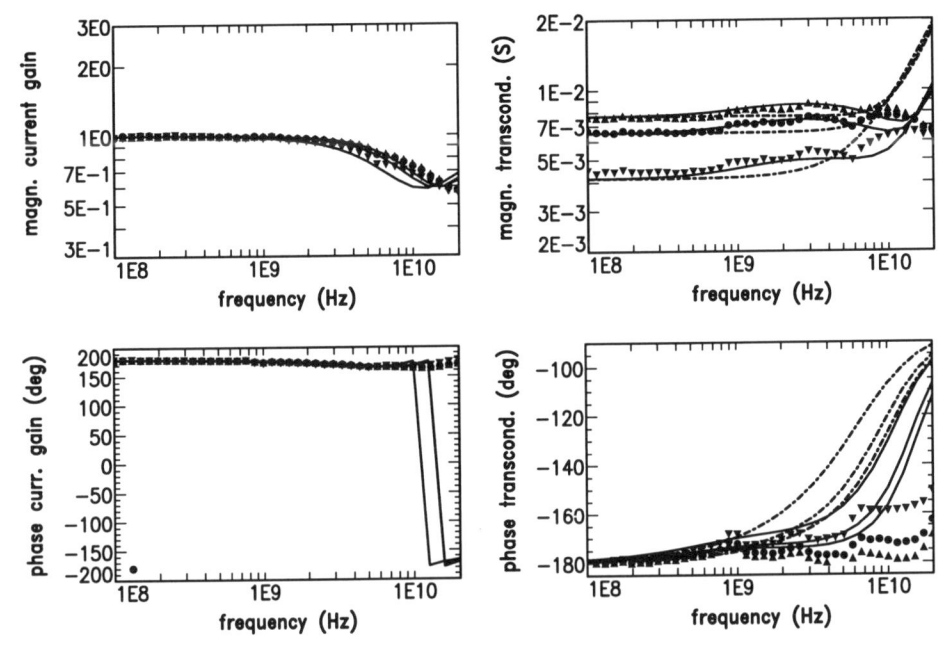

Figure 13: Magnitude (top) and phase (bottom) of the current gain of a 60/0.5 N-channel transistor ($V_{TO} = 0.6\,V$) in common gate configuration. Symbols represent measurements at $V_{ds} = 2.0\,V$ and $V_{gs} = 0.9\,V$ (downwards-directed triangles), $1.2\,V$ (solid circles) and $1.5\,V$ (upwards-directed triangles). Lines represent MOS MODEL 9 simulations.

Figure 14: Magnitude (top) and phase (bottom) of the transconductance of a 60/0.5 N-channel transistor ($V_{TO} = 0.6\,V$) in common gate configuration. Symbols represent measurements at $V_{ds} = 2.0\,V$ and $V_{gs} = 0.9\,V$ (downwards-directed triangles), $1.2\,V$ (solid circles) and $1.5\,V$ (upwards-directed triangles). Dashed and solid lines represent MOS MODEL 9 simulations without and with bulk resistance, respectively.

see that the input impedance in common gate configuration depends on g_{ds}, which stems from a dependency on the ac output conductance. The frequency behaviour of the ac output conductance is quite sensitive to the bulk resistance. Consequently, also the input impedance depends on the bulk resistance (see Figure 11). In Figure 13 the current gain, i_{out}/i_{in}, as a function of frequency is shown for the device and bias conditions from Figure 11. The current gain is almost independent of frequency and bias condition, which is described correctly by MOS MODEL 9. Neglecting again gate and bulk resistance, one finds for the current gain

$$\frac{i_{out}}{i_{in}} \approx -\left(1 + j\omega \frac{C_{ds}^{eff}}{g_m + g_{ds}}\right)\left(1 + j\omega \frac{C_{ss}^{eff}}{g_m + g_{ds}}\right)^{-1}, \quad (11)$$

where C_{ds}^{eff} is the effective capacitance

$$C_{ds}^{eff} = -\frac{\partial Q_d}{\partial V_s} - \frac{\partial Q_d}{\partial V_b} + C_{jun,d}. \quad (12)$$

From Figure 13 it is clear that the pole in the denominator of Eq. 11 causes the decrease in current gain, while at still higher frequencies non-quasi static effects become visible. The transconductance, i_{out}/v_{in}, of the 60/0.5 N-channel device is shown in Figure 14 as a function of frequency. From Figures 5 or 6 and Figure 14 it can be seen that below f_T again the description of MOS MODEL 9 for frequency and bias dependence is quite accurate. Neglecting gate and bulk resistance, the expression for the transconductance yields

$$\frac{i_{out}}{v_{in}} \approx -(g_m + g_{ds})\left(1 + j\omega \frac{C_{ds}^{eff}}{g_m + g_{ds}}\right). \quad (13)$$

As with the input impedance, the appearance of the dc output conductance in the expression for the ac transconductance coincides with a sensitivity for the bulk resistance.

Cascade configuration

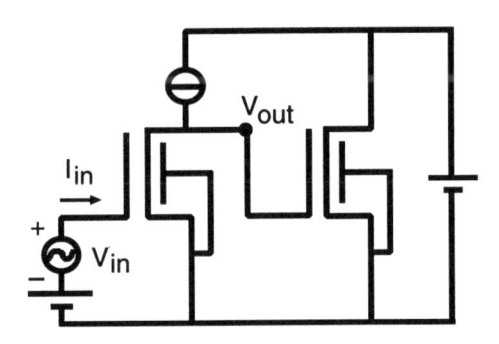

Figure 15: MOSFETs in cascade configuration.

In Figure 15 two MOSFETs in cascade configuration are shown. The gate of the lefthand transistor is voltage driven, while the drain of the righthand transistor is ac short-circuited. If both transistors have the same dc gate and drain bias, small-signal quantities such as input impedance and voltage gain (of the lefthand transistor) can be calculated from the admittance parameters of the individual transistors. This procedure has been followed for both measurements and simulations. The input impedance, $Z_{in} = v_{in}/i_{in}$, of two MOSFETs in cascade configuration is shown in Figure 16 as a function of frequency. For both transistors the drain bias has been chosen at 2 V, being 60% of the supply voltage of the process, while for the gate bias three values equally spaced between the threshold voltage, V_{TO}, and about one Volt above V_{TO}, have been used. Due to the fact that the experimental data are obtained by combining several measured Y-parameters the experimental uncertainty increases, especially at low frequencies. The frequency dependence of both magnitude and phase of the input impedance, Z_{in}, is rather complicated and no simple analytical expression can be obtained. MOS MODEL 9 describes this frequency dependence reasonably well.

In Figure 17 the voltage gain, v_{out}/v_{in}, as a function of frequency is shown for the lefthand transistor in the cascade configuration of Figure 15. Neglecting gate and bulk resistance, one finds for the voltage gain

$$\frac{v_{out}}{v_{in}} \approx -\frac{g_m}{g_{ds}}\left(1 - j\omega\frac{C_{dg}^{\text{eff}}}{g_m}\right)\left(1 + j\omega\frac{C_{gg}^{\text{eff}} + C_{dd}^{\text{eff}}}{g_{ds}}\right)^{-1}. \quad (14)$$

Herein C_{dg}^{eff} is given by Eq. 5; C_{gg}^{eff} is given by Eq. 2; and C_{dd}^{eff} is given

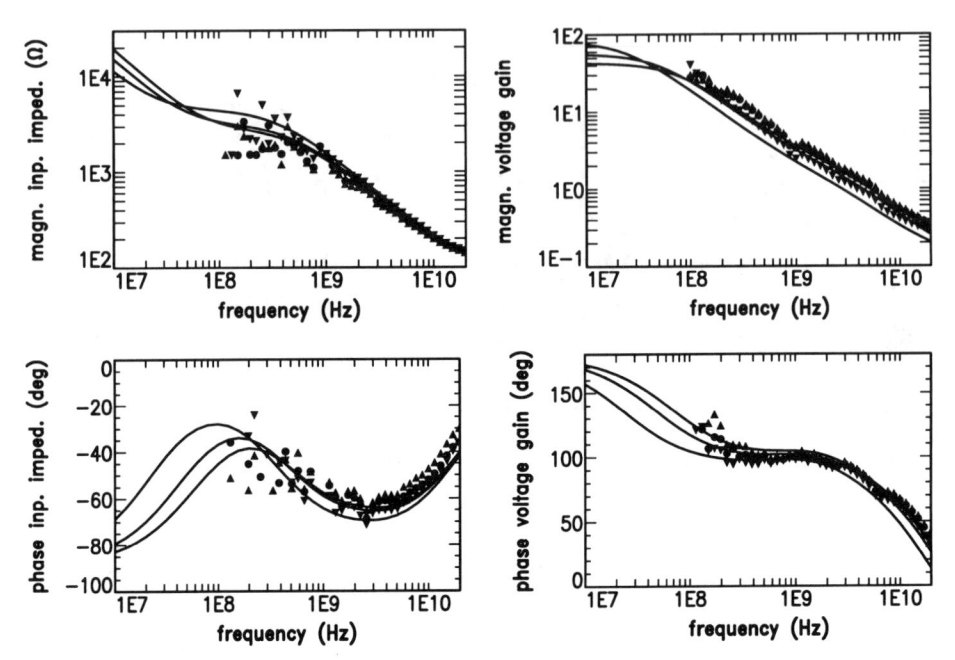

Figure 16: Magnitude (top) and phase (bottom) of the input impedance of two 60/0.5 N-channel transistors ($V_{TO} = 0.6\,V$) in cascade configuration. Symbols represent measurements at $V_{ds} = 2.0\,V$ and $V_{gs} = 0.9\,V$ (downwards-directed triangles), $1.2\,V$ (solid circles) and $1.5\,V$ (upwards-directed triangles). Lines represent MOS MODEL 9 simulations.

Figure 17: Magnitude (top) and phase (bottom) of the voltage gain, $v_{\text{out}}/v_{\text{in}}$, of two 60/0.5 N-channel transistors ($V_{TO} = 0.6\,V$) in cascade configuration (see Figure 15). Symbols represent measurements at $V_{ds} = 2.0\,V$ and $V_{gs} = 0.9\,V$ (downwards-directed triangles), $1.2\,V$ (solid circles) and $1.5\,V$ (upwards-directed triangles). Lines represent MOS MODEL 9 simulations.

by
$$C_{dd}^{eff} = \frac{\partial Q_d}{\partial V_d} + C_{gdo} + C_{jun,d} \quad . \tag{15}$$

From Figure 17 we see that only at the lowest frequencies the magnitude of voltage gain equals g_m/g_{ds}. The decrease of the voltage gain with increasing frequency is well-described by the model.

Cascode configuration

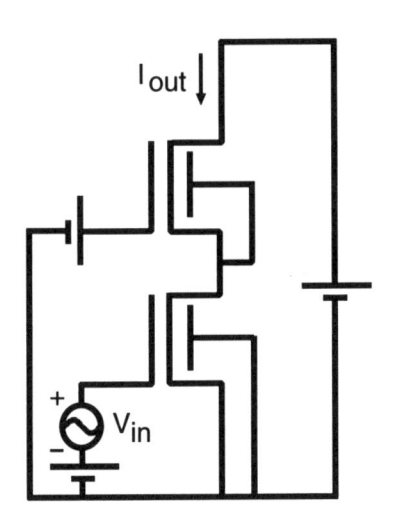

In Figure 18 two MOSFETs in cascode configuration are shown. The gate of the bottom transistor is voltage driven, while the drain of the upper transistor is ac short-circuited. If both transistors have the same dc gate and drain bias, small-signal quantities for the cascode configuration can be calculated from the admittance parameters of the individual transistors. For the transconductance, i_{out}/v_{in}, i.e. the ac drain current from the upper transistor divided by the ac gate voltage of the lower transistor, this procedure has been followed for both experimental and simulated data. In Figure 19 resulting experimental data for this transconductance are compared with calculations using MOS MODEL 9. No simple analytical expression for the transconductance can be derived. However, at low frequencies the transconductance of the cascode configuration should be about equal to that of the single device (cf. Figure 8). The agreement between measurements and calculations in Figure 19 for both frequency and bias dependence is quite good, except for the lowest gate voltage at the highest frequencies.

Figure 18: MOSFETs in cascode configuration.

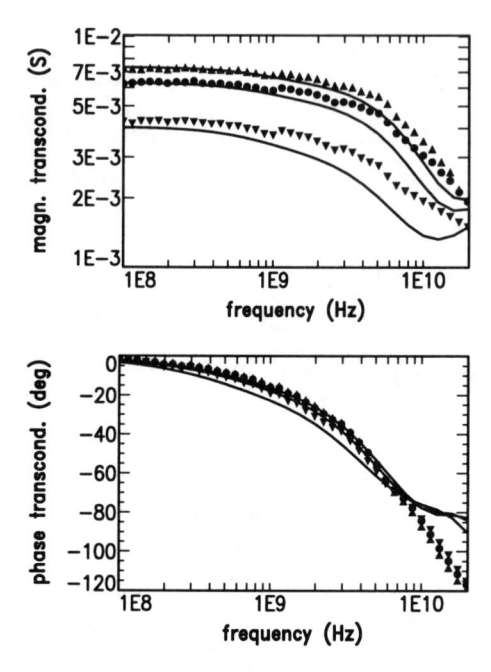

Figure 19: Magnitude (top) and phase (bottom) of the transconductance of two 60/0.5 N-channel transistors ($V_{TO} = 0.6\,V$) in cascode configuration. Symbols represent measurements at $V_{ds} = 2.0\ V$ and $V_{gs} = 0.9\ V$ (downwards-directed triangles), 1.2 V (solid circles) and 1.5 V (upwards-directed triangles). Lines represent MOS MODEL 9 simulations.

Summary and conclusions

The Philips compact MOS model, MOS MODEL 9, has been developed especially for analog circuit design [3]. This model combines a good description of the geometry dependence of the transistor behaviour with continuous derivatives of currents and charges with respect to bias voltages (see Table 1 and [14]).

In the previous sections we have presented a detailed investigation into the accuracy of MOS MODEL 9 for the description of a number of quantities, that are important for RF circuit design. For a number of RF quantities such as impedance, transconductance, current and voltage gain simulations have been compared with measurements on a number of basic transistor configurations. It is of crucial importance to characterize a number of parasitics and include them in the calculations. Not only junction and overlap capacitances should be taken into account, but also the bulk resistance and especially the gate resistance. For both resistive parasitics the effects on the quantities investigated have been demonstrated.

The present implementation of MOS MODEL 9 describes only quasi-static behaviour. This implies that the model is valid up to the cut-off or unity current-gain frequency, f_T, which can reach values as high as 20 GHz for modern mainstream CMOS processes (see Figure 7). It should be noted, however, the formulation of a non-quasi static extension of MOS MODEL 9 is already available [24].

Taking into account the parasitics mentioned above, both the bias and frequency dependence (up to f_T) of impedance, transconductance, current and voltage gain of the transistor configurations investigated, are described with good accuracy by simulations using MOS MODEL 9.

References

[1] F. Najm, *Simulation and Modeling, A New Beginning*, IEEE Circuits & Devices, Vol.12, No.1, 8-10, 1996.

[2] *BSIM3v3 manual*, Department of Electrical Engineering and Computer Science, University of California, Berkeley, CA 94720, U.S.A.

[3] *MOS MODEL 9*, complete model description and documentation for implementation available on request at e-mail address: mm9_mxt@natlab.research.philips.com.

[4] W.R. Eisenstadt, *Low Power RF Technology and Design*, IEDM'95 Short Course on Technologies for Portable Systems, Washington D.C., 1995.

[5] D.C. Shaver, *Microwave Operation of Submicrometer Channel-Length Silicon MOSFET's*, IEEE Electron Device Letters, Vol.6, No.1, 36-39, 1985.

[6] A.E. Scmitz, R.H. Walden, L.E. Larson, S.E. Rosenbaum, R.A. Metzger, J.R. Behnke and P.A. Macdonald, *A Deep-Submicrometer Microwave/Digital CMOS/SOS Technology*, IEEE Electron Device Letters, Vol.12, No.1, 16-17, 1991.

[7] A.L. Caviglia, R.C. Potter and L.J. West, *Microwave Performance pf SOI n-MOSFET's and Coplanar Waveguides*, IEEE Electron Device Letters, Vol.12, No.1, 26-27, 1991.

[8] C. Raynaud, J. Gautier, G. Guegan, M. Lerme, E. Playez and G. Dambrine, *High-Frequency Performance of Submicrometer Channel-Length Silicon MOSFET's*, IEEE Electron Device Letters, Vol.12, No.12, 667-669, 1991.

[9] Y. Mii, S. Rishton, Y. Taur, D. Kern, T. Lii, K. Lee, K. Jenkins, D. Quilan, T. Brown Jr., D. Danner, F. Sewell and M. Polcari, *High Performance 0.1 μm nMOSFET's with 10 ps/stage Delay (85 K) at 1.5 V Power Supply*, Proceedings VLSI Symposium, 91-92, 1993.

[10] J. Hänseler, H. Schinagel and H.L. Zapf, *Test Structures and Measurement Techniques for the Characterization of the Dynamic Behaviour of CMOS Transistors on Wafer in the GHz Range*, Proceedings IEEE IC-MTS, Vol.5, 90-93, 1992.

[11] R. Singh, A. Juge, R. Joly and G. Morin, *An Investigation into the Nonquasi-Static Effects in MOS Devices with On-Wafer S-Parameter Techniques*, Proceedings IEEE IC-MTS, Vol.6, 21-25, 1993.

[12] R.R.J. Vanoppen, J.A.M. Geelen and D.B.M. Klaassen, *The High-Frequency Analogue Performance of MOSFETs*, Proceedings IEDM, 173-176, 1994.

[13] F.M. Klaassen and R.M.D. Velghe, *Compact Modelling of the MOSFET Drain Conductance*, Proceedings ESSDERC, 418-422, 1989.

[14] R.M.D.A. Velghe, D.B.M. Klaassen and F.M. Klaassen, *Compact MOS Modeling for Analog Circuit Simulation*, Proceedings IEDM, 485-488, 1993.

[15] R.M.D.A. Velghe, D.B.M. Klaassen and F.M. Klaassen, *Compact MOS Modelling for Analogue Circuit Simulation*, Proceedings ESSDERC, 833-836, 1994.

[16] M.J. van Dort and D.B.M. Klaassen, *Sensitivity Analysis of an Industrial CMOS Process using RSM Techniques*, Proceedings SISPAD, 432-435, 1995.

[17] R.M.D.A. Velghe and D.B.M. Klaassen, *Prediction of Compact MOS Model Parameters for Low-Power Application*, Proceedings ESSDERC, 565-568, 1995.

[18] M.J. van Dort and D.B.M. Klaassen, *Circuit Sensitivity Analysis in Terms of Process Parameters*, Proceedings IEDM, 941-944, 1995.

[19] H.C. de Graaff and F.M. Klaassen, *Compact Transistor Modelling for Circuit Design*, Springer-Verlag, Wien, 1990.

[20] Y. Tsividis and K. Suyama, *MOSFET Modeling for Analog Circuit CAD: problems and prospects*, Proceedings CICC, p.14.1.1, 1993.

[21] Y.P. Tsividis and K. Suyama, *MOSFET Modeling for Analog Circuit CAD: problems and prospects*, IEEE Journal of Solid-State Circuits, Vol.29, No.3, 210-216, 1994.

[22] *JUNCAP*, complete model description and documentation for implementation available on request at e-mail address: mm9_mxt@natlab.research.philips.com.

[23] Y.P. Tsividis, *Operation and Modeling of the MOS Transistor*, McGraw-Hill, New York, 1988.

[24] T. Smedes and F.M. Klaassen, *An Analytical Model for the Non-Quasi-Satic Small-Signal Behaviour of Submicron MOSFETs*, Solid-State Electronics, Vol.38, No.1, 121-130, 1995.

[25] G. Gonzalez, *Microwave Transistor Amplifiers: Analysis and Design*, Prentice-Hall, Englewoods-Cliffs, N.J., 1984.

High Integration CMOS RF Transceivers

F. Brianti, G. Chien, T. Cho, S. Lo, S. Mehta, J. Ou,
J. Rudell, T. Weigandt, J. Weldon, P. Gray

Department of Electrical Engineering and Computer Sciences,
University of California at Berkeley

ABSTRACT

This paper presents an overview of technical challenges in achieving higher integration levels, lower power dissipation, smaller form factor, lower cost, and multistandard operation in portable battery-powered RF transceivers for personal communications applications.

1. Introduction

Digital radio personal communications devices utilizing the bands between 800MHz and 2.5GHz will play an increasingly important role in the overall communications infrastructure in the next decade. In addition to the wide-area and mid-range transceivers in use today exemplified by cellular telephones and cordless telephones, respectively, requirements will evolve for short-range transceivers for picocell applications at high data rates appropriate for wireless LANs and other applications. In addition, the rapidly proliferating standards for digital RF communications will require development of transceivers that can interface with more than one RF standard and yet be economical to produce. Considerations of power dissipation, form factor, and cost for these new generations of transceivers will dictate that the RF/IF portions of these devices evolve to higher levels of integration than is true at present. The major challenge in RF transceiver design is to more effectively utilize scaled technologies to improve the integration level of RF transceivers, with resulting further improvements in power dissipation, form factor, and cost. A typical example of present practice for the implementation of such a transceiver as might be used in a frequency-hopped wireless LAN application is shown in block diagram form in Fig. 1. This architecture is not particularly amenable to higher levels of integration.

In this paper, some of the key barriers to realizing higher levels of integration are discussed. In particular, the possible approaches for achieving higher integration and multistandard operation using scaled CMOS technology are described. The most promising initiatives involve new architectures, such as direct-conversion or low-IF receiver architectures that eliminate external IF filtering, new approaches to frequency synthesis that eliminate the need for external VCO resonators, and more effective utilization of inductors available in near-standard IC technology to provide the tuning function essential to low-power realizations of RF functions, and power-optimized baseband filtering and A/D conversion techniques. Some preliminary results from an experimental integrated CMOS receiver for 1.9GHz operation are discussed for illustration. A hypothetical single-chip transceiver that might result from success in these areas is illustrated in block diagram form in Fig. 2.

2. Receiver Architectures for High-integration RF Transceivers in CMOS

The vast majority of currently-manufactured transceivers for the applications mentioned above utilize single- or dual-conversion configuration for the receive path. This architecture is not particularly well suited to higher levels of integration because the required high-Q, high-dynamic range IF bandpass filter. Image rejection consideration usually dictate that the center frequency of this filter be on the order of 10% of the carrier frequency in conventional receivers, and as a result it does not lend itself to implementation using monolithic filtering approaches. A second problem is that in conventional architectures the frequency synthesizer which generates first local oscillator must

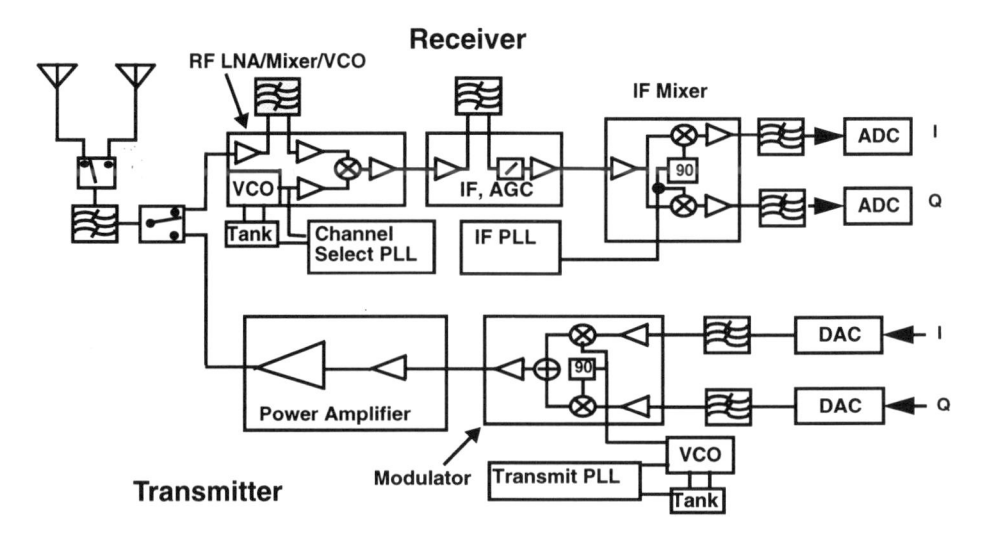

Fig. 1: Block diagram of a typical multi-chip, multi-technology transceiver implementation

simultaneously achieve extremely low phase noise and produce the fine channel spacings required to tune the radio.

A number of promising architectures for realizing higher integration in RF transceivers are under investigation in various laboratories. The most promising of these involve is the use of zero IF, low-IF, or wideband-IF (WBIF) configurations in the receiver. These configurations have been investigated intensively for years (see for example [5][10]) but have made their way into practice in only a few specialized applications[8][9][10][6][7][11][13][14][12][15]. These configurations eliminate the external IF filtering function since the IF filter is replaced by two (I and Q) lowpass filters in the case of zero IF and quasi-IF receivers, or by a low-frequency, low-Q bandpass IF filter in the low-IF case.

The most severe problems in direct conversion receivers result from the fact that the baseband signal often contains low-frequency information that must be distinguished from DC and low-frequency errors that arise in the base-

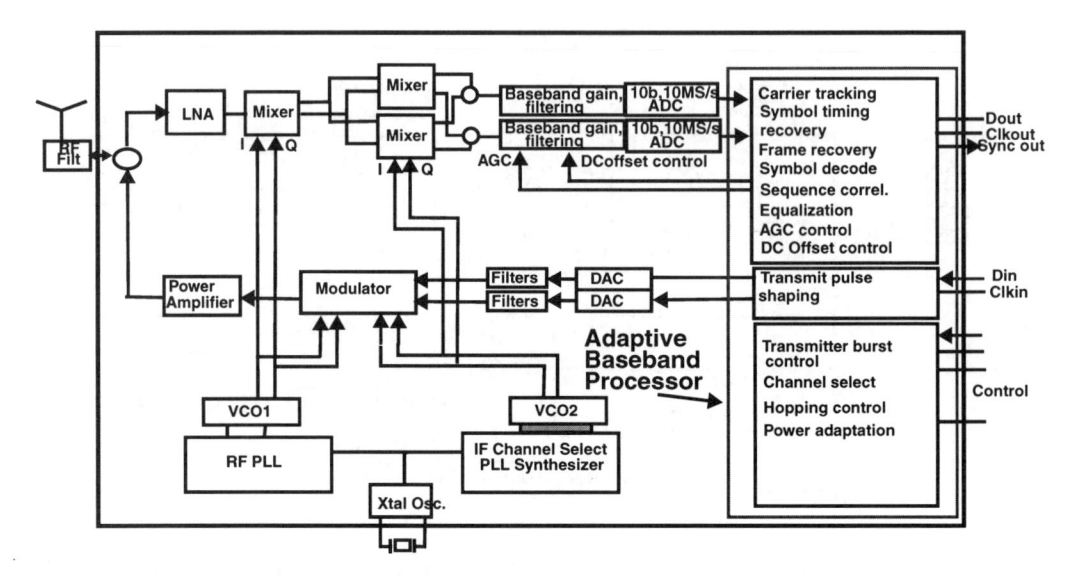

Fig. 2: Block diagram of a possible future high-integration adaptive transceiver

band signal path. One important error source is the device-mismatch-induced DC offset and 1/f noise of the signal path itself. For reasons of large-signal blocking performance, the gain of the LNA is usually restricted to the 20dB range, so that the wanted signal level reaching the mixer under weak signal conditions is on the order of 20-100 microvolts in amplitude. The accumulated DC offset referred to the mixer output can easily be 10mV, 30-50 times larger than the signal. Another important contributor is LO leakage, resulting from the fact that since the LO is at the carrier frequency, any energy from it reaching the RF path demodulates to a DC offset. Because the effects of LO leakage can be a function of the impedance seen at the antenna, these DC offsets can vary with time in an unpredictable manner. The problem is more severe in frequency-hopping receivers because the carrier leakage is different at each hop frequency, giving a time variation to the DC offset that is induced. The cumulative effect of carrier feedthrough and DC offsets is to superimpose large, possibly time-varying additive errors on the small wanted low-frequency AC signal in the baseband signal path.

Numerous approaches have been tried to attack these problems. Most have been attempted within the context of a more conventional, bipolar technology multi-chip receiver implementation using analog baseband filtering and demodulation. However, a closer coupling between the demodulation process and the RF and baseband analog signal path may well allow the separation of the DC offsets to be carried out using an adaptive approach that combines this function with carrier recovery, symbol timing recovery, automatic gain control (AGC), and data detection in a mixed analog-digital implementation. Most TDMA systems, for example, utilize a preamble in the frame structure which when demodulated to baseband has either zero or known DC content, allowing adaptive, frame-by-frame DC offset removal[21]. The problem of 1/f noise can be attacked in a number of ways, one of which is to simply use correlated double sampling or chopper stabilization of the active elements in the baseband signal path. A/D conversion of the baseband signal at high resolution is a requirement for this approach.

Most of the benefits of homodyne receivers accrue if the IF is translated to a low but nonzero value instead of to DC as in homodyne receivers. The IF needs to be low enough that normal monolithic filtering techniques such as g_m/C continuous filters or switched-capacitor filters can be used. The advantage of this over homodyne receivers is that the problems of DC offset and 1/f noise are greatly reduced. However, a new problem of image rejection of the relatively close-in image frequency is introduced. This image energy must be eliminated through the use of an image-rejection mixer configuration following the LNA. Since this mixer will have to provide image rejection on the order of 60-70dB in some applications, phase shift accuracy and path matching accuracy within the mixer must be extremely precise. Progress has been made in this area in recent research [16].

Receiver architecture also has a strong influence on another difficult problem, that of providing a local oscillator with low enough phase noise. As discussed later, the use of fixed LO frequency as opposed to a tunable one allows the use of a high comparison frequency and loop bandwidth, potentially allowing large reductions in close-in phase noise. For this reason an architecture that allows the use of a fixed first LO is advantageous for monolithic integration. This can be provided using a variation on the double heterodyne architecture called wideband IF or WBIF. In this approach the entire band of frequencies to be tuned by the receiver is translated down to IF in the first mixer, and then subsequently translated directly to baseband in the second mixer with little or no IF selective filtering. The channel- select filtering is done at baseband with a lowpass filter following translation to baseband. This technique has several important advantages. The first local oscillator can be implemented as a fixed frequency oscillator, making it easier to realize the required phase noise performance. The second LO, used to tune the desired channel, is at much lower frequency and its phase noise contributions, as well as the spurs associated with the narrow channel spacing and associated low comparison frequency, can be made much smaller. The carrier feedthrough problem is also eliminated. The technique eliminates the IF filter, but retains the image reject problem at RF and also many of the DC offset and drift problems of direct conversion receivers since adjacent channel blocking signals are carried to baseband and as a result most of the gain applied to the desired signal is done at baseband. Image rejection must be realized through a combination of a passive RF filter and an image reject configuration in the double conversion mixer. The example receiver in Fig. 2 has this configuration.

Great benefits could potentially accrue from a unification of approaches to data communications transceiver design. From an applications viewpoint, there is no fundamental reason that a single transceiver device could not provide the functionality of multiple communications standards at multiple frequencies. This might allow-for example, a single hand-held device to perform the functions of cellular phone and cordless phone compatibly with the varying standards for such service in Japan, North America, and Europe. Certainly multiple RF ports and antennas would be required for optimum performance at the different wavelengths, but following this a single transceiver should be able to adapt the different transmit power levels and modulation schemes. Central to achieving this would be an architecture in which a major portion of the IF (baseband in homodyne receivers) signal processing is performed in the digital domain, as well as carrier recovery, symbol timing recovery, equalization, DC offset control, power control, and so forth. A high-integration CMOS or BiCMOS implementation would be essential.

3. Implementation of Narrowband RF LNAs

The implementation of RF low-noise amplifiers in CMOS at power dissipations competitive with corresponding bipolar technologies requires the use of tuned narrowband techniques. The inherently low g_m/I_d ratio of MOSETs at reasonable bias points requires that a relatively high bias current be used to achieve the reasonable noise figure in a 50 ohm environment. The use of on-chip spiral and bond wire inductors to implement impedance transforming networks at the input, to realize input matching, and to tune the LNA output node allow significantly lower power dissipation for a given noise figure and intermodulation performance. Recently results from several experimental LNAs implemented in CMOS technology have shown noise figures in the range of 3dB with input 3rd order intercepts on the order of 4 to 8 dBm at 3 volts with 30mW of power dissipation.[44][45]at 900MHz. Similar results have been obtained at 1.9Ghz.

Perhaps the most difficult problem in high-integration transceiver is the noise coupling through the substrate and power supplies from various sections of the receiver into the critical parts such as the LNA and synthesizers. A differential implementation of the LNA with differential signal path off chip is critically important in order to minimize this problem. A differential implementation carries a power dissipation penalty for a given dynamic range and places a burden on the off-chip coupling circuitry, requiring either a balun or differential ceramic filter. However, it is likely that such an approach will be necessary to control the noise coupling problem.

4. Circuit Approaches for Integrated Synthesizers

The phase noise present in the local oscillator signal generated by the synthesizer contributes directly to phase noise on the IF or baseband signal after frequency conversion, and as a result directly degrades the effective signal-to-noise ratio (SNR) of phase-modulated signals. More importantly, LO phase noise mixes with adjacent channel signal energy, degrading overall receiver SNR and limiting receiver blocking performance. Finally, in the transmitter the phase noise of the LO contributes noise energy outside the band of the channel being transmitted. Spurious transmitted energy at adjacent channel frequencies must be closely controlled in most systems. With proper PLL loop design, the phase noise of the synthesizer is dominated by the phase noise of the crystal reference for frequencies far below the PLL loop bandwidth, and by the inherent phase noise of the VCO itself for frequencies far beyond the PLL loop bandwidth.

In conventional synthesizers realizing fine channel spacing, the phase comparison frequency is low and loop bandwidth is low. As a result, synthesizer phase noise in the regions of interest is dominated by inherent VCO phase noise. For an LC oscillator, the ratio of the internally generated VCO phase noise power to the carrier power can be shown to be directly related to the amount of energy stored per cycle to the thermal energy kT. The energy stored is Q times the energy which must be supplied per cycle, and phase noise is directly related to the inverse of resonator Q, a fact predicted by a number of analyses of phase noise in oscillators.[26] A number of approaches show promise for realizing the VCO function on-chip. These include the use of on-chip spiral inductors[39][46][47][20] the use of bond wire inductance [19][43]and the use of plated-up gold inductors over thick oxide [24]. Even with higher-Q inductors, the realization of a low-resistance, wide-range varactor tuning capacitance using standard IC technology is difficult. The simultaneous realization of high Q and tunability, together with either wide tuning range or high center frequency accuracy, is a very difficult task. The use of synthesizer configurations that allow wide PLL bandwidth so that the phase noise of the VCO is suppressed in the range of interest[22][23] is an attractive option in some receiver architectures. The effects of close-in VCO phase noise can be minimized if PLL loop bandwidth, phase comparison frequency, and loop order can be kept high. The use of a WBIF receiver architectures, in which the first VCO frequency is fixed, also relieves the phase noise problem because the comparison frequency and loop bandwidth can be kept much higher in a fixed-frequency first VCO. The second VCO/synthesizer performs the tuning function, but its impact on receiver phase noise is smaller since it operates at a much lower frequency.

5. Low-Power Baseband Signal Conversion and Processing

Depending on the type of transceiver, baseband operations of IF filtering, equalization, timing recovery, symbol constellation decoding, signal correlation, symbol generation, quadrature modulation, frequency synthesis and so forth are required. A major body of current research is aimed at performing more of these functions in the digital domain than is currently the case, with resultant improvement performance, adaptability, and manufacturability. The principal trade-offs are the incurred penalty in power dissipation and die area of the digital implementation, and the cost and die area of the required A/D converter. Rapid progress has been made in the implementation of these types of functions in VLSI CMOS [33][32][34] [14].

The implementation of baseband processing and A/D conversion for the various types of direct conversion architectures presents special problems. In this type of architecture, most of the channel select filtering is to be implemented at baseband using a combination of on-chip continuous-time filter, switched capacitor filter, and digital signal processing. This allows both a higher integration level and easier adaptability to multiple standards of operation. A key problem in the implementation of such circuits is to achieve sufficiently low levels of power dissipation in the baseband filtering and signal processing circuits at the relatively high value of dynamic range required. Since the composite baseband signal coming from the mixer in a direct conversion receiver contains energy from all of the channels within the approximately 20MHz band passed by the RF filter, exceptionally large dynamic range and an exceptionally low noise level are required.

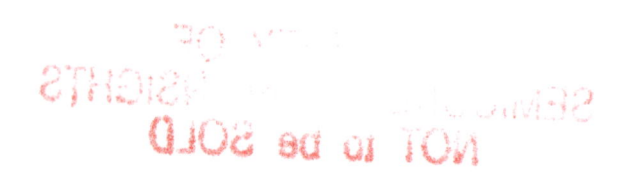

The particular nature of the filtering function required at baseband presents particular problems but also particular opportunities for power minimization. The composite signal at the output of the mixer consists of the desired low-frequency signal at a level of on the order of 200uV RMS, with the adjacent channels energy appearing at higher frequencies and with amplitudes larger by 40 to 60dB depending on the application. The total in-band input-referred noise of the overall filter must be low enough so that this source of noise does not degrade the overall receiver noise figure excessively, which in this case implies an equivalent noise resistance at the input of less than 10K. In addition, input linearity must be sufficient that intermodulation between adjacent channel signals does not degrade in-band SNR. Finally, channel offsets from all sources must be kept low enough to maintain the A/D converter following the channel filter in its linear range for all signal conditions. Because of the lowpass nature of the channel, 1/f noise and channel offsets present particular problems of signal degradation. Special techniques such as data-driven offset cancellation or the use of inherently balanced baseband codes are necessary to control this problem. Two recent experimental prototypes have demonstrated performance at a level needed for cordless telephones at overall power levels much less than 50mW per channel for filtering and A/D conversion at 3 volts in 0.6 micron CMOS [48][49].

Many benefits accrue in pushing baseband signal processing into the digital domain, particularly for multistandard adaptive transceivers. For direct conversion receivers, the composite baseband signal contains all the large adjacent-channel blocking signals, and as a result an all-digital implementation of the baseband signal processing would require two A/D converters of greater than 80dB dynamic range and 20MHz effective sampling rate. Some combination of analog and digital filtering will be optimum. For at least the higher-frequency portions of this set of applications, low-power, high-speed approaches such as pipelining will be required. Finding techniques for reducing the power dissipation of these A/D converters is a key goal. Current state of the art for this class of converters is about 1mW/ MHz of sample rate at 10 bits in 0.6m technology [48].

6. Alternative Technologies for High-Integration RF Transceivers

The alternative technologies for a transceiver at the integration level of Fig. 2 are BiCMOS and CMOS. Bipolar and BiCMOS solutions are attractive because of the inherent capability of bipolar transistors to provide high g_m at low current, and because of the well-developed family of circuit techniques for RF design using bipolar technology[18]. The interest in utilizing CMOS for high-integration transceivers, stems from the potentially lower cost of a CMOS implementation It appears likely that at least for the lower performance applications it will be possible to overcome the poorer characteristics of CMOS for RF by utilizing alternative receiver architectures by using narrowband tuned circuits (taking advantage of the high f_{max} of the NMOS device), and by using more adaptation in the receiver. The continued scaling of CMOS technology, with 0.1 micron devices with f_ts of near 100GHz recently demonstrated,[36][37] should eventually allow this approach in some applications.

7. Summary

Prospects for continued progress in high-integration, low-cost RF transceivers is excellent. A key requirement for progress is close collaboration between transceivers and system designers, RF and digital circuit designers, and device and package modeling engineers, so that opportunities for innovation with new architectural approaches can be identified and exploited.

Acknowledgments

Research sponsored by NSF under grant MIP9101525, ARPA under contract J-FBI-92-150, and ARO under grant DAAHO4-93-G-0200.

References

1. Rapeli, J. "IC Solutions for Mobile Telephones," book chapter in *Design of VSLI Circuits for Telecommunications and Signal Processing,* Kluewer, June 1993.
2. S. Sheng, R. Allmon, L. Lynn, I. O'Donnell, K. Stone, R. W. Brodersen, "A Monolithic CMOS Radio Systems for Wideband CDMA Communications," Wireless '94, Calgary, Canada, July 1994.
3. V. Thomas, et al, "A One-chip 2 GHz Single-Superhet Receiver for 2Mb/s FSK Radio Communications," *Digest of Technical Papers, 1993* International Solid-State Circuits Conference, San Francisco, CA, February 1994.
4. J. Sevenhans, et al, "An Analog Radio Front-end Chip Set for a 1.9 GHz Mobile Telephone Application," *Digest of Technical Papers, 1993* International Solid-State Circuits Conference, San Francisco, CA, February 1994.
5. Cavers, J.K., Liao, M.W. "Adaptive compensation for imbalance and offset losses in direct conversion transceivers," *IEEE Transactions on Vehicular Technology,* Nov. 1993.
6. J. Sevenhans, et al, "An Integrated Si Bipolar RF Transceiver for a Zero IF 900 MHz GSM Digital Mobile Radio Single Chip RF Up and RF Down Converter of a Hand Portable Phone," *Digest of Technical Papers,* 1991 Symposium on VLSI Circuits, Honolulu, June 1991.
7. P. Weger, et al, "Completely Integrated 1.5 GHz Direct Conversion Receiver," Digest of Technical Papers, 1994 Symposium on VLSI Circuits, Honolulu, Hawaii, June 1994.
8. Voudouris, K., Noras, J.M., "Direct conversion receiver for the TDMA mobile terminal," *IEEE Colloquium on Personal Communications: Circuits, Systems and Technology,* London, UK, Jan. 1993.
9. Plessey GP1010 GPS Receiver Preliminary Data Sheet, October 1992.
10. Bateman, A., Haines, D.M., "Direct conversion transceiver design for compact low-cost portable mobile radio terminals," 39th IEEE Vehicular Technology Conference, San Francisco, CA, May 1989.

11. J. Min, et al, "An All-CMOS Architecture for a Low-Power Frequency-Hopped 900 MHz Spread Spectrum Transceiver," *Digest of Technical Papers*, 1994 Custom Integrated Circuits Conference, San Diego, June 1994.
12. Vanwelsenaers, A., Rabaey, D., Vanzieleghem, E., Sevenhans, J., and others, "Alcatel chip set for GSM handportable terminal," *Proceedings of 5th Nordic Seminar on Digital Mobile Radio Communications DMR V*, Helsinki, Finland, 1-3 Dec. 1992.
13. A. Abidi, "Radio Frequency Integrated Circuits for Portable Communications," *Digest of Technical Papers*, 1994 Custom Integrated Circuits Conference, San Diego, June 1994.
14. A. Chandrakasan, S. Sheng, R. W. Brodersen, "Design Considerations for a Future Portable Multi-Media Terminal," *Third-Generation Wireless Information Networks*, Kluewer Academic Publisher, 1992.
15. P. Baltus and A. Tombeur, "DECT Zero-IF Receiver Front-end," *Proceedings of the AACD* (Leuven), pp. 295-318, March 1993.
16. M. Steyaert and J. Crols, "Analog Integrated Polyphase Filters," Proceedings of the AACD (Eindhoven), March 1994.
17. M. Thiriamsut, et al, "A 1.2 Micron CMOS Implementation for a Low-Power 900 MHz Mobile Telephone Radio-Frequency Synthesizer," *Digest of Technical Papers,* 1994 Custom Integrated Circuits Conference, San Diego, CA, June 1994.
18. Meyer, R.G., Mack, W.D., "1- GHz BiCMOS RF front-end IC," *IEEE Journal of Solid-State Circuits,* March 1994.
19. Steyaert, M., Craninckx, J., "A 1.1 GHz oscillator using bondwire inductance," *Electronics Letters*, 3 Feb. 1994.
20. Chang, J.Y.-C., Abidi, A.A., Gaitan, M., "Large suspended inductors on silicon and their use in a 2- mu m CMOS RF amplifier," *IEEE Electron Device Letters*, May 1993
21. Sampei, S., Feher, K., "Adaptive DC-offset compensation algorithm for burst mode operated direct conversion receivers," Vehicular Technology Society 42nd VTS Conference. Frontiers of Technology. From Pioneers to the 21st Century, Denver, CO, 10-13 May 1992.
22. T. Weigandt, et al, "Analysis of Timing Jitter in CMOS Ring Oscillators," *Digest of Technical papers*, 1994 International Symposium on Circuits and Systems, London, June, 1994.
23. B. Kim, et al, "DLL/PLL System Noise Analysis for Low Jitter Clock Synthesizer Design," *Digest of Technical Papers*, 1994 International Symposium on Circuits and Systems, London, June 1994.
24. K. Negus, et al, "A Highly Integrated Transmitter IC with Monolithic Narrowband Tuning for Digital Cellular Handsets," *Digest of Technical Papers,* 1993 International Solid-State Circuits Conference, San Francisco, CA, February 1994.
25. C. Nguyen, "Integrated Filters Using Micro-mechanical Resonators," Ph.D. Dissertation, University of California, Berkeley, California, Nov 1994.
26. J. K. A. Everhard, "Low-Noise Power-Efficient Oscillators, Theory and Design," *IEEE Proceedings*, August, 1986.
27. Chang, G., et al, "A Low-Power CMOS Digitally Synthesized 0-13 MHz Agile Sinewave Generator," *Digest of Technical papers,* 1994 International Solid-State Circuits Conference, San Francisco, CA, February 1994.
28. R. Gharpury, "Modeling of Substrate Interactions in Integrated Circuits," MS Report, University of California, Berkeley, September, 1994.
29. Hull, C.D., Meyer, R.G,, "A systematic approach to the analysis of noise in mixers," *IEEE Transactions on Circuits and Systems I: Fundamental Theory and Applications*, Dec. 1993.
30. T. Cho, et al, "A 10-bit, 20MS/sec, 35mW Pipeline A/D Converter," *Digest of Technical Papers*, 1994 Custom Integrated Circuits Conference, San Diego, June, 1994.
31. Lu, F., Samueli, H., Yuan, J., Svensson, C. "A 700- MHz 24-b pipelined accumulator in 1.2- mu m CMOS for application as a numerically controlled oscillator," *IEEE Journal of Solid-State Circuits*, Aug. 1993.

32. Jain, R., Samueli, H., Yang, P.T., Chien, C., and others. "Computer-aided design of a BPSK spread-spectrum chip set," *IEEE Journal of Solid-State Circuits,* Jan. 1992.
33. Chung, B.-Y., Chien, C., Samueli, H., Jain, R. "Performance analy*sis of an all-digital BPSK direct-sequence spread-spectrum IF receiver architecture,"* IEEE Journal on Selected Areas in Communications, Sept. 1993, vol.11,(no.7):1096-107.
34. Wong, B.C., Samueli, H." A 200- MHz all-digital QAM modulator and demodulator in 1.2- mu m CMOS for digital radio applications," *IEEE Journal of Solid-State Circuits,* Dec. 1991.
35. Loinaz, M.J., Su, D.K., Wooley, B.A., " Experimental results and modeling techniques for switching noise in mixed-signal integrated circuits," *Digest of Technical Papers,* 1992 Symposium on VLSI Circuits. Seattle, WA, USA, 4-6 June 1992.
36. Yan, R.H., Lee, K.F., Jeon, D.Y., Kim, Y.O., and others, "High performance 0.1- micron room temperature Si MOSFETs," *Digest of Technical Papers*, 1992 Symposium on VLSI Technology. *Digest of Technical Papers,* Seattle, WA, USA, 2-4 June 1992.
37. Jian Chen, Parke, S., King, J., Assaderaghi, F., and others, "A high speed SOI technology with 12 ps/18 ps gate delay operating at 1.5V," *Proceedings of IEEE International Electron Devices Meeting,* San Francisco, CA, 13-16 Dec. 1992.
38. Nguyen, N.M., Meyer, R.G., "Si IC-compatible inductors and LC passive filters," *IEEE Journal of Solid-State Circuits,* Aug. 1990.
39. Nguyen, N.M., Meyer, R.G., "A 1.8- GHz monolithic LC voltage-controlled oscillator," *IEEE Journal of Solid-State Circuits,* March 1992.
40. M. McDonald, "A 2.5 GHz BiCMOS Image-Reject Front-End," *Digest of Technical Papers,* 1993 International Solid-State Circuits Conference, San Francisco, CA February, 1994.
41. Riley, T. et al, "Delta-Sigma Modulation in fractional-N Frequency Synthesis," *IEEE Journal of Solid-State Circuits,* May 1993.
42. Su, D.K., Loinaz, M.J., Masui, S., Wooley, B.A., "Experimental results and modeling techniques for substrate noise in mixed-signal integrated circuits," *IEEE Journal of Solid-State Circuits*, April 1993.
43. J. Craninckx, M. Steyart, " A CMOS Low-Phase-Noise Voltage-Controlled Oscillator with Prescaler," *Digest of Technical Papers,* 1995 International Solid-State Circuits Conference, San Francisco, CA, February 1995.
44. Karanicolas, A. N., "A 2.7V 900MHz CMOS LNA and Mixer," *Digest of Technical Papers,* 1996 International Solid-State Circuits Conference, San Francisco, CA, February 1996.
45. A. Rofougaran, J. Y-C Chang, M. Rofougaran, and A. A. Abidi, "A 1Gha CMOS RF Front-End IC for a Direct-Conversion Wireless Receiver," *IEEE Journal of Solid State Circuits,* in press.
46. Soyeur, et al, "A 3V 4Ghz nMOS Voltage-Controlled Oscillator with Integrated Resonator," *Digest of Technical Papers*, 1996 International Solid-State Circuits Conference, San Francisco, CA, February 1996.
47. Rofouragian, J. Rael, M. Rofougaran, and A. Abidi, "A 900Mhz L-C Oscillator with Quadrature Output," *Digest of Technical Papers,* 1996 International Solid-State Circuits Conference, San Francisco, CA, February 1996.
48. Cho, T., G. Chien, F. Brianti, and P. R. Gray, "A Power-Optimized CMOS Baseband Channel Filter and ADC for Cordless Applications," *Digest of Technical Papers,* 1996 Symposium on VLSI Circuits, Honolulu, June 1996.
49. Abidi, A, et al, "A CMOS Channel-select Filter for a Direct-Conversion Wireless Receiver," *Digest of Technical Papers,* 1996 Symposium on VLSI Circuits, Honolulu, June 1996.

2 GHz RF Circuits
in BiCMOS Process

J.M. Fournier and P. Senn
France Télécom - CNET Grenoble

Abstract

RF circuits using BiCMOS architectures for front-end mobile receivers operating at 2 GHz are presented. The advantages and limitations of CMOS and Bipolar technologies are discussed for the development of low noise amplifiers. A comparison between the architectures proposed for the implementation of silicon RF functions and those commonly used is also presented. The main advantages of the proposed new architectures are the significant improvement in integration level and the reduction in power consumption. The future prospects of a low cost BiCMOS one-chip solution are also discussed.

1. Introduction

The fast growing demand for mobile communications has recently led to numerous developments in chip technologies for the RF front-end of these systems. Cellular and cordless telephone systems have gained widespread user acceptance, with an improved quality service compared to previous analog networks, due to the advent of digital systems (CT2, DECT, GSM, PHS, etc.). Cellular radios allow the subscribers to place and receive telephone calls over the wireline network wherever cellular coverage is provided.

New digital cellular systems use many base stations with relatively small coverage radii (of the order of 1 to 10 km, even less in city centers). The system capacity can thus be drastically increased. At the same time, the RF transmission power between the mobile and the base / station can also be reduced. Although each frequency is used by multiple mobile-base station pairs, the ever increasing demand for new subscriptions requires new frequency allocations. As a consequence, the radio frequency bands occupied by cellular systems has continuously increased (Table 1). Existing analog networks are designed around an RF of a few hundred Mhz and the most advanced digital cellular networks use frequencies of around 2 GHz. In spite of this high level of demand, mobile sets still suffer from weaknesses such as: high cost due to low integration, especially for the RF front-ends, and also high power consumption.

However, a high level of integration using advanced submicron CMOS processes has been achieved for the baseband functions including all the channel and source encoders/decoders. Similarly, power supplies have been reduced to 3v, leading to low digital power dissipation: typically by 12mW/MIPS with 5v DSP to 4 mW/MIPS with 3v/0.5µm CMOS DSP (MIPS: Million of Instructions Per Second). In spite of the increasing complexity of the digital parts, the contribution of the RF front-end to the global power consumption, especially the receiver part, remains a critical point.

Many semiconductor technologies are competing today to propose advanced solutions for low cost, high level integration and low power consumption [1,2]. Over the past few years, GaAs has been the only available solution for front-end components operating between 1 and 2 GHz. Recent improvements in bipolar processing and dimensional reduction have raised the silicon operating frequency up to that previously dominated by GaAs. Moreover, in the 900 MHz band, silicon has virtually replaced the III-V technologies [3]. Submicronic CMOS processes have also proved to have high frequency capabilities but still with high power consumption [4].

To date, BiCMOS has always seemed to be a niche technology in spite of the great success in smart-power applications. Mobile applications actually present a good opportunity for such advanced BiCMOS processes. Although bipolar technologies are much faster than CMOS, the latter are much denser and allow practically zero static power consumption. By combining the two advantages together in BiCMOS, it is possible for designers to use bipolar transistors for functions requiring high frequency operations, and CMOS for logic gates and low (or medium) frequency analog signal processing. BiCMOS seems to be a suitable candidate for a one-chip low power consumption RF front-end[5].

Standards	Cellular/ Cordless	Frequency band (MHz)	Carriers spacing (kHz)	Number of carriers	Peak handset power (mW)
CT2	Cordless	864 - 868	100	40	10
DECT	Cordless	1880 - 1900	1728	10	250
PHS	Cordless	1895 - 1918	300	77	80
GSM	Cellular	880 - 960	200	124	2000
DCS1800	Cellular	1710 - 1880	200	374	2000
UMTS	Cellular	1885 - 2200			

Table 1 - Summary of some mobile system parameters

The majority of current transceivers utilize a single or dual intermediate frequency (IF) for the receiver part. A typical example of such a transceiver is shown in Figure 1:

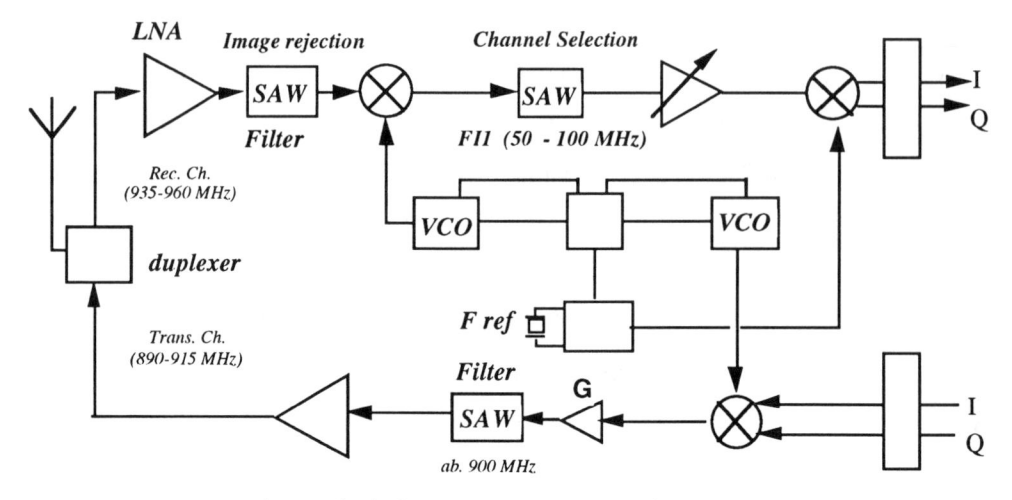

Fig. 1: Block diagram of a GSM transceiver.

This architecture is not easily integratable directly and many discrete elements are generally used in conventional equipment, especially the SAW filter, for which an outband rejection of more than 40 dB is desirable. Generally, the Low Noise and Transmitter Amplifiers are gathered with the duplexer on a GaAs chip. The other functions, such as the mixers, the IF amplifiers, PLL, etc. are usually implemented on bipolar silicon chips. The baseband functions, including the digital I/Q generation, utilize the CMOS process. Previous studies have reported the integration possibilities of RF front-ends at around 1 GHz [6,7]. Recent developments in submicronic BiCMOS processes [8] show a considerable interest in utilizing this technology for high-integration transceivers of up to 2 or 3 GHz.

BiCMOS technologies including 0.8 or 0.5 μm electrical channel length CMOS devices and fast (10 to 15 GHz) vertical NPN bipolar transistors, with a Si-poly emitter, are now being widely developed. These processes were first introduced in order to boost certain digital applications that required high clock and output rates (ATM applications, hard disk drive,...). Very recently, certain analog capabilities (poly-poly capacitors, resistor layers) have been added to expand the field of applications. Although power supply reduction (3.3 v for 0.5 μm) seems to becoming a clear limitation for speed enhancement in fast digital BiCMOS circuits (mainly due to the loss of 2 V_{be} for the gate output dynamic range), there is no direct constraint for analog applications (except for a high voltage level such as the IP3). Moreover, low voltage applications may involve thin epitaxial layers, resulting in low parasitic elements, such as collector resistors and capacitors.

2. BiCMOS process for RF critical parts

A. Low noise amplifier

For a "single chip solution", the LNA is the most critical part of the receiver channel regarding both the noise factor and the third order intercept point (IP3). The LNA must provide enough gain (typ. 15 - 20 dB) in the RF band of interest (typ. 2 GHz for second generation GSM systems) and it must also have a low noise figure (typ. less than 3dB) and sufficient overload IP3 performances (-15 to -10 dBm. for standard receiver specifications). Cascode structures using an integrated LC load have recently been proposed [9]. This structure (Fig.2) is particularly suitable for narrow-band applications.

- f_c^2 (carrier frequency) $= \dfrac{1}{2\pi\ LC}$

- Input impedance matching criteria :

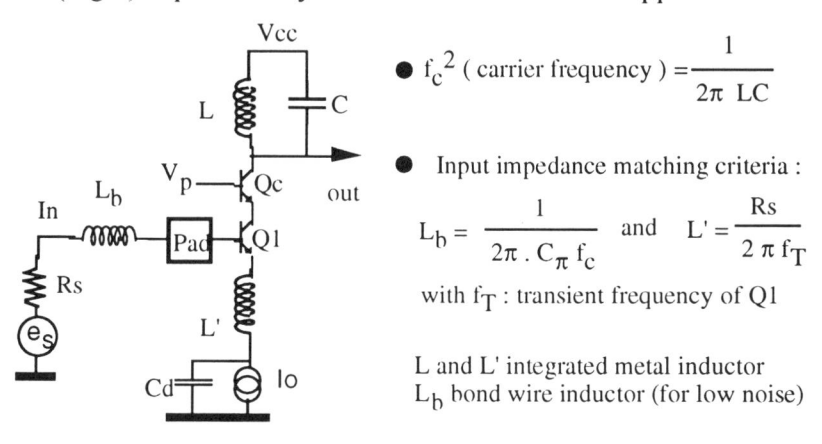

$$L_b = \dfrac{1}{2\pi \cdot C_\pi f_c} \quad \text{and} \quad L' = \dfrac{R_s}{2\pi f_T}$$

with f_T : transient frequency of Q1

L and L' integrated metal inductor
L_b bond wire inductor (for low noise)

Fig 2 Cascode LNA with on chip LC load

This is composed of a large area transistor Q1 designed for low noise and an integrated LC load. The cascode transistor Qc is used for input-output insulation. Integrated inductor L' and bond wire inductor Lb are used to match the input impedance to 50 Ω[6].

Submicron NMOS or vertical bipolar transistors are potentially good candidates for such amplifiers.

For comparison, NF,IP3 and gain simulations versus bias current Io have been performed for a NMOS and a bipolar cascode amplifier in a 0.5μm submicronic BiCMOS process [8]. The same LC load (6nH x 0,8 pF, for a 2 GHz frequency band) is used for both structures and the 50 Ω input impedance matching criteria (fig 2) are satisfied for each current value. The power supply is 3V. The sizes of the devices are respectively :

- M1 : 0.5 X 600 μm2 and MC : 0.5 X 150 μm2 in the NMOS structure
- Q1 : 30 μm2 and Qc : 15 μm2 in the bipolar structure

a. Noise Factor

NF simulation results versus bias current Io are shown in Fig 3.

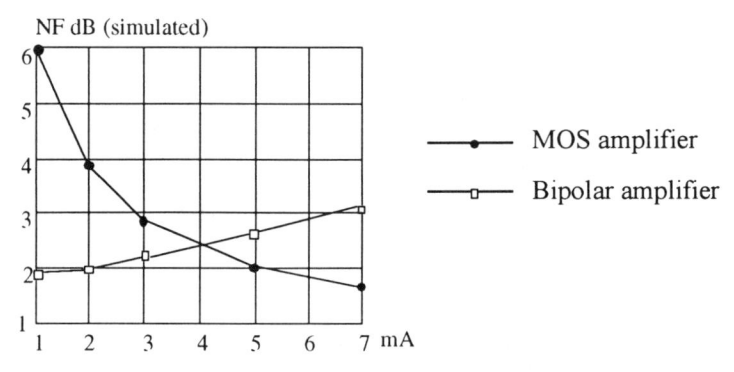

Fig.3: NF versus bias current for the NMOS and bipolar amplifiers
(50 Ω input impedance matching criteria has been satisfied for each current value)

In spite of the narrow-band proposed structure, classical noise models (more suitable for broad-band amplifiers) can however be used to show the effects of devices parameters on noise factor :

$$\text{Bipolar: } F = 1 + \frac{1}{Rs}\left[Rb + \frac{1}{2\,gm_{Bip}}\right] + Rs\,\omega_c^2 \cdot \frac{C_\pi^2}{gm_{Bip}}$$

$$\text{MOS : } F = 1 + \frac{1}{Rs}\frac{2}{3\,gm_N} + \frac{2}{3}\,Rs\,\omega_c^2 \cdot \frac{C_{gs}^2}{gm_N}$$

In the low current region, the bipolar solution is better than NMOS due to its higher gm. The effect of the base resistor is minimized by a careful layout optimization of the bipolar device (folded structure). With increasing current, the noise is lowered for the NMOS amplifier due to the increase in gm_N. On the other hand, in the bipolar solution, the noise current contribution, resulting from: $C_\pi^2/gm_{Bip} = gm_{Bip}/\omega_t^2$ (with $\omega_t/2\pi = F_t$: transient frequency) prevails and is responsible for the slight increase of NF due to saturation of F_t (Fig.4).-

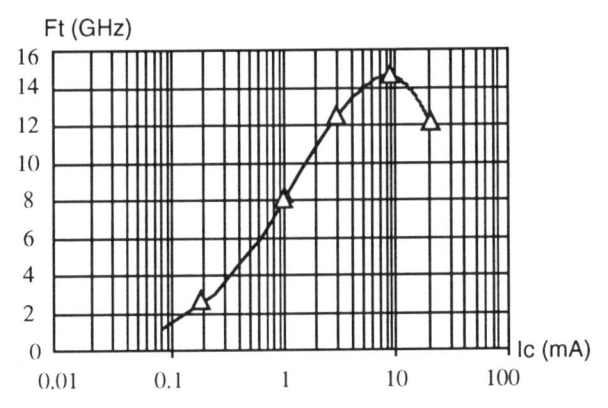

Fig.4: Ft versus collector current for the 30 µm2 optimized bipolar transistor (VCE = 1.5V)

The previous results show that the bipolar solution is less power consuming than NMOS concerning NF. A 2dB NF is obtained with only 2 mA in the bipolar amplifier, whereas 5 mA is necessary in the NMOS solution. Another important feature concerns the sensitivity of the active devices to the problem of cross-talk (which will be discussed later). In the NMOS structure, the sensitive channel region is directly coupled to cross-talk noise via the well region or substrate. In the bipolar structure, the sensitive base -emitter region is isolated by the N+ Buried Layer (BL) and sinker collector path. Furthermore, the larger active region used in the MOS amplifier (450 μm^2 as opposed to 40 μm^2 for bipolar) increases the noise aspect sensitivity. Differential structures for LNAs, currently being experimented [10] can lead to a reduction in the cross-talk effect by subtracting the common mode induced noise.

b. Gain and IP3

Fig.5 and Fig.6 show respectively simulation results concerning the gain and IP3 of the amplifier versus bias current Io.

Fig.5: Simulated Gain versus bias current

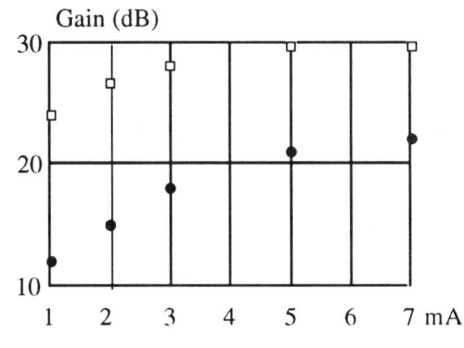

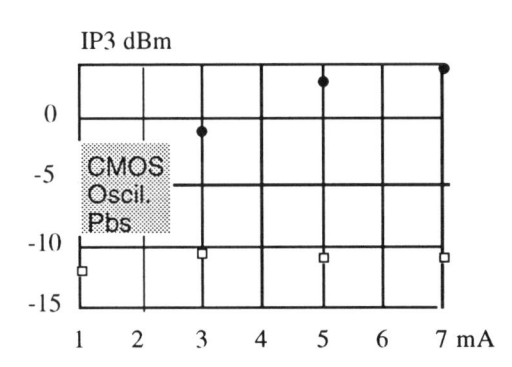

Fig.6: Simulated IP3 versus bias current

In the bipolar configuration, the gain, due to the high gm, is 10 dB higher than the corresponding NMOS amplifier. On the other hand, the IP3 is limited to about -10 dBm (32 mV zero-to-peak on 50 ohm input load) due to the input diode distortion [11]. In the NMOS solution, a high IP3 (10 to 15 dB higher than in bipolar case) is obtained according to the quadratic input characteristic law and the large input dynamic range of MOS devices. Nevertheless, in the low current region, the cascode NMOS device Mc does not provide sufficient input-output insulation (due to low gm) and oscillations may occur for large signals. This effect drastically decreases the IP3 for a low bias current (Fig. 6). In conclusion, taking into account the power consumption, previous simulation results show that the bipolar device is more suitable concerning the NF. On the other hand, MOS devices are better suited to low distortion in the middle power range. BiCMOS processes are potentially promising for new circuit structures with a good tradeoff between low NF and high IP3 for low power consumption applications.

B. On-chip inductors

As shown previously, the use of on-chip inductors for LNA is now being widely adopted. Furthermore, these devices are increasingly being used as LC integrated resonators in low phase noise voltage control oscillators for frequency synthesis. The resonance frequency of such on-silicon inductors, without specific process adaptations, is higher than 5 to 6 GHz, with a Q factor of the order of 5, allowing the use of such devices in the 2 GHz frequency range. CAD tools using standard models [12] [13] [14] and process characteristics have been developed to obtain the equivalent circuit and layout for an inductor. Experimental results show that an absolute accuracy < 5% for the L value has been reached [15]. The CMOS and BiCMOS submicronic process evolution is very promising in terms of performances of the on-chip inductors. Fig.7 shows the simulated Q factor (at 2 Ghz) and resonance frequency Fr versus inductor value L for a standard 0.7 µm BiCMOS process and an advanced 0.5 µm BiCMOS process. The width of the metal lines (third metal level) and the spacing between the lines were respectively 10 µm and 5 µm.

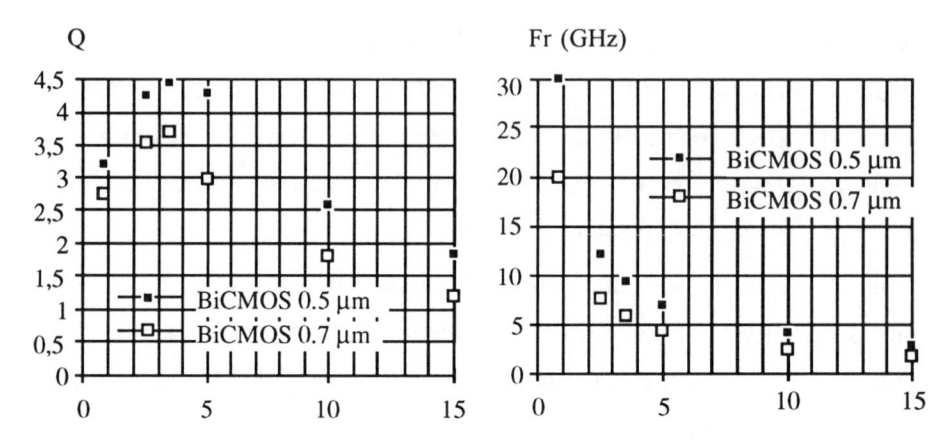

Fig.7: Q factor at 2Ghz and resonance frequency versus inductance value (nH)

The thicker dielectric involved in the 0.5 µm BiCMOS process (4 µm versus 2.5 µm in 0.7 µm BiCMOS) results in slightly enhanced performances concerning the Q factor and resonance frequency.

The Q factor is a very important parameter in the LC integrated resonator for low phase noise voltage control oscillators. In a spiral inductor it is approximated by the expression

$$Q = \frac{L\omega}{R_{dc} + R_{rl}}$$

where Rdc is the DC resistor of the metal path and Rrl represents the high frequency radiative loss in the Si substrate [16]. The Q factor is thus mainly dependent on three parameters:

(i) - the thickness of the metal layer (1 µm in industrial process) which leads directly to Rdc

(ii) - the SIO2 layer thickness (which also affects the resonance frequency).

(iii) - the silicon conductivity under the oxide. The higher the silicon conductivity, the lower the Q factor due to radiative loss in the silicon layers.

Resistor Rrl results from these last two parameters.The first point can be improved by interconnecting several metal levels [17] (to the detriment of the resonance frequency) or by using a very thick metallization (4 µm) and polyimide dielectric (10 µm) [18] not easily compatible with industrial processes. More promising is the use of a copper layer which is planned for future submicron processes.

The second point can be improved in proportion to the number of metallization levels which will grow with future generation processes.

Finally, the third point is the most difficult to solve and requires complex electromagnetic models for accurate loss modeling. In the BiCMOS process, a thin low conductivity nepi layer (for bipolar devices) deposited on a lightly doped P substrate (Fig. 8) is available. This layer arrangement is favorable to low radiative loss. On the other hand, the CMOS (Fig. 9) process using thick and more heavily doped wells deposited on a heavily doped P substrate (to avoid latch-up) seems less efficient in terms of the radiative loss. Experiments are currently running in a CMOS and a BiCMOS process for comparison.

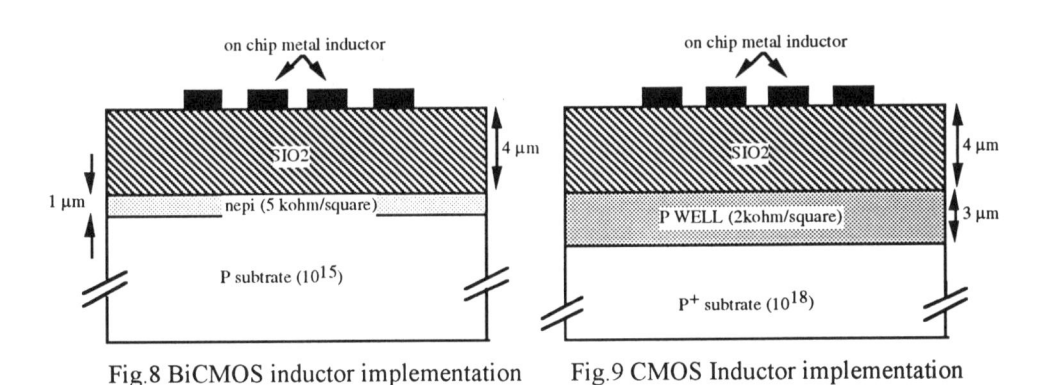

Fig.8 BiCMOS inductor implementation   Fig.9 CMOS Inductor implementation

C. Substrate cross-talk insulation

A potential limitation for a one-chip solution comes from the cross-talk problem in analog/digital mixed-mode applications. The low input level of RF signals (typ. - 90 dBm) may particularly be perturbed by the switching transient effects through the conductive silicon substrate, induced by the digital components. It is generally believed that silicon-on-insulator (SOI) substrates provide a high immunity for cross-talk problems. It has recently been shown that equivalent isolation performances for high frequency cross-talk have been obtained with full junction isolation wells, as compared with a more expensive SOI process [19]

Thus, CMOS and BiCMOS submicron processes involve a capacity of isolation in the high frequency (GHz) domain. The heavily doped P+ substrate involved in the CMOS process serves as an effective conductor path for noise throughout the I.C. This configuration leads to the use of guard rings that are not very effective at suppressing cross-talk [20] [21]. Only backside contacts are efficient at reducing noise by creating a low impedance node [21] at the expense of packaging costs. The BiCMOS process seems to be more effective at suppressing induced noise in three ways (Fig.10):

- (i) : the quiet region can be insulated from the noisy area by including a large resistive (Nepi on P- subtrate) region.

- (ii) : the P+ guard ring connected to the P+ buried layer can provide a better control of the local substrate potential around sensitive devices (Bipolar or NMOS used in LNAs for example) due to the high resistance of the bulk substrate.

- (iii) : a surrounding N+ deep sinker connected to an N-iso B.L. (available in the BiCMOS process) builds an efficient full junction isolation well for the noisy region including MOS devices

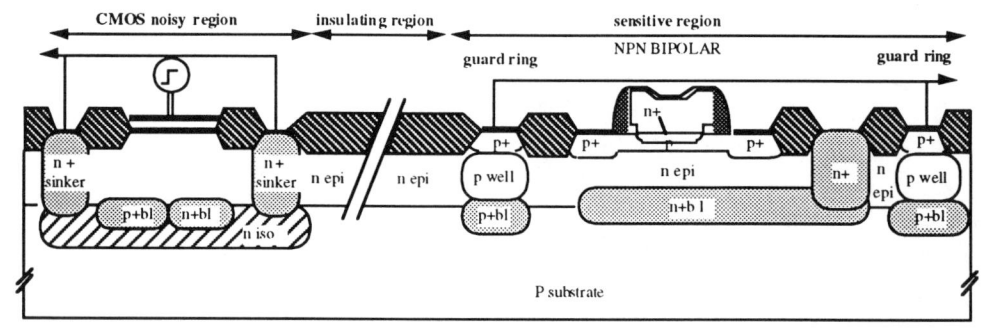

Fig.10 proposed solution for cross-talk H.F reduction in BiCMOS process

To conclude, the BiCMOS process seems a very promising solution to cross-talk problems in RF circuits assuming that a low inductive dedicated connection for guard-ring contacts is provided. Experiments are currently running to evaluate the previous assumptions.

3 - Circuit developments and results

This part is dedicated to experimental results concerning the integration of critical RF functions recently developed [22]. The improved architectures used for the circuits meet the requirements for new portable transceivers : increased integration level, reduced power supply (3V) leading to low power consumption and finally, adaptation to multi-standard. To this effect, two designs have been developed: an improved front-end image reject mixer which allows the image filter to be suppressed, and an LC VCO with internal LC resonators.

The circuits have been implemented in the BiCMOS4 process (from SGS-Thomson) with a typical NPN f_T of 10 GHz and 0.7 µm MOSFET channel length. All precise capacitors are fabricated using two poly layers. The integrated inductors are implemented in the upper (third) level interconnection metal. All decoupling capacitors are fabricated using poly gates and diffused wells.

A. Front-end image reject mixer

a. Classical image cancelling architecture and previous developments.

In a heterodyne receiver, the rejection of the image frequency can be achieved by an image reject mixer. Such a system allows the SAW image reject filter to be suppressed, assuming that the rejection is higher than 40 dB. This system (Fig.11) uses the complex representation of the transposed signals at the intermediate frequency (f_{IF}) to eliminate the image.

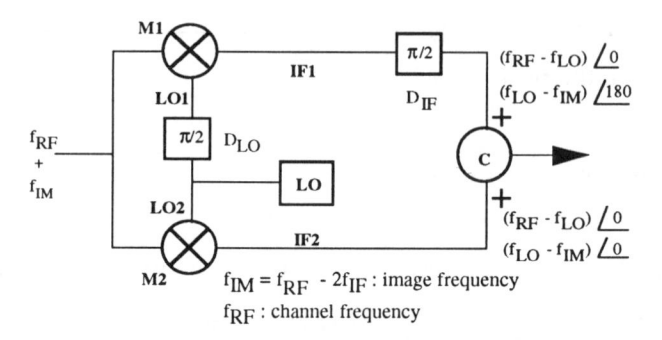

Fig.11: Classical image reject mixer

The 90° phase shifters (PS) D$_{LO}$ and D$_{IF}$ are critical parts of the system. The image rejection is greatly dependent on the phase shift accuracy and the gain matching in the two paths LO1 - IF1 and LO2 - IF2. In recent developments [23][24], the PS D$_{LO}$ is a combination of an RC integrator and a CR differentiator. This network arrangement introduces no phase error but an amplitude error due to the LO frequency variation and process fluctuations. In [24], the D$_{IF}$ PS is implemented as a differential RC/CR bridge in one of the IF paths and is responsible for a phase error depending on the IF adjustment and chip-to-chip process fluctuations. Consequently, the resulting image rejection does not exceed 20 dB.

b. Improved mixer architecture

In the proposed new architecture (Fig.12) two main improvements have been achieved which allow a stronger (45 dB) image rejection :
 - The D$_{IF}$ PS is distributed among the two IF paths as differentiator-integrator networks. As for the D$_{LO}$ PS, this network arrangement introduces a constant 90° phase difference between the two IF paths. Consequently the D$_{IF}$ and D$_{LO}$ PS achieve the appropriate phase shift between the two paths LO1 - IF1 and LO2 - IF2 independently of the LO frequency variation, IF adjustment and process fluctuations.

- Only the magnitudes of the signals between the two paths remain to be matched to obtain a good image rejection. This is performed through two steps as shown in Figure 12. First, an RC (CR) LO network is connected to a CR (RC) IF network in each LO-IF path so that the attenuation in each network is compensated. Second, a voltage controlled gain (VCG) is introduced into the combiner. A feedforward loop can also be used between the VCO and VCG to compensate the gain variation due to D_{LO} throughout the VCO frequency range. In the same way, the VCG can be trimmed to compensate the gain variation due to D_{IF} throughout the IF adjustment range.

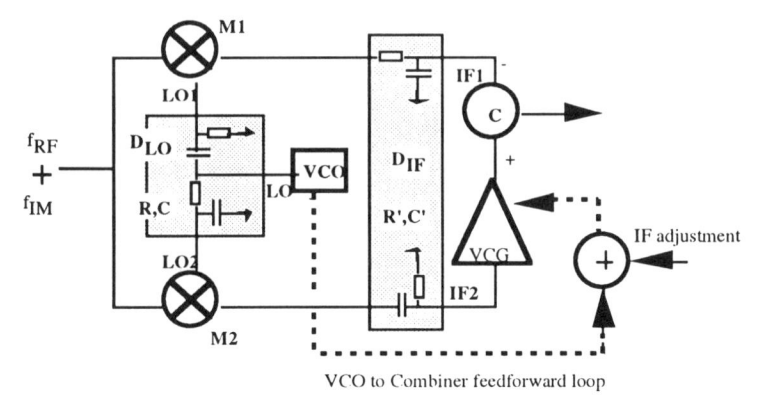

Fig.12: Improved new architecture of the mixer.

The improvement can be quantified by writing the amplitude ratio of the signals A_{IF1} and A_{IF2} at the inputs of the combiner :

$$A_{IF1}/A_{IF2} = RC \cdot f_{LO}/R'C' \cdot f_{IF} \quad (1)$$

from which the amplitude difference of the two signals is deduced :

$$(A_{IF1} - A_{IF2}) = A_{IF} \cdot [\Delta R/R - \Delta R'/R' + \Delta C/C - \Delta C'/C' + \Delta f_{LO}/f_{LO} - \Delta f_{IF}/f_{IF}] \quad (2)$$

The terms $\Delta R/R - \Delta R'/R' = \Delta C/C - \Delta C'/C'$ are cancelled due to the new distributed phase shifters, while $\Delta f_{LO}/f_{LO}$ and $\Delta f_{IF}/f_{IF}$ are cancelled due to the VCG.

c. Cascode low noise amplifier

To obtain a complete image rejection front-end, a low noise amplifier (Fig. 13) was implemented before the mixer. The LNA uses the cascode structure with than LC on-silicon load as described previously

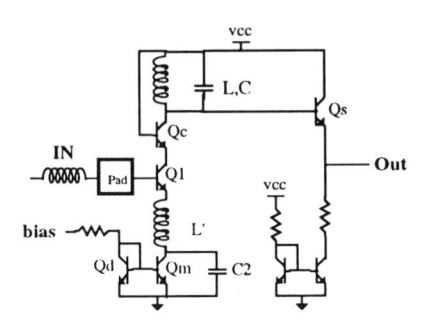

Fig 13: Representation of the low noise amplifier

The current mirror Qm-Qd and internal decoupling capacitance C2 form a low noise bias scheme for Q1. Transistor Qs acts as source follower and level shifter. The NF and IP3 are respectively 4.5 dB and -11 dBm for a 12 dB voltage gain and 5 mA current supply . Simulation results in the advanced BiCMOS 0.5 µm show a great improvement in the Noise factor (2.5 dB).

d. Experimental results

Fig.14 shows the image rejection and conversion gain performances of the front end circuit . A 45 dB image rejection in a 200 MHz bandwidth centered at 2GHz has been obtained for a 200 MHz IF. Furthermore, the gain control system allows a 35 dB minimum image rejection for an IF adjustment from 150 MHz to 250 Mhz. The performances of the mixer and overall front end circuit are summarized in Table 2.

Fig.14 Image rejection (after VCO trimming)

Table 2. Front-end measured results

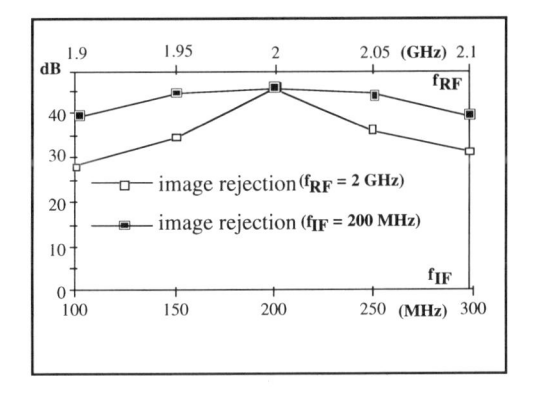

Parameter	Measurement
Supply voltage	3 V
Supply current (LNA + Mixer)	20 mA
S11 (LNA)	-10 dB
NF (LNA+mixer)	19 dB
Gain (LNA + mixer)	18 dB
Image Rejection (200 Mhz IF)	45 dB
Input IP3 (mixer)	-4.5 dBm
LO-RF insulation (LNA+mixer)	27 dB
LO-IF insulation (LNA+mixer)	34 dB

B. VCO-Prescaler

a. Previous developments

The critical parts of a synthesizer are the VCO, which must exhibit a low phase noise level, and prescaler divider, which must run at high frequency. LC sinusoidal oscillators achieve the best performances in terms of phase noise level but require external inductors or varactors at the expense of cost and power consumption. Fully integrated LC VCO's, including metal spiral on silicon inductors, have recently been proposed. The low phase noise level obtained is very promising but the power consumption is rather high [25]. On the other hand, the proposed circuit in [26] exhibits no differential output signal. This last requirement is needed to minimize the LO to RF leakage in a fully differential mixer. Furthermore the power supply must be adapted to a small size 3V battery and recent studies have shown that a 2 GHz operating frequency may be obtained, even with a standard digital submicron CMOS process [29].

b. Proposed VCO architecture.

A simple fully integrated LC VCO exhibiting a low phase noise level, differential output access and low current consumption from a 3V power supply is proposed. Fig. 15 shows the basic topology of the VCO -Prescaler.

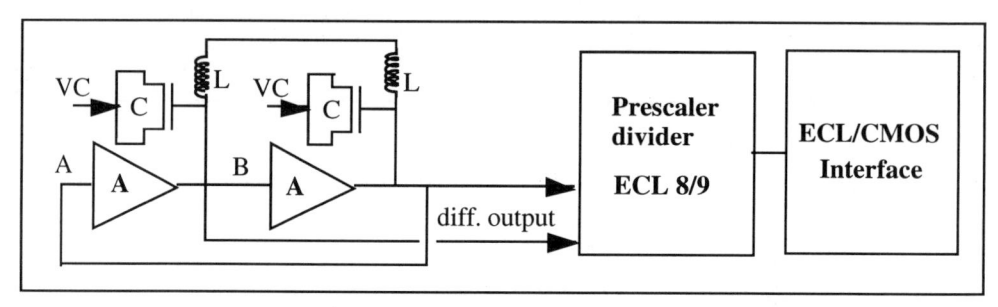

Fig.15 VCO - Prescaler topology

Two identical amplifier stages (using a simple common emitter configuration) with integrated LC loads are arranged in a positive loop. The oscillation occurs at a frequency corresponding to 0° in the phase transfer function. The differential output scheme is obtained by the necessary 180° phase shift between the two output nodes A and B. The loop gain for oscillation is adjusted by the common bias current in each stage. L are integrated metal inductors with a value of 5.6 nH and a Q factor of 4.5. The voltage controlled MOS capacitors C are used as varactors. This arrangement achieves an inherent static insulation between the drain-source access, connected to the external control voltage, and the internal gate sensitive nodes. Consequently, no additional capacitors are needed for this insulation. Furthermore, a 100 MHz frequency tuning range is obtained without capacitor switching. A minimum channel length has been used for MOS capacitors so as to minimize the channel resistor effect on the Q factor.

To obtain a complete circuit including the critical parts of a synthesizer, an ECL prescaler 8/9 and an ECL/CMOS interface were integrated with the VCO on the same chip.

c. Experimental results

Fig.16 displays the measured oscillation frequency of the VCO as a function of the controlling voltage. The output voltage vs. oscillation frequency is illustrated in Fig.17, which shows a maximum voltage variation of about +/- 10mV.

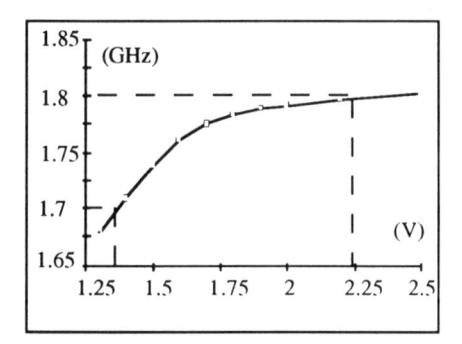

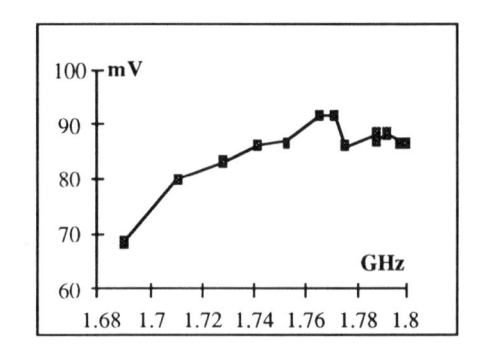

Fig.16: Frequency versus controlling voltage

Fig.17: Output voltage vs. oscillation frequency

The VCO phase noise obtained in a free running condition is -83 dBc/Hz at 100 KHz offset from the carrier (Fig.18).

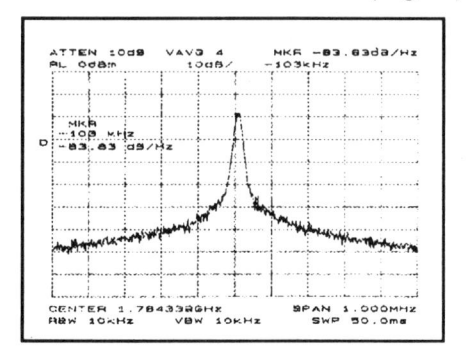

Fig.18 VCO output spectrum (free running)

Parameter	Measurement
Supply voltage	3 V
VCO supply current	4 mA
Prescaler supply current	15 mA
Fmin to Fmax (VCO)	1,7 to 1,8 GHz
Vout (VCO) (diff. output)	80 mV
VCO P.N. (100 KHz offset)	- 83 dBc/Hz
PLL P.N. (100 KHz offset)	- 99 dBc/Hz
PLL P.N (10 KHz offset)	- 97 dBc/Hz

Table 3 Summary of measured results

With an external PLL for synthesizer operation, the phase noise is reduced to -97 dBc/Hz at 10 KHz offset. The overall performances of the circuit are summarized in Table 3. The VCO power consumption is only 12 mW. Microphotographs of the front-end circuit and VCO-prescaler are shown in Fig.19. and Fig 20 respectively.

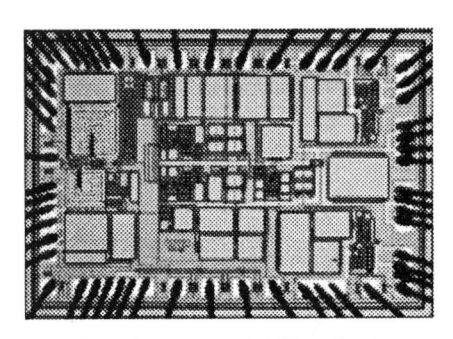

Fig. 19 Front-end 2Ghz circuit

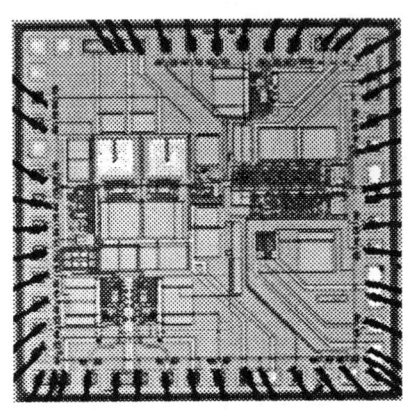

Fig. 20 VCO-prescaler circuit

C. IF Filtering

High Q bandpass SAW filters are widely used in RF conversion schemes. The main limitation lies in the fact that this technology is not compatible with silicon ICs and thus SAW filters remain external components. In addition, power dissipation is needed due to input/output adapters. An alternative solution is the use of a zero IF configuration in the receiver that eliminates the IF bandpass filtering function. This configuration has been investigated for many years and some industrial developments are now available [28].

An alternative solution could be the implementation of a fully integrated 200 MHz IF filter based on a method utilizing the frequency translation technique [27]. The system, shown schematically in Fig.21, consists of two identical channels, each composed of an analog multiplier followed by a low-pass filter whose output is fed into a second multiplier. The two identical channels are summed in order to cancel the unwanted sum-frequency term appearing at the output of the channels.

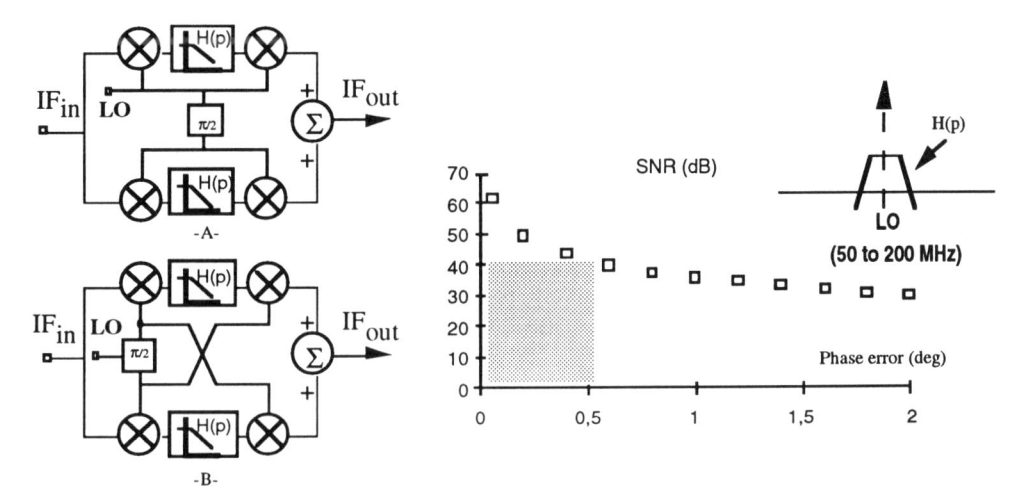

Fig. 21: Block diagram of the selective filter

This technique allows some interesting features such as: quasi-IF (the channel selection may occur at the IF level and consequently allows the use of a fixed VCO at the RF level) and multi-standard applications (different channel width and center frequencies). Bandpass filter adaptations are simply achieved by adjusting the VCO input of the filter and the bandwidth of the corresponding low-pass filters. The drawbacks of this filter architecture are well known, such as the sensitivity to the gain mismatch between the two channels and the phase error of the 90° phase shifter. Figure 21 presents the SNR (ratio of wanted to unwanted signals) versus phase error of the 90° phase shifter (Fig. 21-A). A maximun 0.5 degree error is required in order to guarantee a SNR more than 40 dB (a maximum gain mismatch between the two channels of 1% is also required for the same SNR). The modified architecture proposed in Fig. 21-B allows the gain mismatch and phase error constraints to be reduced by using distributed phase shifters similar to those described in the image reject mixer. Another critical point remains : the leakage between the LO and the channel of interest. This point can be minimized by using highly efficient bipolar mixers.

4. Conclusion

This paper has presented some capabilities of the BiCMOS process for a 2 GHz RF development. New architectural RF front-end solutions have been presented. BiCMOS seems to be a good technology candidate for 2 GHz applications and experimental results using a 0.8 µm BiCMOS process have been described. The three functions (amplifier-mixer, synthesizer and IF filter) have recently been integrated on the same chip for an RF to IF mono-chip development. The concepts used for the circuits are particularly suitable for applications in multi-standard radio terminals. Some recent developments in SiGe devices (F_t up to 40 GHz) indicate the potential of BiCMOS as a low cost single chip solution for RF telecommunications applications.

References

1. Asad A. Abadi, "Low-Power Radio-Frequency IC's for Portable Communications", Proceeding of the IEEE Vol 83, N°4, April 1995
2. P. R. Gray, R. G. Meyer, "Future Directions in Silicon ICs for RF Personal Communications", in proceedings of CICC, May 1995, USA
3. S. Weber, " RF ICs critical to growth in mobile arena", Electronic Engineering Times, October 1995.
4. D. Eggert, W. Barthel, W. Budde, H. J. Jentschel, R. Richter, F. Sawade, " CMOS - Microwave Wideband Amplifiers and Mixers on SIMOX-substrates ", in proceedings of ESSCIRC'95, Lille, France
5. P. Senn " Radio Frequency IC's in BiCMOS Process for Telecom Applications.",in Nomadic Microwave for Mobile Com. and Detect., Novembre 1995, Arcachon - France
6. R.G.Meyer, W.D. Mack, "A 1 GHz BiCMOS RF front-end IC", IEEE Journal of Solid-State Circuits, March 1994.
7. A. Rofougaran, J.Y-C. Chang, M. Rofougaran, S. Khorram, A.A. Abidi, "A 1 Ghz CMOS RF Front-end IC with wide Dynamic range", ESSCIRC'95 - Lille - France
8. A. Greiner, M. Laurens, A. Monroi ,"High Performance of a quasi-self aligned 0.5 µm BiCMOS technology ", IEEE International topical Meeting -November 95 - Arcachon - F
9. D. pache, J.M. Fournier, G. Billiot, P. Senn " An improved 3v 2 GHz BiCMOS Image Rejecter Mixer IC", in proceedings of CICC, May 1995 , USA
10. Paul R. Gray, T.cho, J.OU, T.Weigandt, C. Rudell, F. Brianti, S. Lo, S. MeHta, G. Chien, J. Weldon, " High-Integration CMOS RF Transceivers" AADC , April 96.

11 D.O Pederson and K. Mayaram, "Analog Integrated Circuits for Communication". Kluwer,
12 F. W. Grover "Inductance Calculations", Dover Publications, New York, 1962.
13 H. M. Greenhouse "Design of Planar Rectangular Microelectronic Inductors"IEEE Trans. on Parts, Hybrids, and Packaging, VOL PHP-10, NO 2, June 1974, pp 101-109.
14 D. M. Krafcsik and D. E. Dawson"A Closed-Form expression for representing the distributed nature of the Spiral Inductor", IEEE MTT Monolithic Circuits Symposium Digest 1986, pp.87-92.
15 D. Pache "Etude de nouvelles architectures pour l'intégration de fonctions RF en technologie BiCMOS" PhD Thesis - CNET -Grenoble , France (to be published)
16 R.B. Merril, T. W. Lee, Hong YOU, R, Rasmussen, L.A. Moberly "Optimization of High Q integrated Inductors for Multi-level Metal CMOS" IEDM 95 P 983-986.
17 Joachim N. Burghartz and all "High-Q Inductor in Standard Silicon Interconect Technology and its Application to an Integrated RF Power Amplifier", IEDM 95 ,p;1015
18 Bon-Kee Kim and all " Monolithic Planar RF Inductor And Waveguide Structures on Silicon With   Performance Comparable to Those in GaAs MMIC, IEDM 95 , p.717.
19 K. Joardar " Signal Isolation in BiCMOS Mixed Mode Integrated Circuits", in proceeding of BCTM95, October 1995 - USA
20 Timothy J. Schmerbeck"Minimizing Mixed-Signal Coupling and Interaction" ISCAS 1993
21 D.K. Su, M.J. Loinaz, S. Masui and B.A. Wooley "Experimantal Results and ModelingTechniques For Substrate Noise in Mixed Signal Integrated Circuits", IEEE Journal of Solid-state Circuits, pp 420, April   1993.
22 D. Pache, J.M. Fournier, G. Billot and P. Senn " An improved 3V 2GHz Image Rejecter Mixer and a VCO - Prescaler Fully Integrated in a BiCMOS Process" NomadicMicrowave for Mobile Communications and Detection. - Arcachon 16-17 November 1995 - France
23 M. Steyaert and R. Roovers, "A 1GHz single chip quadrature modulator", IEEE J. Solid State Circuits VOL. 27, N0 8, August 1992, pp 1194-1197.
24 M. D. McDonald, "A 2,5GHz BiCMOS image reject front end", ISSCC 93, PAPER TP 94, pp 144-145.
25 R. Duncan, K. Martin and A. Sedra "A 1 GHz Quadrature Sinusoidal Oscillator" IEEE 1995 CICC Conf. pp.91-94.
26 M. Nguyen and R.G. Meyer "A 1.8-GHz Monolithic LC Voltage-Controlled Oscillator"IEEE J. Solid State   Circuits VOL. 27, N0 3, March 1992, pp 444-450.
27 G. A. Rigby " Integrated Selective Amplifiers Using Frequency Translation", IEEE Journal of Solid-State   Circuits, vol. sc-1, N°1, September 1966
28 C. Berland, J. Dulongpont, P. Genest, E. Laurent « Radios in Mobile Communication Equipments », in Nomadic Microwave for Mobile Communications and Detection, 16-17 Novembre 1995, Arcachon - France
29 J. Craninckx and M. Steyaert, " A 1.75 GHz/3V Dual-Modulus Divide-by- 128/129 Prescaler in 0.7 µm   CMOS ", ESSCIRC'95, Lille - France

RF CMOS Design, Some Untold Pitfalls

Michiel Steyaert, Marc Borremans, Jan Craninckx,
Jan Crols, Johan Janssens and Peter Kinget

*Katholieke Universiteit Leuven, ESAT-MICAS,
Kardinaal Mercierlaan 94, B-3001 Heverlee, Belgium*

Abstract

Since several years research has been carried out on the design of RF circuits in CMOS technologies. Since then, the usability of CMOS for RF design has been demonstrated by several research groups. However, there are still some fundamental problems and limitations which may not be overlooked. The purpose of this work is to present some of those 'untold pitfalls' in the design of RF CMOS circuits for fully integrated transceivers for telecommunication applications.

1. Introduction

A few years ago the world of wireless communications and its applications started to grow rapidly. The main cause for this event was the introduction of digital coding and digital signal processing in wireless communications. This digital revolution is driven by the development of high performance, low cost, CMOS technologies which allow for the integration of an enormous amount of digital functions on a single die. This allows on its turn for the use of sophisticated modulation schemes, complex demodulation algorithms and high quality error detection and correction systems, resulting in high performance, lossless communication channels.

Today, the digital revolution and the high growth of the wireless market bring also many changes to the analog transceiver front-ends. The front-ends are the interface between the antenna and the digital modem of the wireless transceiver. They have to detect very weak signals (µV's) which come in at a very high frequency (1 to 2 GHz) and, at the same time, they have to transmit at the same high frequency high power levels (up to 2 Watt). This requires high performance analog circuits, like filters, amplifiers and mixers which translate the frequency bands between the antenna and the A/D-conversion and digital signal processing. Low cost and a low power consumption are the driving forces and they make the analog front-ends the bottle neck in future RF design. Both low cost and low power are closely linked to the trend towards full integration. An ever further level of integration renders significant space, cost and power reductions. Many different techniques to obtain a higher degree of integration for receivers, transmitters and synthesizers have been presented over the past years [1],[2],[3]. This paper introduces and analyses some advantages and disadvantages and their fundamental limitations.

Parallel to the trend to further integration, is there the trend to the integration of RF circuitry in CMOS technologies. The mainstream use for CMOS technologies is the integration of digital circuitry. The use of these CMOS technologies for high performance analog circuits yields however, if possible, many benefits. The technology is of course, if used without any special adaptions towards analog design, cheap. This is especially true if one wants to achieve the ultimate goal of fully integration : the complete transceiver system on a single chip, both the analog front-end and the digital demodulater implemented on the same die. This can only be achieved in either a CMOS or a BiCMOS process. BiCMOS has better devices for analog design, but its cost will be higher, not only due to the higher cost per area, but also due to the larger area that will be needed for the digital part. Plain CMOS has the extra advantage that the performance gap between devices in BiCMOS and nmos devices in deep submicron CMOS, and even nmos devices in the same BiCMOS process, is becoming smaller and smaller due to the much higher investments in the development of CMOS than bipolar. The f_T's of the nmos devices are getting close to the f_T's of the npn devices.

Although some research had been done in the past on the design of RF in CMOS technologies [4], it is only since a few years that real attention has been given to its possibilities [5],[6]. Today several research groups at universities and in industry are researching this topic [7],[2],[3],[9]. As bipolar devices are inherently better than CMOS devices, RF CMOS is by some seen as a possibility for only low performance systems, with reduced specification (like ISM) [10],[8], or that the CMOS processes need adaptions, like substrate etching under inductors [7]. Others feel however that the benefits of RF CMOS can be much bigger and that it will be possible to use plain deep submicron CMOS for the full integration of transceivers for high performance applications, like GSM, DECT and DCS 1800 [2],[3]. First, this paper analyses some trends, limitations and problems in technologies for high frequency design. Secondly, the downconverter topologies and implementation problems are addressed. Thirdly, the design and trends towards fully integrated low phase noise PLL circuits are discussed. Finally, the design of fully integrated upconverters is studied.

2. Technology

A. Active Devices

Due to the never ending progress in technology and the requirement to achieve a higher degree of integration for DSP circuits, submicron technologies are nowadays considered standard CMOS technologies. The trend is even towards deep submicron technologies, e.g. transistor lengths of 0.1 µm. Using the square law relationship for MOS transistors to calculate the f_t of a MOS device does not longer hold, due to the high electrical fields. Using a more accurate model which includes the mobility degradation due to the high electrical fields, results into

$$f_t = \frac{g_m}{2\pi C_{gs}}$$

$$= \frac{\mu}{2\pi 2/3 L^2} \frac{(V_{GS}-V_T)}{(1 + 2(\theta + \frac{\mu}{v_{max}L})(V_{GS}-V_T))} \quad (1)$$

Hence by going to deep submicron technologies, the square law benefit in L for speed improvement drastically reduces due to the second term in the denominator of the equation above. Even for very deep submicron technologies the small signal parameter g_m has no square law relationship anymore:

$$gm = \frac{\mu C_{ox} W (V_{GS}-V_T)}{L(1 + 2(\theta + \frac{\mu}{v_{max}L})(V_{GS}-V_T))} \qquad (2)$$

with transistor lengths smaller than approximately

$$L < \frac{\mu}{v_{max}} \frac{1}{\frac{1}{2(V_{GS}-V_T)} - \theta} \approx 0.12 \mu m \qquad (3)$$

with $\mu/v_{max} = 0.3$, Vgs-Vt =0.2 (boundary of strong inversion) and $\theta = 0.06$, the transistor has only the weak inversion and the velocity saturation area. This will result in even higher biasing currents in order to achieve the required g_m and will result in higher distortion and intermodulation components, which will be further discussed in the trade-off of low-noise amplifier designs.

Furthermore, the speed increase of deep submicron technologies is reduced by the parasitic capacitance of the transistor, meaning the gate-drain overlap capacitances and drain-bulk junction capacitances. This can clearly be seen in fig. 1 in the comparison for different technologies of the f_t and the f_{max} defined as the 3dB point of a diode connected transistor [11]. The f_{max} is more important because it reflects the speed limitation of a transistor in a practical configuration. As can be seen, the f_t rapidly increases, but for real circuit designs (f_{max}) the speed improvement is only moderate.

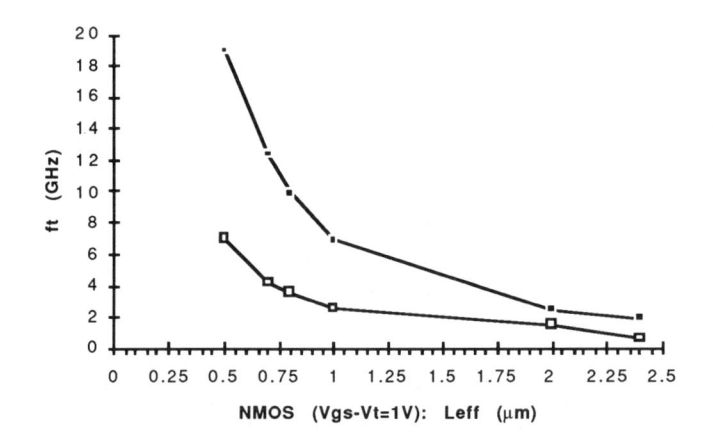

Fig. 1 : Comparison of ft and fmax

B. Passive Devices

The usability of a process for RF design does not only depend on the quality of the active devices, but also, more and more, on the availability of good passive devices. The three passive devices (resistors, capacitors and inductors) will be discussed.

Low-ohmic resistors are today available in all CMOS technologies and their parasitic capacitance is such that they allow for more than high enough bandwidth (i.e. more than 2 to 3 GHz). A more important passive device is the capacitor. In RF circuits capacitors can be used for AC-coupling. This allows DC-level shifting between different stages, resulting in a more optimal design of each stage and in the ability to use lower power supply voltages. The quality of a capacitor is mainly determined by the ratio between the capacitance value and the value of the parasitic capacitance to the substrate. A too high parasitic capacitor loads the transistor stages, thus reducing their bandwidth, and it causes an inherent signal loss due to a capacitive division. Capacitors with ratio's lower than 8 are as a result difficult to use in RF circuit design as coupling devices.

The third passive device, the inductor, is gaining more and more interest in RF circuit design on silicon. The use of inductors allows for a further reduction of the power supply voltage and for a compensation of parasitic capacitances by means of resonance, resulting in higher operating frequencies. The problem is that the conducting silicon substrate under a spiral inductor reduces the quality of the inductor. Losses occur due to capacitive coupling to the substrates and eddy currents induced in the substrate will also result in losses and in a reduction of the effective inductance value. This problem can be circumvented by using extra processing steps which etch away the substrate under the spiral inductor [23], having the large disadvantage that it eliminates all the benefits of using a standard CMOS process. It is therefore important that in CMOS spiral inductors are used without any process changes and that there losses are accurately modelled. In [12] it is shown that spiral inductors can be accurately modelled and that they can be used in CMOS RF circuit design. As an example section 4 discusses all the different possibilities for the use of inductors in the design of VCO's. It shows that high performance VCO's can be integrated with spiral inductors, even on lossy substrates. without requiring any external component.

3. The Receiver

A. Receiver Topologies

The heterodyne or IF receiver is the best known and most frequently used receiver topology. In the IF receiver the wanted signal is downconverted to a relatively high intermediate frequency. A high quality passive bandpass filter is used to prevent a mirror signal to be folded upon the wanted signal on the IF frequency. Very high performances can be achieved with the IF receiver topology, especially when several IF stages are used (e.g. 900 MHz to 300 MHz, 300 MHz to 70 MHz, 70 MHz to 30 MHz, 30 MHz to 10 MHz). The main problem of the IF receiver is the poor degree of integration that can be achieved as every stage requires going off-chip and requires the use of a discrete bandpass filter. This is both costly (the cost of the discrete filters and the high pin-count for the receiver chip) and power consuming (often the discrete filters have to be driven by a 50 Ω signal source).

The homodyne or zero-IF receiver has been introduced as an alternative for the IF receiver that can achieve a much higher degree of integration. The zero-IF receiver uses a direct, quadrature, downconversion of the wanted signal to the baseband. The wanted signal has itself as mirror signal and sufficient mirror signal suppression can therefore be achieved, even with a limited quadrature accuracy (e.g. 3° phase accuracy and 1 dB amplitude accuracy). Theoretically, there is thus no discrete high frequency bandpass filter required in the zero-IF receiver, allowing in this way the realization of a fully integrated receiver. A limited performance of the LNA and the mixers makes however that, although not for mirror signal suppression, a high frequency bandpass filter is still required. The reason why low performance LNA's and mixers require bandpass filtering and how this can be prevented, is explained further on.

In the zero-IF receiver downconversion can be performed in a single stage (e.g. direct from 900 MHz to the baseband), giving large benefits towards full integration, low cost and low power consumption [13]. The problem of the zero-IF receiver is its poor performance compared to IF-receivers. The zero-IF receiver is intrinsically very sensitive to parasitic baseband signals like DC-offset voltages and crosstalk products caused by RF and LO selfmixing. It are these drawbacks that have kept the zero-IF receiver from being used on large scale in new wireless applications. The use of the zero-IF receiver has therefore been limited to either low performance applications like pagers and ISM [10] or as only a second stage in a combined IF - zero-IF receiver topology [14],[15]. It has however been shown that with the use of dynamic non-linear DC-correction algorithms, implemented in the DSP, the zero-IF topology can be used for high performance applications like GSM and DECT [1],[16].

In recent years new receiver topologies, like the quasi-IF receiver [3] and the low-IF receiver [2] have been introduced for the use in high performance applications. The quasi-IF receiver uses a quadrature downconversion to an IF frequency, followed by a further quadrature downconversion to the baseband. The channel selection is done with the second local oscillator on the IF frequency, giving the advantage that a fixed frequency first local oscillator can be used. The disadvantages of the quasi-IF receiver are that, with a limited accuracy of the first quadrature downconverter (e.g. a phase error of 3°), the mirror signal suppression is not good enough and an HF filter which improves the mirror signal suppression is still necessary. A second disadvantage is that a high IF is required in order to obtain a high enough ratio between the IF frequency and the full band of the application. Otherwise will the tunability of the second VCO have to be to large. A high IF requires a higher power consumption. The first stage of mixers can not be true downconversion mixers in the sense that they still have to have a relatively high output bandwidth and multistage topologies inherently require more power.

The low-IF receiver performs a downconversion from the antenna frequency directly down to, as the name already indicates, a low IF (i.e. in the range a few 100 kHz) [2]. Downconversion is done in quadrature and the mirror signal suppression is performed at low frequency, after downconversion, in the DSP. The low-IF is thus closely related with the zero-IF receiver. It can be fully integrated (it does not require a HF mirror signal suppression filter) and uses a single stage direct-downconversion. The difference is that the low-IF does not use baseband operation, resulting in a total immunity to parasitic baseband signals, resolving in this way the main disadvantage of the zero-IF receiver. The drawback is that the mirror signal is different from the wanted signal in the low-IF receiver topology, but by carefully choosing the IF frequency an adjacent channel with low signal levels can be selected for which the typical mirror signal suppression (i.e. a 3° phase accuracy) is sufficient.

B. Full Integration

With newly developed receiver topologies as the zero-IF receiver and the low-IF receiver the need disappears for the use of external filters which suppress the mirror signal (see previous section). This does not mean however that there would not be any HF filtering required anymore. Filtering before the LNA is, although not for mirror signal suppression, still necessary to suppress the blocking signals. Between the LNA and the mixer filtering can be necessary in order to suppress 2nd and 3rd harmonic distortion products which are introduced by the LNA. Due to either the use of a switching downconverter or the non linearities of the mixer and the local oscillator these distortion products will be downconverted to the same frequency as the wanted signal. The latter problem can be eliminated by using either a very good blocking filter before the LNA (resulting in small signals after the LNA) or by using a highly linear LNA. The use of linear downconverters, i.e. based on the multiplication with a sinusoidal local oscillator signal, reduces of course the problem as well.

Very high, out of band, signals can be present. In order to prevent saturation of the LNA, these signals must be suppressed with a HF filter which only passes the signals in the band of the application. For e.g. the GSM system is the range between the largest possible out-of-band signal and the lowest detectable signal is 107 dB. Without blocking filter, the LNA and the mixer must be able to handle this dynamic range. For the LNA this means that the input should be able to handel an input signal of 0 dBm (i.e. the -1 dB compression point P_{-1dB} should be about 0 dBm), while having a noise figure of 6 dB. Consequently, this means that the IP3 value should be +10 dBm ($IP3 \approx P_{-1dB} + 10.66 dB$). The IMFDR3 of an LNA or mixer for a given channel bandwidth is given by :

$$IMFDR3 = \frac{2}{3} \cdot [IP3 + 174dB - 10\log(BW) - NF] \qquad (4)$$

The required IMFDR3 for an LNA is thus (for a 200 kHz bandwidth) 80 dB. CMOS downconverters can, by using MOS-transistors in the linear region, be made very linear [6],[2],[17], much more linear than the bipolar cross coupled multipliers. IP3 values of +45 dBm and noise figures of 18 dB have been demonstrated for CMOS realizations [2],[6]. This results is an IMFDR3 for a 200 kHz bandwidth of more than 95 dB. The consequence is that the IMFDR3 spec of 80 dB (i.e. without blocking filter) is achievable for the mixer. In this manner CMOS opens the way to the development of a true fully integrated single-chip receiver for wireless systems which does not require a single external component, not even a blocking filter. In order to achieve this goal highly linear mixers which operate by multiplication with a single sine must be used. However, the mixer noise performance is intrinsically more worse than the noise of an amplifier and the use of an LNA will still be necessary. In order to be able to cope with the blocking levels, the LNA will have to be highly linear and its gain will have to be reduced from a typical value of e.g. 18 dB to e.g. 12 dB. The mixers noise figure will than have to be lowered with about 6 dB too. This will require a higher power consumption from the downconversion mixer, but the benefit would be that the receiver can then be fully integrated.

C. The Downconverter

The most often used topology for a multiplier is the multiplier with cross coupled variable transconductance differential stages. The use of this topology or related topologies (e.g. based on square law) in CMOS is limited for high frequency applications. Two techniques are used in CMOS : the use of the MOS transistor as a switch and the use of the MOS transistor in the linear region.

The technique often used in CMOS downconversion for its ease of implementation is subsampling on a switched capacitor amplifier [5],[18],[19]. The MOS transistor is here used as a switch with a high input bandwidth. The wanted signal is commutated via these switches. Subsampling is used in order to be able to implement these structures with a low frequency opamp. The switches and the switched capacitor circuit run at a much lower frequency (comparable to an IF frequency or even lower). The clock jitter must however be low so that the high frequency signals can be sampled with a high enough accuracy. The disadvantage of subsampling is that all signals and noise on multiples of the sampling frequency are folded upon the wanted signal. The use of a high quality HF filter in combination with the switched-capacitor subsampling topology is therefore absolutely necessary.

Fig. 2 shows the block diagram of a fully integrated quadrature downconverter realized in a 0.7 µm CMOS process [2]. The proposed downconverter does not require any external components, nor does it require any tuning or trimming. It uses a newly developed double quadrature structure, which renders a very high performance in quadrature accuracy (less than 0.3° in a very large passband). The topology used for the downconverter is based on the use of nMOS transistors in the linear region [6],[2]. By using capacitors on the virtual ground a low frequency opamp can be used for downconversion. The MOS transistors in the linear region result is very high linearity (input IP3 is 27 dBm) for both the RF and the LO input. The advantages of such a high linearity on both inputs are, as explained in the previous section, that the mixer can handle a very high IMFDR3, resulting in no need for any kind of HF filtering. This opens the way to the implementation of a fully integrated receiver.

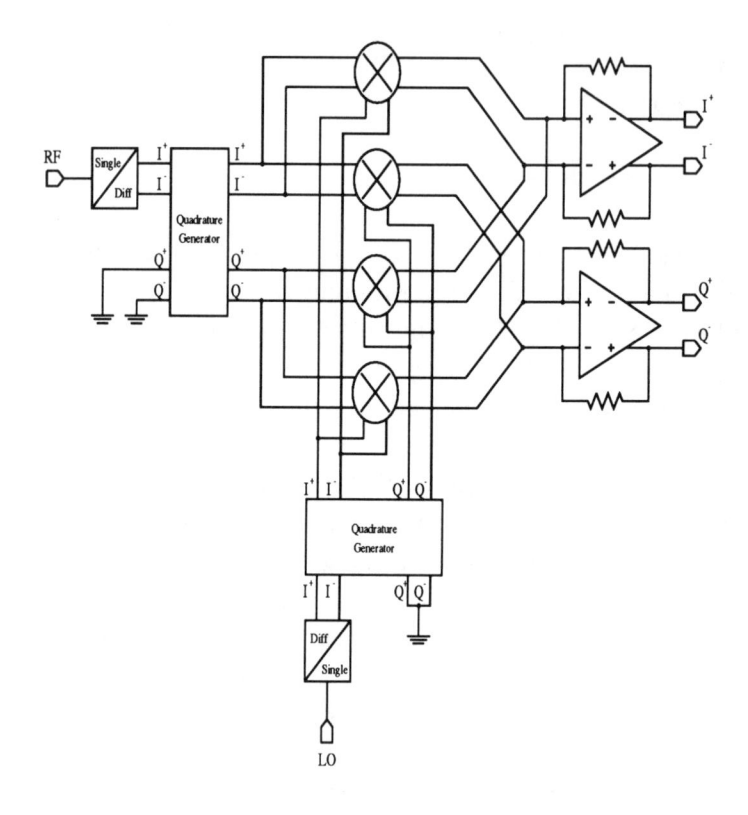

Fig. 2 : A double quadrature downconversion mixer

D. The LNA

The noise figure (NF) of a LNA embedded in a 50Ω system is generally dominated by the noise of the first device in the amplifier. It is defined as :

$$NF = \frac{4kT50 + equivalent\ input\ noise\ power}{4kT50} \quad (5)$$

Figure 3 and 4 compare some common input structures regarding noise. As can be seen from the NF equations and the plotted noise figure as function of the gm of the transistor for the different topologies, the non-terminated common source input stage and the (terminated) transimpedance stage are superior as far as noise is concerned. For those circuits the NF can be approximated as :

$$(NF-1) = \frac{1}{50 \cdot gm} = \frac{(Vgs-Vt)I}{50 \cdot 2 \cdot I} \quad (6)$$

indicating that a low noise figure needs a high transconductance in the first stage. In order to generate this transconductance with high power efficiency, a low $Vgs-Vt$ is preferred. However, this will result in a large gate capacitance. Together with the 50 Ohm source resistance in a 50 Ohm antenna system, the achievable bandwidth is limited by :

$$f_{3dB} \cong \frac{1}{2\pi \cdot 50\Omega \cdot Cgs} \quad (7)$$

Together with equation (6) this results in (f_T is the cutoff frequency of the input transistor (equation 1)) :

$$(NF-1) = \frac{f_{3dB}}{f_T} \quad (8)$$

Due to overlap capacitances and miller effect, this relationship becomes approximately (f_{max} is the 3dB point of a transistor in diode configuration [11]) :

$$(NF-1) = \frac{f_{3dB}}{f_{max}} \quad (9)$$

This means that a low noise figure can only be achieved by making a large ratio between the frequency performance of the technology (f_{max}) and the working frequency (f_{3dB}). Because for a given technology f_{max} is proportional to $Vgs-Vt$, this requires a large $Vgs-Vt$ and associated with it a large power drain. Only by going to real deep submicron technologies, the f_{max} will be high enough to achieve GHz working frequencies with low $Vgs-Vt$ values. Only then the power drain can be reduced to an acceptable value.

Nevertheless, there are some drawbacks in using short channel devices for low noise. The large electric field at the drain of a submicron transistor may produce hot carriers, having a noise temperature significantly above the lattice temperature [20]. This indicates that a good LDD (Lightly Doped Drain) is as crucial for low noise as it is for device reliability.

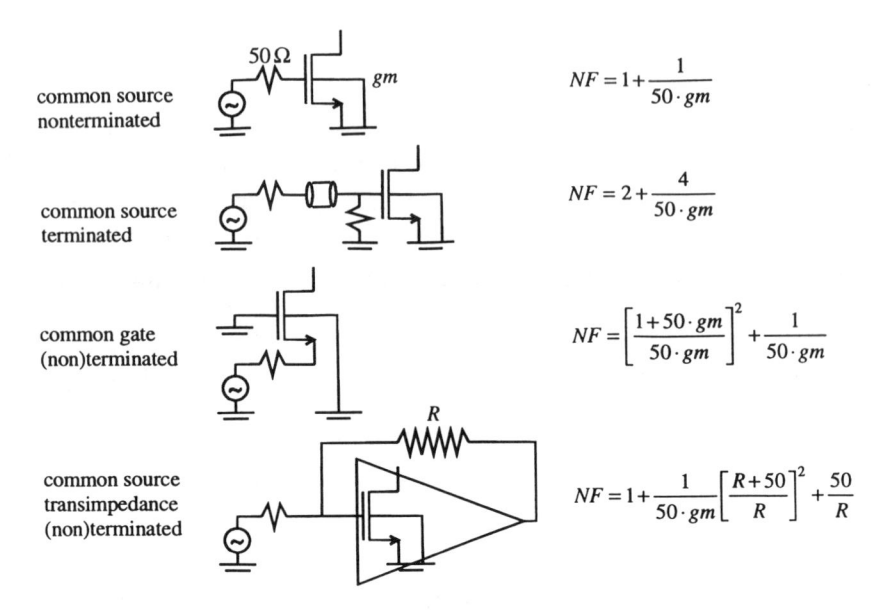

common source nonterminated		$NF = 1 + \dfrac{1}{50 \cdot gm}$
common source terminated		$NF = 2 + \dfrac{4}{50 \cdot gm}$
common gate (non)terminated		$NF = \left[\dfrac{1 + 50 \cdot gm}{50 \cdot gm}\right]^2 + \dfrac{1}{50 \cdot gm}$
common source transimpedance (non)terminated		$NF = 1 + \dfrac{1}{50 \cdot gm}\left[\dfrac{R+50}{R}\right]^2 + \dfrac{50}{R}$

Fig. 3 : Noise and amplification efficiency of some common input structures.

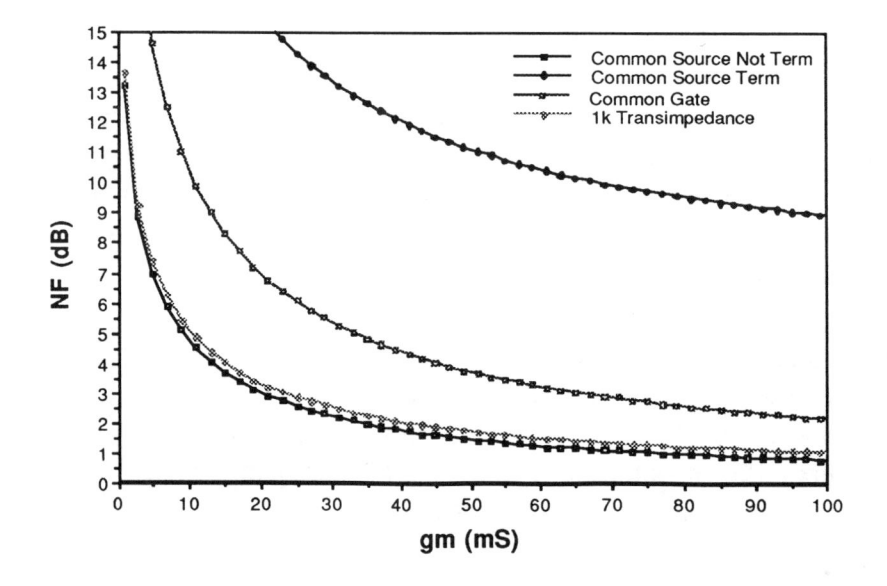

Fig.4 : Performance comparison.

A far more serious problem in using deep submicron technologies is the introduction of higher order intermodulation products. Long channel transistors are described by a quadratic model. Consequently, a one transistor common source amplifier ideally suffers only from second order distortion and produces no third order intermodulation products. As a result high IP3 values should easily be achieved. In fact :

$$IM2 = \frac{1}{2}\frac{v}{Vgs-Vt} \quad \text{and} \quad IM3 = 0 \qquad (10)$$

where v denotes the input amplitude of the amplifier. For short channels, the quadratic model is no longer valid : mobility degradation due to the longitudinal electric field causes the current to eventually become linear in $Vgs-Vt$. To evaluate the effect on intermodulation products, a more accurate transistor model is used:

$$I_{ds} = \frac{\mu C_{ox}}{2}\frac{W}{L}\frac{(Vgs-Vt)}{1+\Theta\cdot(Vgs-Vt)} \quad \text{with} \quad \Theta = \theta + \frac{\mu}{L_{eff}\cdot V_{max}} \qquad (11)$$

where θ is the mobility degradation caused by surface scattering and the $\mu/(L_{eff}\cdot V_{max})$ term is the degradation caused by the longitudinal electric field. The θ term can often be neglected with respect to the $\mu/(L_{eff}\cdot V_{max})$ term : in a typical 0.7 micron CMOS process θ is 0.06 whereas the other term is 0.43. As a result of the short channel effects, the intermodulation relations become

$$IM2 = \frac{v}{Vgs-Vt}\cdot\frac{1}{(1+r)(2+r)} \quad \text{and}$$

$$IM3 = \frac{3}{4}\frac{v}{Vgs-Vt}\cdot\frac{v}{Vsv}\cdot\frac{1}{(1+r)^2(2+r)} \qquad (12)$$

Vsv is the transit voltage between strong inversion and velocity saturation, calculated as

$$Vsv = \frac{1}{\Theta} \qquad (13)$$

and r is the relative amount of velocity saturation, given by

$$r = \frac{Vgs-Vt}{Vsv} \qquad (14)$$

The transit voltage Vsv depends only on technology parameters and is about 2 volt for a 0.7 µm CMOS technology. For deep submicron processes, e.g. a 0.1 micron technology, this voltage becomes even smaller than 300 mV, which is very close to the $Vgs-Vt$ at the boundary of strong inversion.

From the intermodulation equations above, one can see that for sufficiently large values of Vsv, causing r going to zero, the distortion formulas reduce to (10). However, as a consequence of the low values of Vsv in deep submicron FETs, third order intermodulation products start appearing whereas the second order intermodulation becomes slightly smaller than expected due to the linearization effect. Because *Vsv* is inversely proportional to the effective channel length, the generated amount of third order distortion increases roughly with 6 dB every time the transistor length is halved. This may pose a problem when very small transistor lengths are required to reduce the power drain and a high linearity is necessary to avoid the blocking filters.

4. The Synthesizer

A. *The Oscillator*

The local oscillator is responsable for the correct frequency selection in the up- and downconverters. Since the frequency spectrum in modern wireless communication systems must be used as efficiently as possible, channels are placed very close together. The signal level of the desired recieve channel can be very small, whereas adjacent channels can have very large power levels. Therefore the phase noise specifications for the LO signal are very high, which makes the design of this frequency synthesizer very critical.

Meanwhile, mobile communication means low power consumption, low cost and low weight. This implies that a completely integrated synthesizer is desirable, where integrated means a standard CMOS technology without any external components or processing steps. Usually, the LO is realized as a phase-locked loop as shown in figure 5. The very hard specs are reflected in the design of the two high-frequency building blocks present, i.e. the Voltage-Controlled Oscillator (VCO) and the Dual-Modulus Prescaler (DMP).

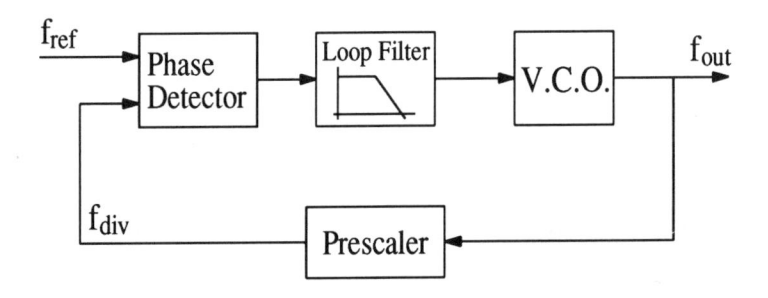

Fig. 5 : PLL-based frequency synthesizer

For the realization of a gigahertz VCO in a submicron CMOS technology, two options exist : ring oscillators or oscillators based on the resonance frequency of an LC-tank. The inductor in this LC-tank can be implemented as an active inductor or a passive one. It has been shown that for ring oscillators [21] as well as active LC-oscillators [22], the phase noise is inversely related to the power consumption.

$$\text{Ring osc. [21]} : L\{\Delta\omega\} \sim kT.R.\left(\frac{\omega}{\Delta\omega}\right)^2 \text{ with } g_m = \frac{1}{R}$$

$$\text{Active-LC [22]} : L\{\Delta\omega\} \sim \frac{kT}{2\omega C}\cdot\left(\frac{\omega}{\Delta\omega}\right)^2 \text{ with } g_m = 2\omega C \quad (15)$$

Therefore, the only viable solution to a low-power, low-phase-noise VCO is an LC-oscillator with a passive inductor. In this case the phase noise changes proportionally with the power consumption :

$$\text{Passive-LC [22]} : L\{\Delta\omega\} \sim kT.R\cdot\left(\frac{\omega}{\Delta\omega}\right)^2 \text{ with } g_m = R\cdot(\omega C)^2 \quad (16)$$

As could be expected, the only limitation in this oscillator is the integrated passive inductor. Equation (16) shows that for low phase noise, the resistance R (i.e. the equivalent series resistance in the LC-loop) must be as small as possible. A low resistance also means low losses in the circuit and thus low power needed to compensate for these losses. Capacitors are readily available in most technologies. But since the resistance R will be dominated by the contribution of the inductors' series resistance, the inductor design is critical. Three solutions exist.

Spiral inductors on a silicon substrate usually suffer from high losses in this substrate, which limit the obtainable Q-factor. Recently, techniques have been developed to etch this substrate away underneath the spiral coil in a post-processing step [7],[23]. The cavity created by such an etching step can clearly be seen in figure 6. However, since there is an extra etching step required after normal processing of the IC's, this technique is not allowed for mass production.

Fig. 6 : Etched spiral inductor

For extremely low phase noise requirements, the concept of bondwire inductors has been investigated. Since a bondwire has a parasitic inductance of approximately 1 nH/mm and a very low series resistance, very-high-Q inductors can be created. Bondwires are always available in IC technology, and can therefore be regarded as being standard CMOS components. Two inductors, formed by four bondwires, can be combined in a enhanced LC-tank [22] to allow a noise/power tradeoff. A microphotograph of the VCO is shown in fig. 7 [25]. The measured phase noise is as low as -115dBc/Hz at an offset frequency of 200kHz from the 1.8-GHz carrier. The power consumption is only 8mA at 3V supply. Although chip-to-chip bonds are used in mass commercial products [28], they are not characterized on yield performance for mass production. Therefore, the industry is reluctant to this solution.

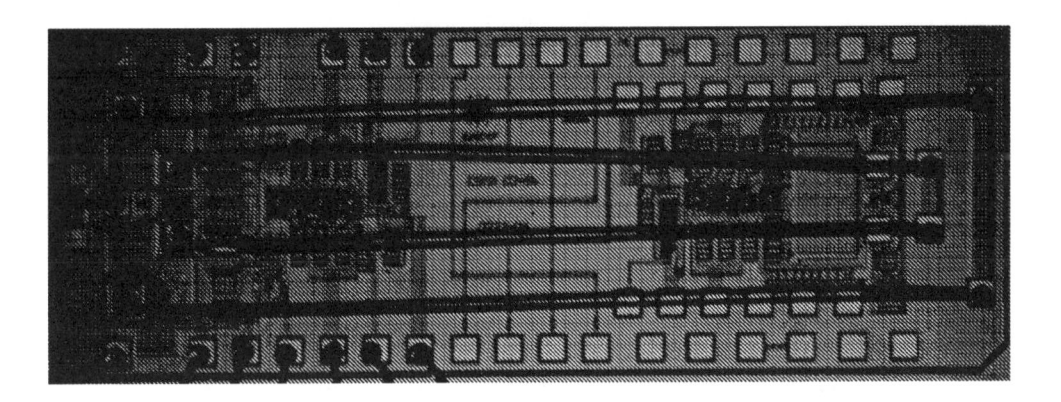

Fig. 7 : Microphotograph of the bondwire LC-oscillator

The most elegant solution is the use of a spiral coil on a standard silicon substrate, without any modifications. Bipolar implementations do not suffer from substrate losses, because they usually have a high-ohmic substrate [24]. Most submicron CMOS technologies use a highly doped substrate, and have therefore large induced currents in the substrate, which is responsible for the high losses. The effects present in these low-ohmic substrates can be investigated with finite-element simulations. This analysis can lead to an optimized coil design, which has been used in a spiral-inductor LC-oscillator. Only two metal layer are available, and the substrate is highly doped. With a power consumption as low as 6mW, a phase noise of -116 dBc/Hz at 600kHz offset from the 1.8-GHz carrier has been obtained [29].

B. The Prescaler

To design a high-speed dual-modulus prescaler, a new architecture has been developed that is based on the 90-degrees phase relationship between the master and the slave outputs of a M/S toggle-flipflop [26]. This architecture is shown in fig. 8. That way a dual-modulus prescaler have been developed that is as fast as an asynchronous fixed divider. A 1.75-GHz input frequency has been obtained at a power consumption 24mW and 3V powert supply. At 5V power supply input frequencies above 2.5 Ghz can even be processed.

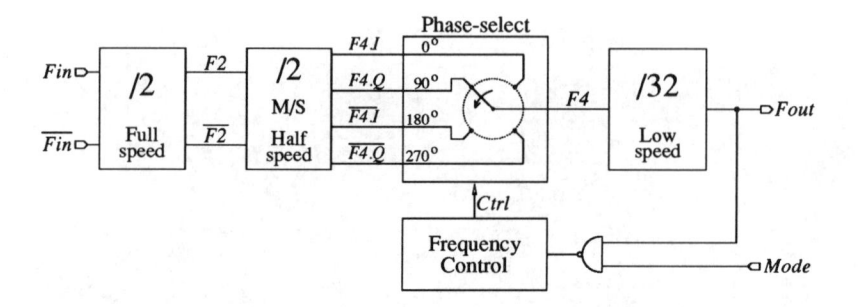

Fig. 8 : New dual-modulus prescaler architecture

The fully integrated VCO and dual modulus prescaler make it possible to integrate a complete LO synthesizer in a standard CMOS technology, without tuning, trimming or post-processing, that achieves modern telecom specs.

5. The Transmitter

Until now in CMOS, only downconversion mixer circuits for receivers have been reported in the open literature. For communication systems like GSM, a two way communication is required and a transmitter circuit must be implemented to achieve a full transceiver system. This implies that still a lot of research for the development of CMOS transmitter circuits has to be done.

A. Downconversion vs. upconversion

The modulation of the baseband signals on the local oscillator carrier frequency require an upconversion mixer topology. In classical bipolar transceiver implementations, the up- and downconverter mixer use typically the same four-quadrant topology. There are, however, some fundamental differences between up- and downconverters, which can be exploited to derive optimal dedicated mixer topologies (see table 1).

	Downconversion	Upconversion
Input 1	High	High
Input 2	High	Low
Output	Low	High

Table 1 : Comparison of the signal frequencies in up- and downconversion

In a downconversion topology the two input signals are at a high frequency (e.g. 900 MHz for GSM systems) and the output signal is a low frequency signal of maximum a few MHz for low-IF or zero-IF receiver systems. This extra degree of freedom has been used in the design of very successful downconverter-only CMOS mixer topology [6].

For upconversion mixers, the situation is totally different. The high frequent local oscillator (L.O.) and the low frequent baseband (B.B.) input signal are multiplied to form a high frequent output signal. All further signal processing has to be performed at high frequencies, which is very difficult and power consuming when using current sub-micron CMOS technologies. Furthermore, all unwanted signals like the intermodulation products and L.O.-leakage, have to be limited to a level below -30 dB of the signal level.

B. CMOS Mixer Topologies

<u>Switching Modulators</u>

Many published CMOS mixer topologies are based on the traditional variable transconductance multiplier with cross coupled differential modulator stages. Since the operation of the classical bipolar cross coupled differential modulator stages is based on the translinear behaviour of the bipolar transistor, the MOS counterpart can only be effectively used in the modulator or switching mode. Large L.O.-signals have to be used to drive the gates and this results in a huge L.O.-feedthrough. Already in CMOS downconverters this is a problem; in [9] e.g. the output signal level is -23 dBm with a L.O. feedthrough signal of -30dBm, which represents a suppression of only -7 dB. This gives rise to very severe problems in direct upconversion topologies. Moreover, by using a square wave modulating LO signal, 30 % of the signal power is present at the third order harmonic. This unwanted signal can only be filtered with an extra external output blocking filter.

In CMOS the variable transconductance stage is typically implemented using a differential pair biased in saturation region. To avoid distortion problems large V_{GS}-V_T values or a large source degeneration resistance have to be used, which results in large power drain and noise problems, especially compared to the bipolar converter circuits. This can be avoided by replacing the bottom differential pair by a pseudo differential topology with MOS transistors in the linear region [17].

Linear MOS mixers.

These problems can be overcome in CMOS by linearly modulating the current of a MOS mixing transistor biased in its linear region (fig.9). For a gate voltage of V_1+v_{in1}, a drain voltage of $V_2+v_{in2}/2$ and a source voltage of $V_2-v_{in2}/2$ the current through the transistor is given by :

$$i_{DS} = \beta \cdot (v_{in1} \cdot v_{in2}) + \beta \cdot (V_1 - V_T - V_1) \cdot v_{in2} \qquad (17)$$

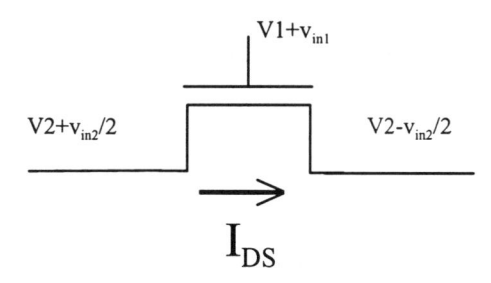

Fig. 9 : Mos mixing transistor biased in the triode region

When the LO signal is connected to the gate and the baseband signal to v_{in2}, the current contains frequency components around the LO due to the first term and components of the baseband signal due to the second term in equation (17). In order to implement this configuration, a high frequency current buffer circuit with a source follower voltage input must be realized which is shown in figure 10 [27]. Transistor M1 is the modulating transistor and transistors M2 and M3 act as source followers and copy the differential baseband signal across the modulating transistor. The nMOS only unity-feedback current buffers M2-M4-M6 and M3-M4-M7 copy the signal current to the output load. The gm of M2 and M3 are boosted by the operation of the feedback loop so that they provide a low impedance input and act as good source followers with low distortion.

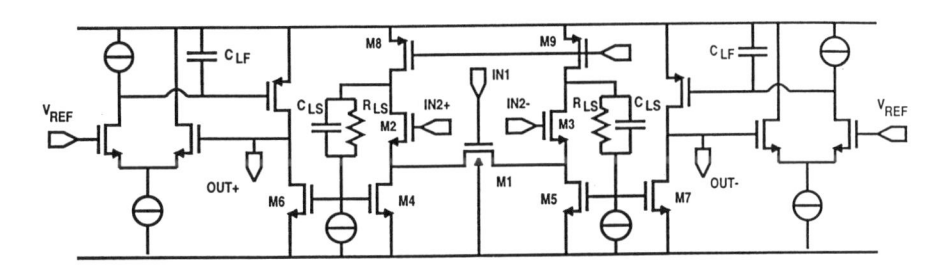

Fig. 10 :nMOS only mixer upconverions topology

At the output an active coil circuit is added to suppress the low-frequency signal components and avoid further intermodulation products in the subsequent stages. Thanks to the differential topology, the L.O. feedthrough signals are common mode signals and are suppressed. This rejection is limited by the matching in the two paths; for a typical mismatch of 10 % it is theoretically possible to achieve more than 39 dB carrier suppression at 900 MHz.

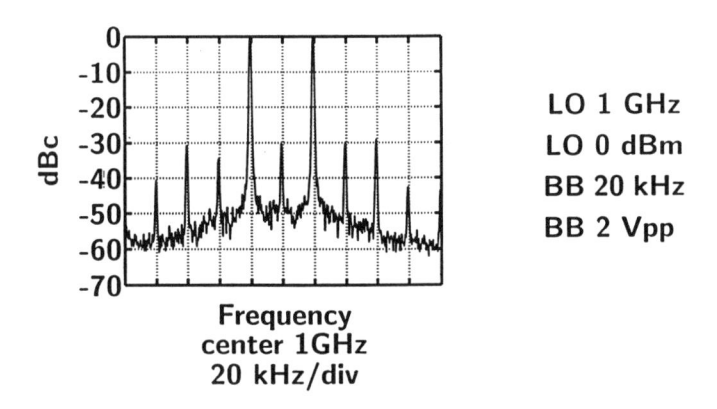

LO 1 GHz
LO 0 dBm
BB 20 kHz
BB 2 Vpp

Fig. 10 : Measured double side band output spectrum for a sine modulation signal.

Figure 11 represents a measured double side band modulation diagram of this mixer circuit. All unwanted signals are below -30 dBc. A conversion gain of -10 dB is achieved for an on-chip 500 Ω load and a 0 dBm LO signal. However, classical RF building block interconnects use a characteristic impedance of 50 Ω. This implies that the CMOS transceiver function requires an extra power preamplifier to drive the input impedance of the external high efficiency power amplifier block. The implementation of this pre-amplifier block is a pitfall in the design of fully integrated CMOS transceivers for present sub-micron technologies. The mixer of figure 10 e.g. is implemented in a 0.7 μm CMOS technology which achieves an fmax of only 6 GHz for a gate overdrive of 1 V or a gm/I ratio of 2. Typical bipolar technologies used for the implementation of 900 MHz fully integrated transceivers have cut-off frequencies of over 20 GHz [14]. Due to the low gm/I of present sub-micron technologies for high frequency operation, the power consumption in CMOS pre-amplifiers will be up to 20 times higher than in bipolar. However, thanks to the rapid down-scaling of CMOS technologies, the present CMOS building block realizations show that full CMOS transmitters with an acceptable power consumption will be feasible in very deep submicron CMOS.

6. Conclusions

The trends towards deep submicron technologies have resulted in the exploration by several research groups of the possible use of CMOS technologies for the design of RF circuits. Especially the developement of new receiver topologies, such as quasi-IF and low-IF topologies, in combination with highly linear downconverters, has opened the way to fully integrated downconverters with no external filters or components. However, due to the moderate speed performance of the present submicron technologies, lower noise circuits in combination with less power drain have to be worked out. The trends towards deep submicron technologies will allow to achieve those goals as long as the short channel effects will not limit the performance concerning linearity and intermodulation problems.

Concerning synthesizers, the last few years high performance low phase noise, low power drain, fully integrated VCO circuits have been demonstrated. Starting with difficult post processing techniques research has resulted in the use of standard CMOS technologies by using bondwires as inductors. Today, even low phase noise peformances with optimizd integrated spiral inductors in standard CMOS technologies without any post-processing, tuning, trimming or external components have been announced. This opens the way towards fully integrated receiver circuits.

However, telecommunication systems are usually two-way systems, requiring transmitter circuits as well. It is only recently that in open literature CMOS upconverters are announced, with moderate output power. Again, thanks to the trends towards deep submicron technologies fully integrated CMOS transmitter circuits with an acceptable power consumption will be feasible. This opens the way towards fully integrated transceiver circuits in standard CMOS technologies.

References

[1] J. Sevenhans, A. Vanwelsenaers, J. Wenin and J. Baro, "An integrated Si bipolar transceiver for a zero IF 900 MHz GSM digital mobile radio front-end of a hand portable phone," *Proc. CICC*, pp.7.7.1-7.7.4, May 1991.

[2] J. Crols and M. Steyaert, "A Single-Chip 900 MHz CMOS Receiver Front-End with a High Performance Low-IF Topology," *IEEE J. of Solid-State Circuits*, vol.30, no.12, pp.1483-1492, Dec. 1995.

[3] P.R. Gray and R.G. Meyer, "Future Directions in Silicon ICs for RF Personal Communications," *Proc. CICC*, May 1995.

[4] Bang-Sup Song, "CMOS RF Circuits for Data Communications Applications," *IEEE J. of Solid-State Circuits*, vol. SC-21, no.2, pp.310-317, April 1986.

[5] P.Y. Chan, A. Rofougaran, K.A. Ahmed and A.A. Abidi, "A Highly Linear 1-GHz CMOS Downconversion Mixer," *Proc. ESSCIRC*, pp.210-213, Sevilla, Sept. 1993.

[6] J. Crols and M. Steyaert, "A 1.5 GHz Highly Linear CMOS Downconversion Mixer," *IEEE J. of Solid-State Circuits*, vol. 30, no.7, pp.736-742, July 1995.

[7] J. Y.-C. Chang, A. A. Abidi and M. Gaitan, "Large Suspended Inductors on Silicon and Their Use in a 2-um CMOS RF Amplifier", IEEE Electron Device Letters, vol. 14, no. 5, pp. 246-248, May 1993

[8] C.H. Hull, R.R. Chu and J.L. Tham, "A Direct-Conversion Receiver for 900 MHz (ISM Band) Spread-Spectrum Digital Cordless Telephone," *Proc. ISSCC*, pp.344-345, San Francisco, Feb. 1996.

[9] A.N. Karanicolas, "A 2.7 V 900 MHz CMOS LNA and Mixer," *Proc. ISSCC*, pp.50-51, San Francisco, Feb. 1996.

[10] A.A. Abidi, "Radio Frequency Integrated Circuits for Portable Communications," *Proc. CICC*, pp.151-158, San Diego, May 1994.

[11] M. Steyaert and W. Sansen, "Opamp Design towards Maximum Gain-Bandwidth," *Proc. of the AACD workshop*, pp.63-85, Delft, March 1993.

[12] J. Crols, P. Kinget, J. Craninckx and M. Steyaert, "An Analytical Model of Planar Inductors on Lowly Doped Silicon Substrates for High Frequency Analog Design up to 3 GHz," *to be published in Proc. VLSI Circuits Symposium*, June 1996.

[13] D. Rabaey and J. Sevenhans, "The challenges for analog circuit design in Mobile Radio VLSI Chips," *Proc. of the AACD workshop*, vol. 2, pp.225-236, Leuven, March 1993.

[14] T. Stetzler, I. Post, J. Havens and M. Koyama, "A 2.7V to 4.5V Single-Chip GSM Transceiver RF Integrated Circuit," *Proc. ISSCC*, pp.150-151, San Francisco, Feb. 1995.

[15] C. Marshall et al., "A 2.7V GSM Transceiver ICs with On-Chip Filtering," *Proc. ISSCC*, pp.148-149, San Francisco, Feb. 1995.

[16] J. Sevenhans et al., "An Analog Radio front-end Chip Set for a 1.9 GHz Mobile Radio Telephone Application," *Proc. ISSCC*, pp.44-45, San Francisco, Feb. 1994.

[17] A. Rofougaran et al., "A 1GHz CMOS RF Front-End IC with Wide Dynamic Range," *Proc. ESSCIRC*, pp.250-253, Lille, Sept. 1995.

[18] D.H. Shen, C.-M. Hwang, B. Lusignan and B.A. Wooley, "A 900 MHz Integrated Discrete-Time Filtering RF Front-End," *Proc. ISSCC*, pp.54-55, San Francisco, Feb. 1996.

[19] S. Sheng et al., "A Low-Power CMOS Chipset for Spread Spectrum Communications," *Proc. ISSCC*, pp.346-347, San Francisco, Feb. 1996.

[20] A. A. Abidi, "High-frequency noise measurements on FETs with small dimensions," *IEEE Trans. Electron Devices*, vol. 33, no. 11, pp. 1801-1805, Nov. 1986.

[21] B. Razavi, "Analysis, Modeling, and Simulation of Phase Noise in Monolithic Voltage-Controlled Oscillators", *Proc. CICC*, pp. 323-326, May 1995.

[22] J. Craninckx and M. Steyaert, "Low-Noise Voltage Controlled Oscillators Using Enhanced LC-tanks", *IEEE Trans. on Circuits and Systems - II : Analog and Digital Signal Processing*, vol. 42, no. 12, pp. 794-804, Dec. 1995.

[23] A. Rofourgan, J. Rael, M. Rofourgan, A. Abidi, "A 900-MHz CMOS LC-Oscillator with Quadrature Outputs", *Proc. ISSCC*, pp. 392-393, Febr. 1996.

[24] N. M. Nguyen and R. G. Meyer, "A 1.8-GHz Monoithic LC Voltage- Controlled Oscillator", *IEEE Journal of Solid-State Circuits*, vol. 27, no. 3, pp. 444-450, March 1992.

[25] J. Craninckx and M. Steyaert, "A 1.8-GHz Low-Phase-Noise Voltage-Controlled Oscillator with Prescaler", *IEEE Journal of Solid-State Circuits*, vol. 30, no. 12, pp. 1474-1482, Dec. 1995.

[26] J. Craninckx and M. Steyaert, "A 1.75-GHz/3-V Dual Modulus Divide-by-128/129 Prescaler in 0.7-um CMOS", *Proc. ESSCIRC*, pp. 254-257, Sept. 1995.

[27] P. Kinget and M. Steyaert, "A 1 GHz CMOS Upconversion Mixer", *to be published in Proc. CICC*, session 10.4, May 1996.

[28] -, "AD 7886, a 12-Bit, 750 kHz, Sampling ADC," *Analog Devices data sheet*, Apr. 1991.

[29] J. Craninckx and M. Steyaert, "A 1.8-GHz Low-Phase-Noise Spiral-LC CMOS VCO," *to be published in Proc. VLSI Circuits Symposium,* June 1996.

SILICON INTEGRATION FOR
DIGITAL CELLULAR COMMUNICATION

Jan Sevenhans, Jacques Wenin, Damien Macq
Alcatel Belgium
Jacques Dulongpont
Alcatel Paris

ABSTRACT

Mobile cellular terminals built for new cellular communication systems rely heavily on advanced semiconductor parts. This paper gives an overview of the various implementations currently in use in commercial GSM products. It discusses the recently announced evolutions and offers conclusions on longer term perspectives both at ASIC and Terminal product standpoints.

1. INTRODUCTION

In Europe, digital cellular communication is a reality. A reality that extends to many non–European nations. At the time of writing, there is a total of 70 signatories of the GSM MoU (Memorandum of Understanding) from 66 countries, and 14 organizations hold the observer status. The sum total of the population covered approaches 3 billion people.

This digital reality is even more of a commercial success, with tremendous market growth and drastic competition between all the major telecommunications companies: network installation and supply of end user equipments being the most attractive playgrounds.

In particular, competition on tiny and light handportable terminals started at the very beginning of GSM deployment and is gaining momentum every month with the release of new smaller and lighter equipments. Today, 160 cm^3 handportables are in commercial service.

To achieve such a performance (that 160 cm^3 box contains roughly 5 to 7 million transistors!), it is unanimously acknowledged that ASICs were the enabling technology. This is the subject of the first part of the paper.

But the story is ongoing, and ASIC (r)evolution, together with battery and MMIC (Microwave Monolithic Ic) technologies are preparing the future discussed in the second and last part of this paper.

2. IMPLEMENTING GSM TERMINALS: THE ASIC ENABLER

2.1. The GSM system

GSM represents a pan–European standard for a digital cellular mobile radio system. Two frequency bands of 25 MHz width have been reserved for the 900 Mhz GSM system. The mobile radio equipment transmits at a frequency between 890 and 915 MHz and receives at a frequency exactly 45 MHz higher, i.e. between 935 and 960 MHz. Access to one of the 200 kHz wide radio channels occurs through a combination of frequency and time division multiplexing (TDMA/FDMA).

2.2. Mobile terminal design

A GSM terminal consists of the following functional units :
- RF transmit and receive sections
- Baseband A/D and D/A converters
- Digital baseband processing system
- Device controller
- User interfaces (e.g. microphone, speaker, keypad, display).

Fig.1 shows the functional arrangement of a GSM mobile set belonging to Alcatel's first generation encapsulating each single basic GSM function into a VLSI component (leaving all standard functions to standard parts) [1, 11]. The following paragraphs review succinctly those basic GSM functions. The next sections give a detailed comparison of the various implementations to be found in commercial products.

<u>From User Voice To The Air</u>

The audio signal from the handset microphone is converted into a digital bitstream in the analog front end codec [12]. The speech encoding algorithm compresses the data stream to 13 kbit/s [13]. The algorithms for speech encoding and decoding and for error protection are precisely defined by GSM. The output signal is passed to a channel encoder that provides suitable error protection coding, ensuring reliable transmission even in the presence of radio channel interference [14]. Additional interleaving of data signals helps to convert bundled errors into readily correctable individual errors.

Channel encoding increases the data rate to 22.8 kbit/s. The encoded speech signal is then interleaved. Afterwards, portion by portion, it is packed together with synchronization information into data bursts that are transmitted every 4.625 ms during a 577–µs time slot to a GMSK modulator [16].

Gaussian minimum shift keying (GMSK) is a bandwidth–efficient type of modulation enabling the resulting bitstream of 270.83 kbit/s to be transmitted via one channel of only 200 kHz in width. The result of GMSK modulation is a complex, digital, baseband signal (I, Q). After digital/analog conversion, it is passed to a 900 MHz transmitter and then via the radio channel to the base station [17,18].

From The Air To User's Ear

Due to reflections from various distant objects (e.g. houses, trees, hills), the signal transmitted from the base station is broken up along the way into a large number of signal components whose delay and amplitude differ.

At the mobile terminal, these partial signals are superimposed on one another resulting in a highly distorted signal. The mobile terminal's 900 MHz receiver amplifies the signal, performs the necessary channel filtering and converts it back into the baseband [18,17,16]. After analog/digital conversion, the signal is transmitted to digital signal processing. To equalize the severely distorted signal, an equalizer evaluates a training sequence sent with each burst to determine the channel properties. With this information, the original bit sequence is recovered from the original signal. There are no guidelines for equalizing or synchronizing the signal, leaving equipment manufacturers to develop their own solutions [5,15].

All the measures described for the transmit direction are again reversed in the channel decoder [14]. In this way, bit errors created by faults on the radio path are detected and, wherever possible, corrected. Thereafter, speech decoding and D/A conversion to audio band take place [13,12].

2.3. Radio Implementations on Commercial Products

The low noise radio receiver

A high performance radio receiver basically has the following functionalities [11] :

Low Noise Amplifier

A low noise amplifier (LNA) to bring the femto Watt radio input signal up to an adequate level to cope with the thermal noise of the mixers. The challenge here for analog radio design is to build low cost low noise amplifiers with very low power consumption, 2...5mA to provide +/− 20dB switchable gain and a 3dB or better noise figure and an input third order intercept point above − 14dBm to handle the high intermodulation radio input signals and the blocking levels specified by GSM. Most of the implementations are based on discrete MOSFET/MESFET circuitry although standard component manufacturers promote low noise Silicon bipolar transistors [8].

Frequency conversion

One or two cascaded mixer stages bring the radio signal down from RF to baseband with sufficient gain to further process the analog receiver signal in noisy standard CMOS technology. Even if GaAs is theoretically superior, all reported implementations are silicon bipolar based (f_T ~10..15 GHz) the main reason being the cost detriments of GaAs SSI chips [3,8].

In general we can say that to build a radio receiver with an acceptable Bit Error Rate, we need a low noise, low power (<25mA, 3 to 5V) analog radio front end with 40dB total switchable gain in the low noise amplifier and the mixers before we can add 50nV$Hz^{-1/2}$ CMOS thermal noise to the baseband radio receive signal.

Baseband/IF Implementation

Analog filtering, AGC and analog to digital conversion technique cover all the basic needs for baseband signal processing of a high performance radio receiver. Several options are open to compromise between filtering, AGC and A/D dynamic range.

The most popular solution in commercial products uses simple 8 bit A/D conversion and a conventional rational AGC algorithm in combination with dedicated filtering between all the gain stages [4,16]. The idea is to provide sufficient gain on the wanted signal to overcome the noise of the filters that must suppress the blocking signals and adjacent channels to protect the linear performance of the gain stage to follow, in other words to provide a dB of filtering for each dB of gain until the receive signal reaches the A/D convertors input.

For accurate analog filtering switched capacitor filters are the best. All commercial implementations are based on 1.0...2.0 µm Analog CMOS technology.

Receiver low frequency circuitry is, in all reported implementations, combined with its Transmit counterpart GMSK modulation. 3 to 4 DAC on 8 bits for control functions and power amplifier ramping control are often integrated on the same die. In the majority of the cases, the analog part of the voice coder is also integrated in the same ASIC [4,7].

Receiver Architecture

For the receiver architecture the choice was between heterodyne and homodyne demodulation.

Heterodyne receivers offer the advantage that the local oscillator frequency is different from the radio signal frequency on the antenna. Good design practice has led to the choice of 71 MHz as a standard for the IF (intermediate frequency) in GSM heterodyne receivers, because the 900 MHz GSM band has a width of 70 Mhz : 890–915 MHz for the base station receiver and 935 to 960 MHz for the mobile terminal receiver. A second advantage of heterodyne radio receivers is the opportunity to filter the radio signal at 71 MHz IF. The IF filters can suppress the neighboring channels and blocking signals to optimize the use of the available linear range in the rest of the receive chain. AGC amplifiers at IF have the inherent advantage that offset problems are easily solved by AC–coupling through small (on chip) capacitors.

However, heterodyne receivers have a serious problem due to IF filtering :
Passive filtering (SAW) is the solution for high Q filtering in a low noise receiver but, SAW filters are rather costly, of course are not integrable and their drivers are very power hungry, Other alternatives are to be rejected :
Time continuous or sampled analog filtering at 71 MHz is quite feasible in RF bipolar or GaAs biquads but the noise figure of a high Q active filter increases with the ratio (quality factor / gain) of one filter section.

In a **homodyne** receiver, as the signal goes directly to baseband in one mixer stage, all the filtering is well–known baseband filtering with low Q–factors, easy to integrate in CMOS time continuous or sampled data filters: switched capacitor or switched current circuits. Also the use of a homodyne receiver eliminates the need for an image filter preventing an IF receiver from creating an unwanted response to a spurious signal at 2 times the intermediate frequency away from the wanted signal frequency.

The homodyne receiver, simple and compact as it is, with all its advantages for low Q low pass filtering has one major drawback : OFFSET.

Demodulation of local oscillator leakage in the radio front end and self detection of high blocking levels are a major problem for radio communication systems like GSM, that use a baseband signal spectrum with an important DC and low frequency content. In those systems the high pass action of an AC–coupling is drawing too much energy away from the signal spectrum. For this reason a homodyne receiver for GSM cellular radio mobile communication needs a sophisticated offset cancelling algorithm to cope with 3 types of offset : static offset resulting from transistor and resistor matching errors, dynamic offset resulting from the mixer demodulating the local oscillator leakage to the antenna signal and a second dynamic offset resulting from the high blocking signals self–mixing as the antenna signal leaks to the local oscillator input of the mixers.

Over the past few years elegant solutions have been developed to overcome this offset problem by monitoring the offset in the digital baseband signal, low passing it and feeding the correct offset subtraction back to the input of the CMOS AGC [10,17].

The offset problem has been the dominant reason for many radio telecom companies not using the homodyne receiver for GSM and other radiocom receivers in terminals and base stations. The implementation of sophisticated offset algorithms in the receiver DSP, coupled with D/A converters to subtract the correct offset value have overcome this problem.

At the time of writing, this homodyne receiver architecture is an exclusivity of the Alcatel GSM terminal range.

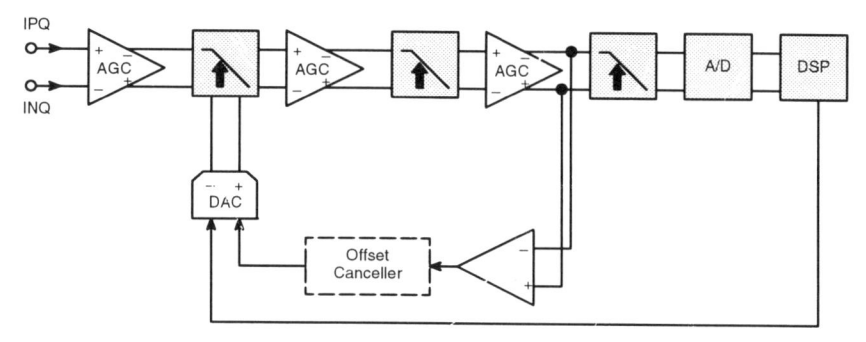

Figure 2 At RF the homodyne receiver by its simplicity and compactness is certainly an advantageous solution for battery powered pocket radio telephones

The low noise radio transmitter

Transmitter architecture

The noise constraints in the transmitter are set 25 dB above the -174 dBmHz^{-1} thermal noise floor, to prevent mobile transmitters from jamming each other and the receiver of the base stations for distant users.

From this point of view the homodyne transmitter was widely adopted : only one mixer stage contributes to the noise of the transmitter signal. Besides Alcatel, Siemens for example, is also promoting the direct conversion solution in Transmit direction [3].

GMSK modulator

The baseband signal generator in a GSM transmitter basically consists of a ROM containing the Gaussian shaped quadrature I and Q signals and a D/A convertor.

The CMOS–based function is combined in all implementations with some DAC for control and with its receive counterpart [3,4,16].

Quadrature modulator

The quadrature phase shifter on the 900 MHz local oscillator signal is a delicate aspect of the homodyne transmitter where a phase accuracy of 5° is required over the 70 MHz range of the GSM receive and transmit band.

Together with the double balanced mixer, the phase shifter is implemented in silicon bipolar, with or without (the 2 exist) its receiver counterpart [3,18].

Power amplifier

The real question today is in the efficiency of the class AB power amplifier using discrete bipolar transistors or FET [9]. Today expensive hybrid modules are used for this application in all commercial products. Efficiency of 35% to 40% is common.

The control of the output power as a function of temperature and aging is another function to measure the output power with an RF peak detector and adjust the gain of the power amplifier in a stable feedback loop. Note that the loop must have sufficient bandwidth to follow the burst ramp up of the time multiplexed GSM transmit signal.

2.4. Digital Baseband Implementation in Commercial Products

Digital Baseband, relatively to Radio, came much more to the forefront. Major manufacturers, some Silicon Vendors disclosed ambitious plans which are "difficult" to find back in commercial products ! (The situation is, however, evolving rapidly as chapter 3 will discuss). Here the text reviews commercial product implementations at time of writing.

Baseband processing

No commercial product has single–chip digital baseband processing. But all are, quite clearly, on their way to achieving this.

From the original 3–4 chip solutions (Equalizer, channel encoder/decoder, speech coder/decoder, glue for timing) all manufacturers diverged onto 2 different routes.

The first gang went to single state–of–the–art DSP (US sourced), associated to an assistant ASIC (signal preprocessing, Viterbi treatments,...).

The second gang followed another route, using 2 existing DSP's, in their core versions. The first DSP running the Transceiver functions (equalization, channel coding), the second one dealing with the user's voice.

Production technologies of the 2 chips generation are 0.8...1.0 µm CMOS.

Micro processor integration ?

At time of writing, very few commercially available implementations integrate the microprocessor as a core within one of the baseband processing chips.

This microprocessor takes up a great deal (if not all) of layer 1 real time processing software, of the GSM protocol. In opposition to the heralded microprocessor integration trend, GSM dedicated micros are becoming available on the open market, featuring a lot of prerequisite functions as GSM timers, deep power down modes, a few 8 bits DAC/ADC for various control tasks... As chapter 3 discusses, the debate is not closed.

3. BACK TO THE FUTURE

A lot of small and powerful products are currently on the market. However, the mass market is heating up competition. Extrapolation on current evolution calls for such targets as –20%/year on volume, –20%/year on weight, –30%/year on energy consumption.

Three technology evolutions are able to bring product development to such ambitious targets : ASIC and Silicon technology, GaAs Monolithic Microwave Technology and finally Battery Technology.

Just as ASIC Technology was the enabler of the 1st generation GSM handportables, these 3 technologies will be the enablers towards the personal communicator of the year 2000. After reviewing the most probable technological steps, the following sections will draw the guidelines of the most probable development paths towards ever better products.

3.1. ASIC Technological Trends

From following tables, presenting the key features of CMOS digital (table 1) and analog (table 2) technology evolutions, 3 determinant advances become evident. With the technological evolution, of course, silicon area (and hence component cost) is dropping rapidly, opening perspectives for higher integration levels. Moreover, partly for reliability reason, partly due to the intrinsic speed advantage, the supply voltage can go down (and hence the battery size or number of elements). A direct consequence of the 2 previous points is that the global power consumption needed to operate the chipset drops drastically.

It is widely believed that standard components (µP, RAM, ROM) will follow the same technological route opening perspectives for coherent product development.

TIME	1993	1994	1995	→	
Feature size	1.2μ	0.8μ	0.5μ	0.35μ	0.25μ
Relative Silicon Area	100%	70%	26%	13%	8%
Typical Supply Voltage	5 V	5 V	3.0 V	3.0 V	2.5 V
Target Supply Voltage	5 V	3.3–3 V	3.0 V	2.4 V	2.0 V
Typical Power Consumption (@ typical voltage)	100%	53%	17%	13%	8%
Target Power Consumption (@ target voltage)	100%	35%	17%	10.5%	6%

Table 1. Digital CMOS Roadmap

TIME	1993	1995	→
Feature size	1.5..1.2μ	0.8..0.5μ	0.35μ..0.25μ
Relative Silicon Area	100%	Need for denser implementation and new analog circuit techniques	
Typical Supply Voltage	5 V	3.0 V	2.5 V..2.0 V
Typical Power Consumption (@ typical voltage)	100%	<60%	<50%
		Need for low power analog and new analog circuit techniques	

Table 2. Analog CMOS Roadmap
(excluding digital part in mixed mode)

The tremendous impact of technological progress on CMOS digital is quite noticeable. This is of prime importance if we know that more than 50% of the energy in today's products is used to activate the digital parts.

In fact, the differences in evolution between technologies are just bringing, at the end of the decade, the power consumption shares to 1/3 for all major contributors: analog, digital, power amplifier.

3.2. Other Important Technological Trends

The March 94 CEBIT exhibition coincided with the first introduction of the 2W power amplifier for 900 MHz and 1800 MHz using GaAs MMIC. This announcement is the first wave of future progress : the MMIC part is much smaller than today's hybrids and the global efficiency is boosted from 35..40% to 50..55%. Further progress is yet to be seen at efficiency and minimum supply voltage level standpoints.

Comparable breakthroughs are likely to be announced at battery technology levels in which new packing techniques for NiCd and NiMH are each year improving the energy per cm^3 density by 10 to 20%, and in which new electrochemical couples are in development.

3.3. Radio Implementation : Which Routes for the Future ?

Mobile radio telephony is becoming a driving application for analog circuit design using silicon CMOS, RF bipolar and GaAs technologies.

Similar things are happening for several wireless personal communication systems. Basically, the cellular radio telephone, the wireless PABX and the wireless SLIC are bringing the same challenges to analog circuit design : i.e. maximum integration of the basic radio functions into 1 or 2 silicon chips, CMOS, Bipolar or BICMOS or GaAs. The analog circuit designer for radio telephone applications will need all the state–of–the–art analog design know–how available today, from RF–mixers and GHz range low noise amplifiers and local oscillator synthesizers over baseband 100 kHz CMOS analog to low frequency speech analog to digital conversion. For all these circuits the message is : minimum power consumption for battery lifetime, minimum silicon area for maximum functional integration per die to obtain a small, low cost pocket size radio telephone.

Radio Front End

Leaving aside the original "One ASIC per transceiver function" approach, the next generation receiver and transmitter functions operating at the radio frequency will be integrated in one RF–bipolar ASIC and the baseband circuitry in a CMOS mixed analog/digital ASIC doing the analog filtering, automatic gain control and A/D conversion.

Analog Baseband

The most difficult task for the analog baseband part is the supply voltage reduction, in a first step to 3.0 Volt. For Mixed Analog and Digital CMOS, the 3V power supply is a technological maximum rating that analog design will have to live with. The use of time continuous filters (OTA–C or MOSFET–C filters) can offer a solution as we go into 3V CMOS analog radio: at least no switches threshold voltage consumes any linear range between the supply rails. And CMOS transconductors with 60 dB linear range for noise and distortion on 5V supply voltage will have to improve in the next few years to make them applicable for 3V CMOS radio analog filter design.

Power amplifiers

As already mentioned above, as the mobile radio telephone market will expand in the near future, RF MMIC design houses will spend the effort to further develop monolithic solutions in GaAs, with improved efficiency and reduced supply voltage.

The local oscillator synthesizer

State-of-the-art mobile radiotelephones use phase modulation on a carrier in the lower GHz spectrum : 0.9 GHz for GSM, 1.8 GHz for PCN and DECT up to 3.5 GHz for other wireless telephone systems in the near future. The radiotelephone terminals rely on temperature controlled reference crystal oscillators (TCXO) to provide the local oscillator signal with a frequency accuracy of 5 ppm. These reference oscillators, operating at 13 or 26 MHz for GSM are not the first candidate for further integration. The challenge for radio analog monolithic integration is in the VCO and the prescaler.

The VCO is very difficult to integrate in an RF bipolar or GaAs front end because of the resonator. Resonators integrated on silicon have been reported with Q factors not higher than 10, which is far too low to obtain a VCO with −100 dBc phase noise at 10 kHz away from the carrier. The Q-factor of the spiral inductors on silicon is physically limited to very low values because in the substrate the electromagnetic field of the spiral inductor is damped in substrate currents.

This damping of RF electromagnetic fields in the bulk is instrumental in providing isolation between functional blocks on a Silicon MMIC, but for an LC resonator it kills the Q-factor. Clever designers will probably one day come up with high Q gyrator solutions or biquad resonators in 20 GHz RF bipolar or GaAs technology, but until then we are stuck with ceramic resonators for high Q VCO's and monolithic BICMOS synthesizers can only integrate the RF divide by 64/65 two modulus prescaler in the bipolar part and the low frequency (15 to 50 MHz) CMOS two modulus divider, the charge pump and the phase comparator.

Another challenge for CMOS analog radio design is the integration of the RF prescaler in submicron CMOS. As the f_T submicron CMOS is well beyond a GHz, a full CMOS 0.9 GHz and 1.8 GHz synthesizer is potentially the next step for analog radio design to further reduce the cost of a mobile radio telephone.

Technological aspects

Radio analog design is in a turbulent evolution today because of the availability of RF-bipolar technology up to 20 GHz f_T. The emphasis is moving from strip line and radio board design towards single chip full radio integration. Single chip radio is still an overstatement for mobile radio telephones today and the economic aspects are not yet proven, but experts now know that in less than 100 mm^2 submicron BICMOS it is possible to integrate analog radio for GSM including the synthesizer with prescaler, the RF-mixers, the baseband filters and AGC as well as the receive and transmit A/D and D/A converters.

For the low noise amplifier, the VCO and the power amplifier there is still some reluctance to go for silicon monolithic integration. Micromodules are certainly a valid alternative and an intermediate stage on the route towards monolithic.

The trade off between GaAs and Silicon is also a moving compromise, 3 years ago monolithic microwave integrated circuits were a GaAs monopoly, today silicon has proven to be more cost effective, at least for applications below 3 GHz. Recent publications show 73 GHz f_T for SiGe polysilicon emitter heterojunction bipolar transistors [2].

Submicron CMOS developments are more important for the digital part of the radio system because in analog the total silicon area will not benefit from the minimum dimensions to the same extent: Matching and other requirements in accurate analog circuits prevent us from using minimum gate–length transistors and minimum width poly for resistors and capacitors.

3.4. Digital Baseband Processing : Which Routes for the Future ?

As stated under section 2.4, all major competitors will probably have a one–chip baseband digital solution. The migration path is already clearly defined in the 1994 products. The 2 routes will end up, taking advantage of 0.6..0.5µm packing density, the first one on a single powerful DSP, in core version integrated with its companion glue, the second one on a dual DSP solution, direct merger of the 2 current components.

A third route, looks promising, even if it must still be introduced in commercial product : Philips announced in 1992 they were working on the KISS16 which is based on a dedicated DSP core optimized, constructed around the GSM basic functions [6].

The open questions for the future will be the amount of layer 1+ software ported onto hardware, the compatibility with half rate speech and finally the target technology for the merge with the microprocessor and (part of) its memories. This merge was announced by SIEMENS [3], when presenting the GOLD.

4. CONCLUSIONS

Starting from existing material coming from commercial products, the text disclosed the most probable development paths based on realistic technology progress.

The way to 'personal communicator' in the year 2000, could be the one illustrated herein. But many technical experts, or marketing forces are working hard to render this vision obsolete ! Companies are investing dozens of manyears each year to go faster, to do better, to use more adequate technologies ... in short, to win the competition race.

Figure 1 Functional Block Diagram of a GSM Mobile Set

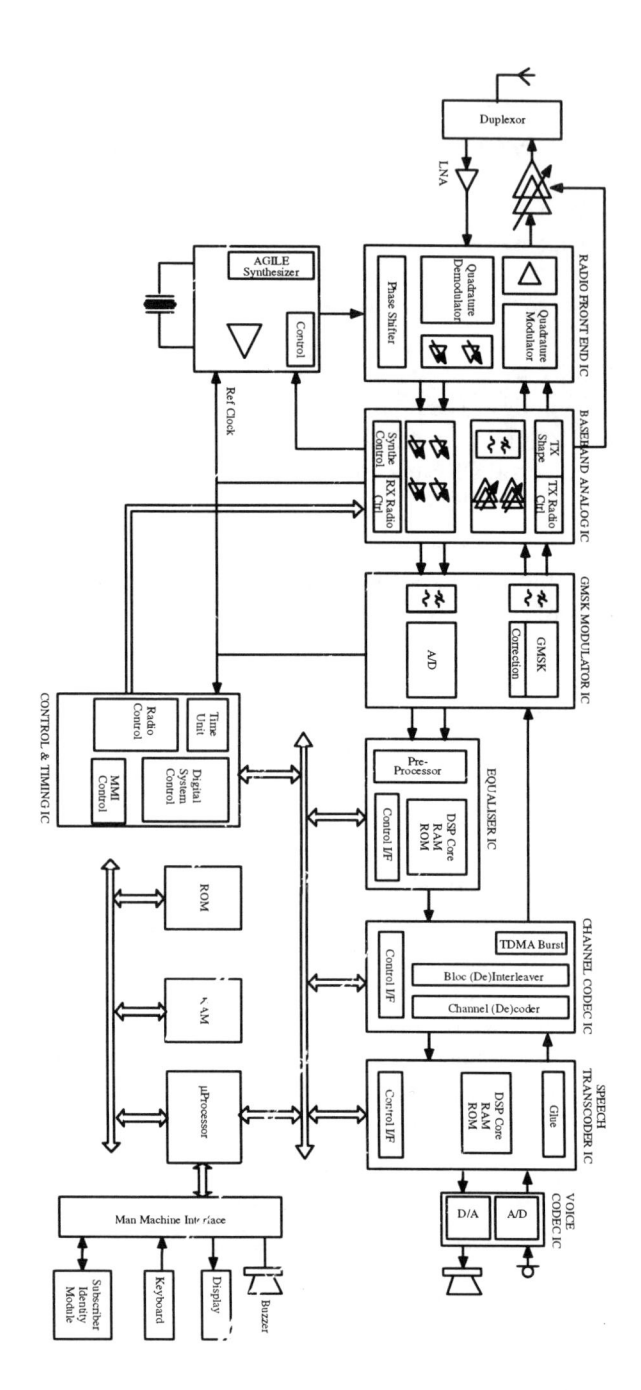

A Monolithic 900 MHz Spread-Spectrum Wireless Transceiver in 1-µm CMOS

Asad A. Abidi

with

Ahmadreza Rofougaran   Glenn Chang
Jacob Rael   James Chang
Maryam Rofougaran   Paul Jinyun Chang
Shahla Khorram

Electrical Engineering Department
University of California
Los Angeles, CA 90095-1594 (USA)

abidi@icsl.ucla.edu   http://www.icsl.ucla.edu/aagroup

INTRODUCTION

The RF and IF sections of wireless communications devices have traditionally comprised a collection of discrete active and passive components, while IC technology has made an impact on the baseband sections. The needs for low power operation and greater miniaturization impel RF and IF circuit technology towards greater levels of integration [1]. The single-chip radio has yet to be realized in the 1 to 2 GHz frequency band, where most of the digital cellular and other widespread data communications take place today. Such a chip would connect to the antenna on one end, and on the other end provide ports for the input and output of baseband data. At the current level of interest in wireless ICs, though, such an integrated radio is expected soon.

There are some historical hurdles to be overcome on the way to realizing the single-chip radio, others which are technological. Until recently, the small group of RF circuit practitioners was mainly trained in the discrete art, and they would normally implement circuits to the specifications of another group of practitioners, the system or radio architects. There was usually little exchange between the two groups, not least because they may have lacked a common language of communication. Miniaturization of the radio was usually a result of packaging the discrete components into smaller form-factors, as was demonstrated by the early generations of the Sony Walkman. The one significant break from this RF tradition appeared in the development of the integrated radio paging receiver in the 1970s, where it may be said that the style of baseband analog IC design was, in its full sense, first applied to a radio-frequency device [2-4]. The technological hurdle arises from the mostly perceived, although sometimes real, inability of well-established, high-volume IC technologies capable of mixed analog-digital integration, namely silicon CMOS, bipolar, and BiCMOS, to implement RF functions.

This is a progress report on one of the first coordinated, large-scale research efforts towards realizing a single-chip 900 MHz digital radio. Several factors are responsible for what has been achieved so far. First, this was an example of a *simultaneous evolution* of the systems architecture with the enabling circuits components, thus allowing for a joint optimization. This luxury is seldom afforded to circuit designers working to the established specifications of an industry standard, but because our transceiver operates in one of the three Industrial, Scientific, and Medical (ISM) frequency bands opened up by the Federal Communications Commission for unlicensed spread-spectrum use, we could design a unique system architecture. A user of these bands is required to spread the transmitted spectrum by a minimum amount, and to transmit no more than a specified power from the antenna, but is otherwise free to use any modulation scheme or spread-spectrum strategy. Second, it was decided at the outset that the *entire transceiver* would be integrated in *CMOS*, which affords an unprecedented flexibility in choosing between an analog or digital implementation of the various receiver blocks. Third, and its importance cannot be underemphasized, the project evolved from a close collaboration between analog and digital circuit designers, communication system engineers, and antenna designers.

SYSTEM ARCHITECTURE

The key transceiver specifications are listed in Table 1, and the rationale for their choice are then described below [5].

Table 1: Transceiver Specifications			
Operating Band	902-928 MHz	Data Rate	Up to 160 kb/s
Modulation Scheme	Binary or Quaternary FSK	Spreading Scheme	Frequency-Hop

The data rate is sufficiently high to accommodate voice and ISDN-type data, without requiring power-hungry equalization of typical delay-spreads in the receiver. A frequency-hopped spectrum spreading method is used over the more well-known direct-sequence, because it is possible with this method to cover an arbitrarily wide frequency range, without the attendant increase in rate of transmitted symbols, or chips. This results in a lower power receiver front-end. Frequency-hopping also enables the use of FSK modulation, which in turn makes it possible to use a direct-conversion receiver, whose virtues of high integration and low power are by now well-known [6].

A block diagram of the transceiver is shown in Figure 1. This shows a single transmitter, and two-branch diversity with two re-

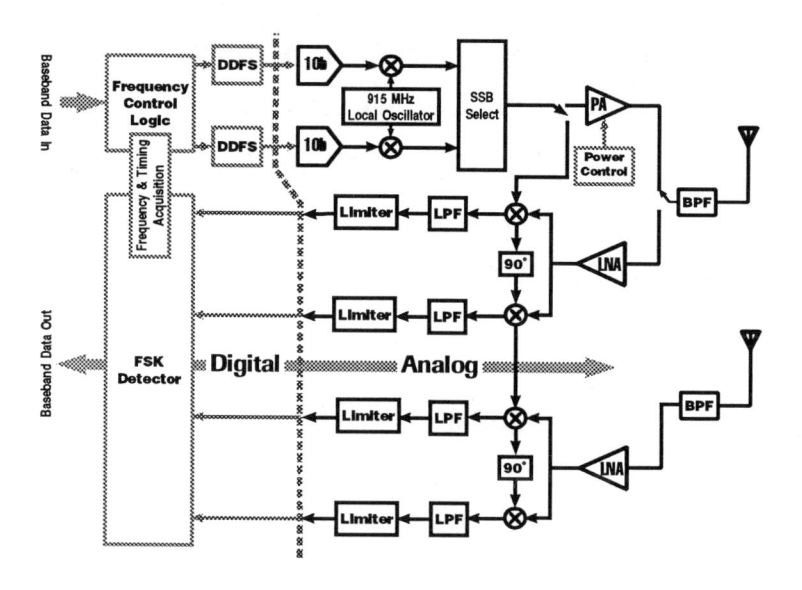

Figure 1: Transceiver Architecture

ceive channels whose outputs combine in the detector. As is characteristic of communications by spread-spectrum, transmission and reception take place in the same frequency band, and therefore one RF preselect bandpass is used bidirectionally. This dielectric resonator filter with a passband from 902 to 928 MHz is the only discrete component shown on this block diagram; not shown is the only other precision discrete component, a crystal with a resonant frequency at about 10 MHz which fixes the 915 MHz frequency of the local oscillator.

At the core of the transceiver is an *agile frequency source* based on direct-digital frequency synthesis (DDFS). This is a table lookup technique, whereby an accumulator which ramps at a rate set by an input frequency control word addresses a sinewave ROM to produce a discrete-time digital sinewave at that frequency. The absolute accuracy of the output frequency is then set by the fixed DDFS clock, its resolution by the input wordlength, and the wordlength of the internal arithmetic as well as the ROM construction determine the spectral purity of the digital sinewave. This approach gives instant frequency agility with continuous phase, but it requires a linear D/A converter, and an analog filter to suppress the images around the clock frequency. The spurious tones in the DDFS output are at least 76 dB below the main tone, so to convert it into a discrete-time analog sinewave with commensurate spectral purity, the DDFS is followed by a 10b DAC. The DDFS also provides an accurate quadrature output, which is converted to the analog domain by a second identical DAC.

This frequency synthesizer operates on the principle of producing a 26 MHz wide spread-spectrum modulated signal at *baseband*, and then upconverting that spectrum as a block with a fixed 915 MHz LO centered at in the middle of the ISM band. Quadrature phases of the LO upconvert the quadrature baseband spectrum into either the upper or lower half of the ISM band, depending on whether the two arms add or subtract. Thus, a DDFS output tone lying in the interval 0 to 13 MHz may be upconverted into either the upper or lower 13 MHz of the ISM band, depending on the sign of the summation, which is itself implemented as a digital phase reversal in one of the two DDFS quadrature channels.

Data modulates the carrier with a binary or quaternary *frequency-shift keyed* (FSK) *modulation*. The DDFS also serves as the data modulator, because a data 1 or 0 imposes a small, fixed offset corresponding to ±160 kHz (as well as ±80 kHz when quaternary FSK is used) on the frequency control word. The output spectrum is therefore constant-envelope, and amenable to *nonlinear power amplification* for a high conversion efficiency. Harmonics created by such a nonlinear element in the large-signal path lie out of the ISM band, but intermodulation distortion between multiple closely-spaced tones may create spurious in-band products. A frequency-hopped system with FSK modulation produces, in principle, only one frequency at a given time, but other effects such as upconversion by harmonics of the LO may cause spurious in-band products. A *passive, on-chip polyphase filter* after the upconversion mixer suppresses these undesirable harmonics, and also improves the extent to which the unwanted sideband is suppressed.

Owing to the choice of FSK modulation, it is feasible to use a *direct-conversion*, or zero-IF, *receiver* [6]. The preselect filter passes a received signal at the antenna comprising all users of the ISM band in the cell, which then passes through the on-chip RF *low-noise amplifier* (LNA). The amplified ISM band is downconverted and dehopped into two channels by mixers driven by quadrature phases of a frequency-hopped local oscillation, which is synchronized to the hopping pattern of the sought user. The desired channel is thus centered at DC, while all other users in the ISM band lie farther away, up to +13 MHz or −13 MHz. An on-chip active *lowpass filter* selects the desired channel, and suppresses all adjacent users. It also provides baseband gain. With the desired channel isolated, the modulation is recovered by driving the signal into a *limiting amplifier* which asynchronously quantizes the received waveform to 1-bit. A digital *correlating FSK detector* then determines which of the two (or four) possible frequency offsets is being received, and makes a data decision.

The hopping local oscillation for the downconversion mixer is derived from the transmitter's frequency synthesizer, which is switched away from the power amplifier during reception. A digital *frequency and timing acquisition loop* connected to the detector synchronizes the DDFS hopping sequence to the incoming hopping pattern, and aligns the symbol clock in the detector, as well as correcting small errors between the receiver and transmitter crystal references.

DETAILED CIRCUIT DESCRIPTION

The various sub-sections below describe the 1-μm CMOS circuits in each block of the transceiver, following the same order as in the previous section. References in all the sub-sections direct the interested reader to other publications devoted to each circuit.

Digital Frequency Synthesizer

A block diagram of the direct-digital frequency synthesizer is given

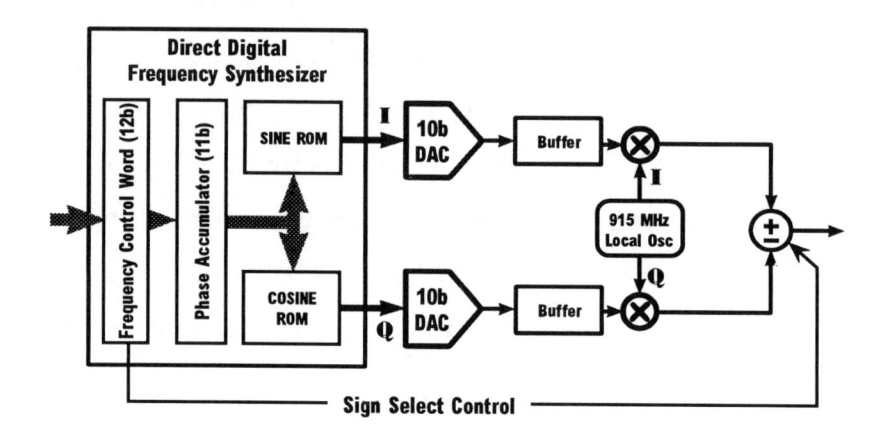

Figure 2: Agile Digital-based Frequency Synthesizer

in Figure 2. The DDFS frequency control word provides a frequency resolution of 2^{-11} of the clock rate. A single quarter-wave ROM stores the difference between amplitude and phase, and comprises coarse and fine tables. Phase-shifted addresses generate both in-phase (I) and quadrature (Q) outputs from the same ROM. These simplifications [7] together yield a 32 times reduction in size compared to a straightforward approach, and this is mainly why a static CMOS implementation dissipates only 60 mW when operating at 80 MHz [8].

A quasi-passive, pipelined charge-redistribution architecture, shown in Figure 3, implements the glitch-free DAC [8]. It consists of ten stages of equal-valued unit capacitors, which operate on a

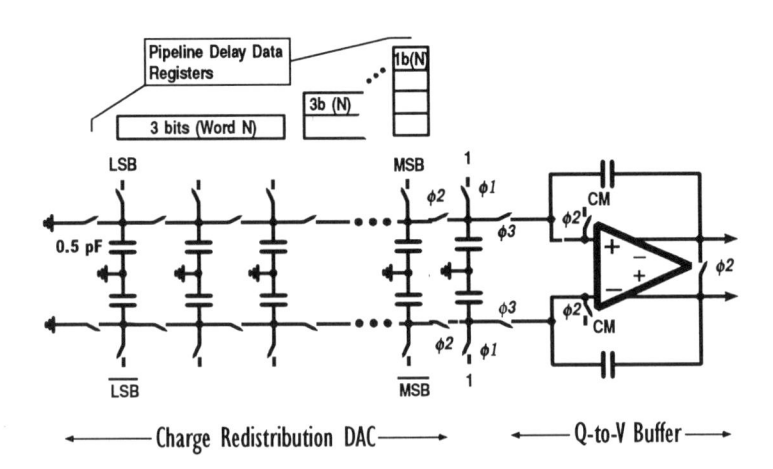

Figure 3: 10b, 80 MHz Pipelined D/A Converter

three-phase clock to successively bisect charge according to decreasingly significant bits of a 10-bit input word. The final charge is converted to voltage in a reset op-amp based integrator. The op amp is a gain-boosted cascode differential pair. The DAC core dissipates only 8 mW from 3 V at 80 MHz, principally in the clock drivers. This simple arrangement yields very pure synthesized output tones, with the highest spurious level of −62 dBc at low synthesized frequencies, rising to -57 dBc worst case at high synthesized frequencies (Figure 4) [8]. The output voltage swing is 0.5V ptp differential.

This high sinewave purity is sacrificed in the buffer following the DAC, which drives the power amplifier through the series switches of the upconversion mixer. An op amp-based solution to this would be too power-hungry, so a simple differential pair is used, resistor-degenerated at the sources to accept the large DAC output with relatively low distortion. The buffer adds a 3rd harmonic of −45 dB, which will upconvert to within the ISM band for synthesized frequencies of less than 4.33 MHz.

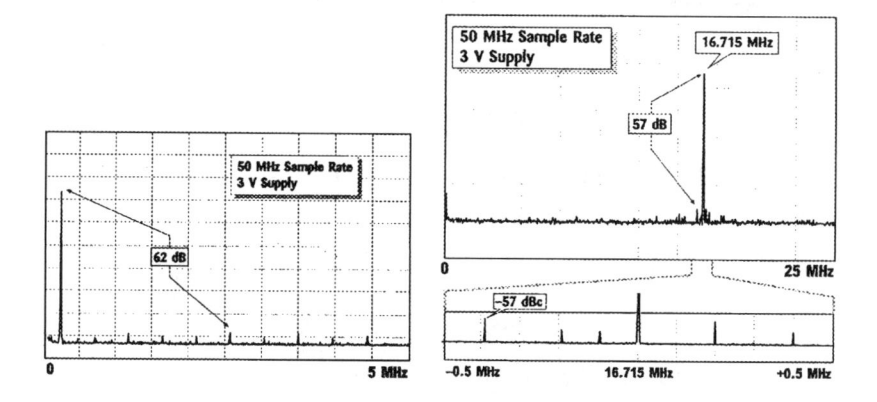

Figure 4: Measured spectral outputs of DDFS/DAC

Upconversion Mixers and Single-Sideband Selection

Following frequency synthesis at baseband, two four-FET switch mixers upconvert the large baseband sinewaves in the quadrature channels to 915 MHz (Figure 5). This simple mixer is very well-suited to MOS implementation, and when driven with large LO signals at the gate, it upconverts with high linearity and efficiency. However, full commutation by the switches also upconverts by the 3rd and 5th harmonics of the LO, and these additional frequencies may produce aliases through the nonlinear power amplifier which lie in-band.

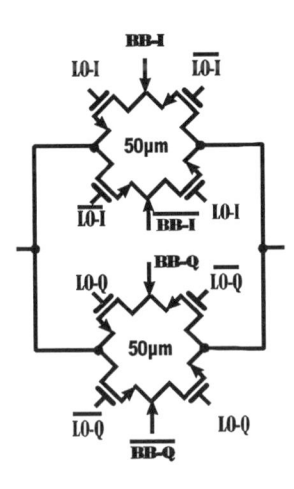

Figure 5: Upconversion FET-switch mixers

Although the mixers are schematically shown here selecting one sideband by summing together the outputs in current-mode, they actually drive separate ports of an RC polyphase filter which further suppresses the unwanted sideband. The polyphase filter (Figure 6) is a generalization of the simple RC-CR filter, and uses phase lag and lead to reinforce one rotational sequence of quadrature phases, say clockwise, while suppressing the other, say counterclockwise [9]. This is a particularly useful addition to quadrature upconverters, because the desired sidebands will assume the opposite rotational sequence to the unwanted sidebands. Fur-

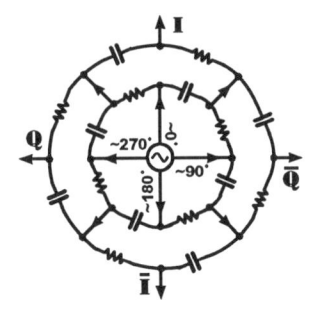

Figure 6: RC Polyphase filter: cascade of two sections

thermore, this is a broadband circuit, and does not require great accuracy in the R and C to be effective. A single polyphase filter section in the transmitter path suppresses the unwanted sideband by another 10 dB after upconversion, and compensates for residual phase errors in quadrature phases of the LO. However, the baseband gains of the two DACs and their buffers in the frequency synthesizer must match to better than 0.1 dB if the unwanted sideband is to be suppressed by more than 45 dB (Figure 7), and the polyphase filter cannot compensate for this source of error.

Owing to its broadband properties, the polyphase filter exerts a fortuitous suppression on the frequencies upconverted by the 3rd harmonic of the LO. When it is driven so as to supplement the inherent selection of the quadrature upconverter, the polyphase filter appears contrary to the 3rd harmonic components, because all the relative angles are now rotated by 3×. Thus, the quadrature upconverter inherently cancels one upconverted sideband at the 3rd harmonic, and this polyphase filter as above cancels the other sideband. The combined effect is to suppress *both* sidebands around the 3rd LO harmonic, which neatly substitutes for an LC notch filter.

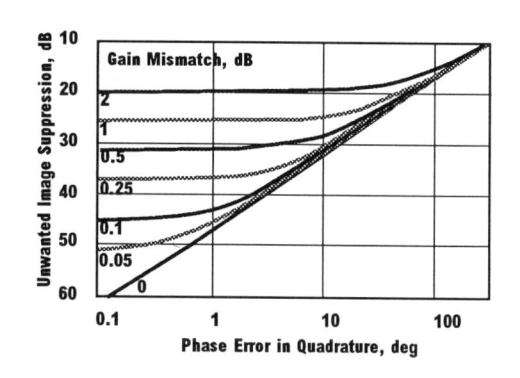

Figure 7: Unwanted sideband suppression with gain and LO phase errors

It is appropriate here to discuss how the single-sideband upcon-

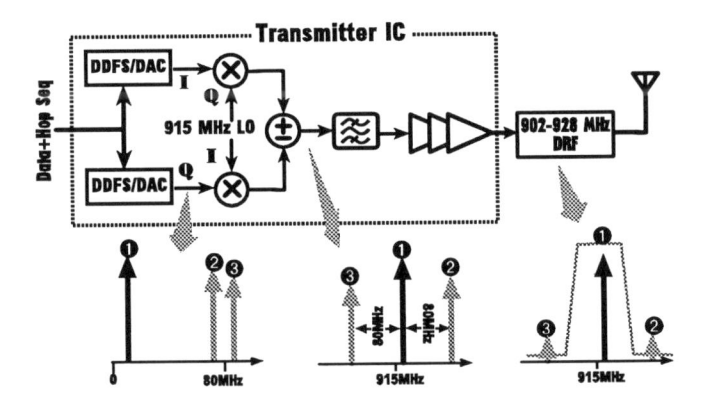

Figure 8: Image frequencies in transmitter

verter treats the image frequencies around the DDFS/DAC clock frequency. After quadrature upconversion, the circuit selects the upper sideband of the lower image, and the lower sideband of the upper image, so that the upconverted images are all separated by exactly the clock frequency (Figure 8), 80 MHz in this implementation. Owing to the *sin x/x* rolloff in the spectrum of the sampled waveform, and the large oversampling between the DAC output of 13 MHz or less with an 80 MHz clock, the images are at least 15 dB lower than the main tone. They are left to propagate through the power amplifier, whose nonlinearity only redistributes energy between the various equally-spaced tones but does not create any new tones. Finally, the off-chip RF preselect filter suppresses the images by an additional 40 dB or so, as the choice of DAC clock frequency forces them to lie in the filter's stopband. Image suppression is therefore accomplished with this passive filter, rather than on-chip filters which would require a prohibitively large power dissipation to handle the large signal levels with sufficiently low distortion.

RF Power Amplifier

Some specifications on the power amplifier in a microcell-based wireless system are relaxed compared to large-cell based systems, while others are more stringent. This power amplifier is integrated

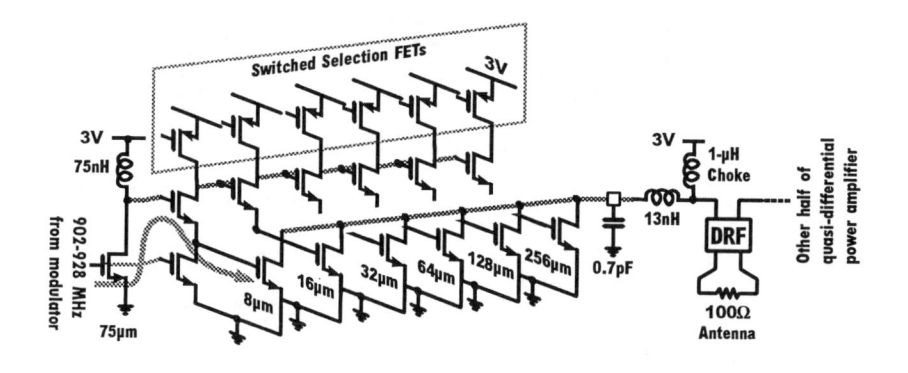

Figure 9: RF Power Amplifier

on the same 1-μm CMOS substrate as the rest of the components in this transceiver, as it is required to deliver a maximum of only 30 mW to the output. However, for maximum user capacity, the output power must also be controllable in 1 dB increments over a range of 30 dB, which corresponds to a minimum of 30 μW.

The power amplifier derives from baseband CMOS circuit design. A binary-weighted array of FETs is connected together at the drains to supply the antenna current (Figure 9) [10]. The FETs are normally biased close to threshold, and are driven into conduction by a large voltage swing at the gates. The requirement for maximum power delivery is met with a 3V supply by providing a push-pull voltage into the antenna with the two halves of the quasi-differential circuit. In turn, a scaled array of source followers drives the gates of the output FETs, and their common gate receives a large swing from a preamplifier with a large on-chip inductor load. This load enables the preamplifier drain voltage to swing up to 5.5 V, thus overcoming the large V_{GS} drop across the source followers. The preamplifier also sets the gate bias of the output FET array through a DC feedback loop.

The output power is controlled by selectively enabling the source followers in the array through PMOS switches in series with the drains. These switches are outside the RF signal path. The effective width of the output FET is thus under direct control of a digital word, while the input voltage to the power amplifier, derived from the upconversion mixer, always stays constant. This reduces output-power dependent pulling of the upconversion LO.

An off-chip inductor matches the capacitive impedance of the power amplifier to 50Ω. This capacitance is dominated by the output bonding pad, and the sum of the drain junction capacitances of the FET array is the next contributor. The latter capacitance does not change much with the power-control word, but the net output resistance contributed by the r_{ds} of the ON FETs in the array does depend on this word. This causes the output reflection coefficient, s_{22}, to vary somewhat but at maximum power it attains a minimum value of –30 dB by design. The amplifier is unconditionally stable.

The power amplifier yields a 0.7 dB droop across the 902 to 928 MHz in the ISM band, provides a power to a balanced 50Ω load of up to +13 dBm with a 35 dB lower range, and attains a peak conversion efficiency of 42% at maximum power, including the preamplifier. The 1 dB compression point in the amplifier input-output

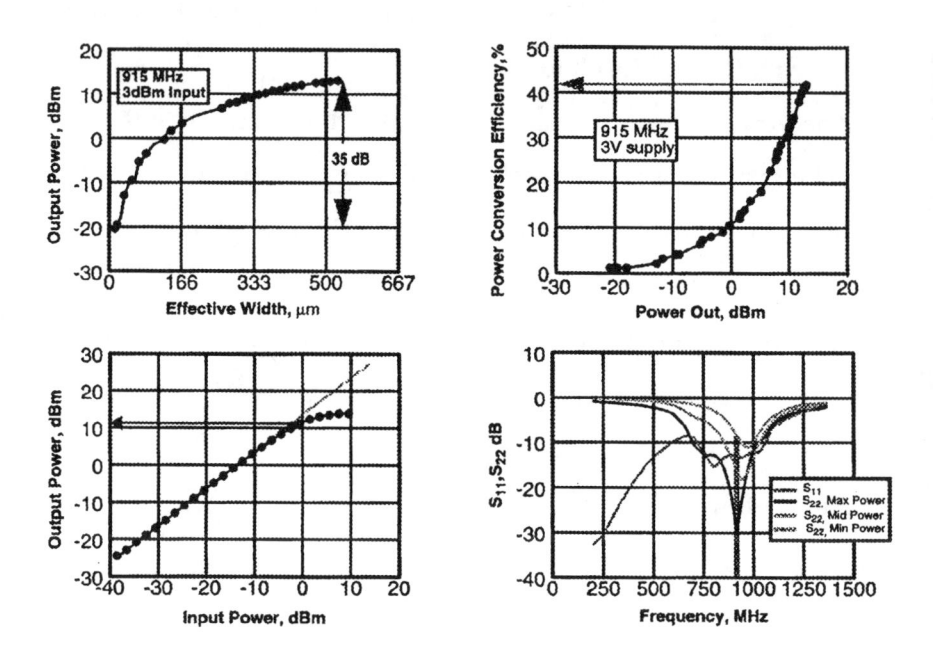

Figure 10: Measured RF Power Amplifier Characteristics

characteristic almost coincides with the peak power (Figure 10).

Local Oscillator with Quadrature Outputs

In keeping with the desire for maximum integration, an on-chip fixed-frequency local oscillator was sought which does not require an external resonator. Aside from the challenge of making a 1μm CMOS oscillator operate at all at 900 MHz, there is the additional requirement that the oscillation must be accompanied by a low phase noise, and that the oscillator should produce a large swing to drive the four-FET switch mixer. Incidentally, the last two requirements are in accord with each other, rather than in conflict.

The oscillator core comprises a cross-coupled, common-source FET pair providing a negative resistance to inductor loads (Figure 11). The inductors are fabricated on-chip. Parallel resonance sets in between the inductors and the net FET capacitance, part of which is a drain junction capacitance with a voltage coefficient. This latter is used as a means of coarse frequency tuning, by varying the top-rail voltage with a FET resistor carrying the common-mode current (Figure 12). However, the voltage swing at the oscillator and its current drain will now change with the frequency. The core circuit drains 7 mA from a 3V supply at 915 MHz, and spans the range from 860 to 958 MHz.

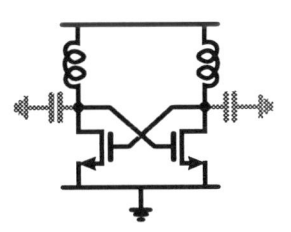

Figure 11: LC Oscillator Core

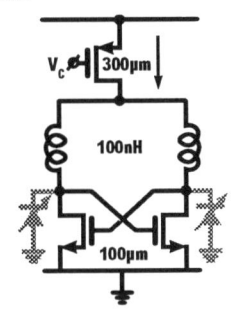

Figure 12: Tuning method

Clipping in the FET characteristics determines the amplitude of oscillation, and this in turn depends on the top-rail voltage. Unlike transistors with a junction at the input which clips by turning ON, the swing at the MOSFET gate may be arbitrarily large, limited only by breakdown. Thus, the oscillator produces a 6.5V ptp differential output at 915 MHz, and for a given transistor noise and Q (=5.5) of the resonant circuit, the corresponding output phase noise is lowered in inverse proportion [11]. An SSB phase noise of −100 dBc/Hz is measured at a 100 kHz offset from the 915 MHz oscillation (Figure 14). The slope of the phase noise vs frequency is

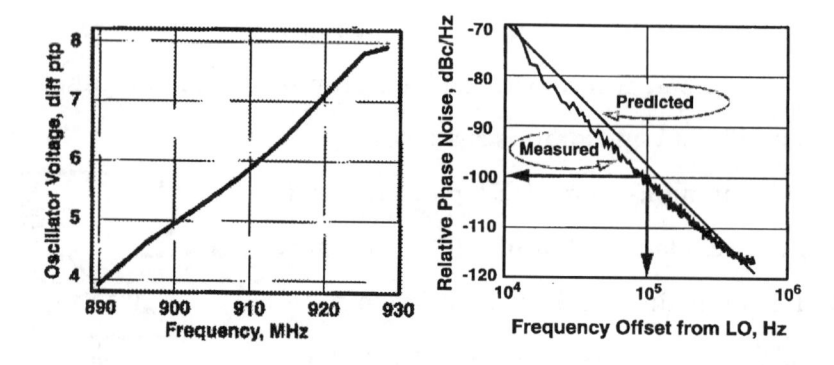

Figure 14: Measured LO characteristics

30 dB/decade up to a 1 MHz offset, which shows that MOSFET flicker noise dominates. We are able to predict this very well using Leeson's model [11], and our own measurements of flicker noise in MOSFETs [12, 13].

Single-sideband modulators require quadrature phases of the lo-

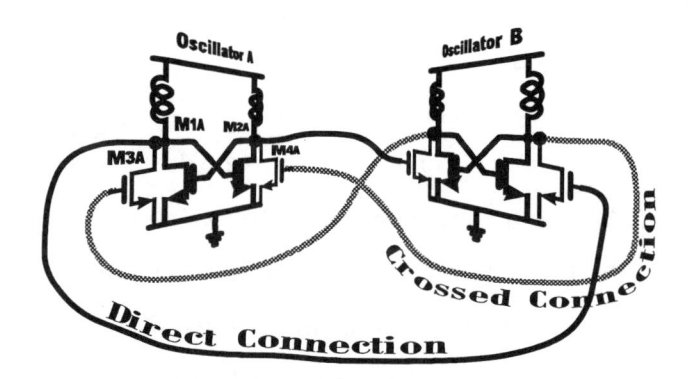

Figure 13: LC Oscillators Synchronized in Quadrature

cal oscillator, usually in the form of balanced signals. Quadrature must further be attained with considerable accuracy for adequate sideband suppression (Figure 7). Most local oscillators produce an unbalanced single-phase oscillation, and various methods, which by and large are marginally satisfactory, are used to derive quadrature phases [6]. This LO uses a new topology to inherently produce balanced outputs in precise quadrature (Figure 13) [14]. It consists of two identical oscillators tightly coupled through FETs so that they injection lock to exactly the same frequency, but the coupling topology suppresses both in-phase and anti-phase synchronization, and permits only synchronization in quadrature with a unique sequence of phases. This negative feedback suppresses errors from quadrature caused by asymmetrical capacitive loading. The accuracy of quadrature is measured with an on-chip single-sideband upconversion (Figure 15), where it is found that the unwanted sideband is suppressed by more than 46 dB, and LO leakage is 49 dB lower than the wanted sideband. This implies a phase error of less than 1°.

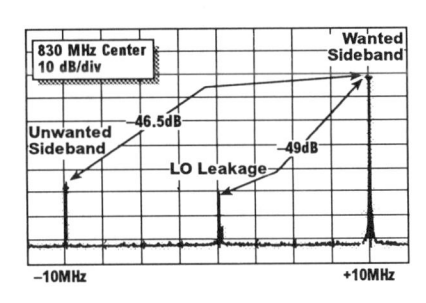

Figure 15: SSB upconversion to measure quadrature accuracy

Integrated Transmitter

The various blocks described above have been integrated into a monolithic transmitter, containing all functions from baseband data in to antenna drive at the output (Figure 16) [15]. Overall performance of the transmitter is evaluated, first, by measuring the spectrum of a single tone produced in the ISM band (Figure 17), and, second, by measuring the spectrum with frequency-hopping (Figure 18). In both cases the unwanted sideband

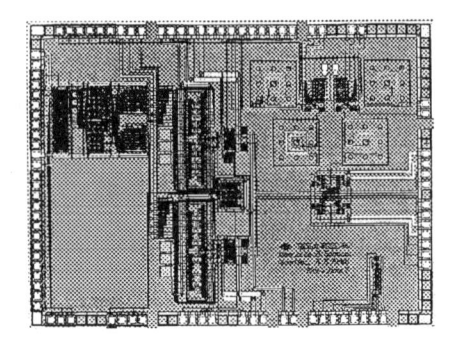

Figure 16: Monolithic Transmitter

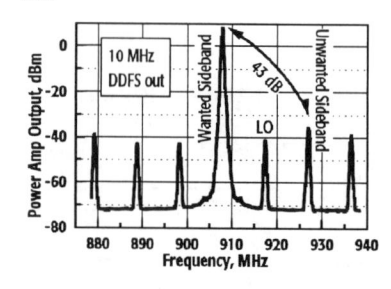

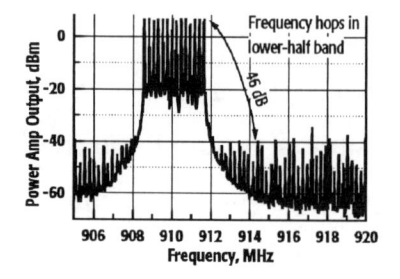

Figure 17: Measured spectrum with single-tone output (med power)

Figure 18: Measured spectrum with frequency hopping (med power)

and LO leakage are at least 43 dB below the desired parts of the transmitted spectrum.

Technology for Large-Value On-Chip Inductors

An enabling component in realizing RF blocks in the transmitter and receiver is a unique method to realize on-chip spiral inductors, 50 to 100 nH in value, and with a self-resonance beyond 2 GHz. Large-value spiral inductors are conventionally thought to be incompatible with silicon IC technology, because the large parasitic capacitance they inherit through oxide to the semiconducting substrate causes self-resonance at a few tens or hundreds of MHz. We have overcome this limitation by using a simple, post-process technique that selectively removes the silicon substrate with a gas-phase etchant, leaving the inductor encased in a membrane of oxide attached to the rest of the substrate, but suspended above an air cavity (Figure 19) [16]. In the MOSIS 1-μm CMOS process, this does not require any extra masks, and the etching is carried out on fully fabricated die. The inductor Q of 4 to 5 is limited by the sheet resistance of the 2nd-level Al metallization.

In many circuits, the larger the available inductor, the lower the bias current required to achieve a certain specification. The main advantage in realizing the inductor on-chip is that no power is wasted in driving the pad, package, and board stray capacitances at radio frequencies which an off-chip, discrete inductor would entail. A simple, three-element model suffices for circuit simulations [16], where the inductance, resistance, and parasitic capacitance to the far-off ground plane may be manually calculated using well-known formulas.

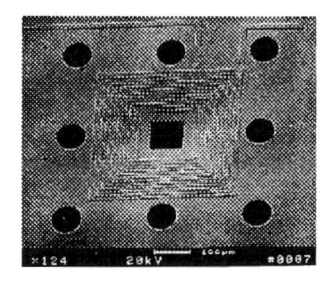

Figure 19: Large spiral inductor suspended over cavity

Low-Noise Amplifier and Downconversion Mixer

A receiver front-end which meets the quartet of requirements of low noise, high gain, wide dynamic range, and a good input match to 50Ω at 900 MHz may well be thought to be the greatest design challenge in a 1-µm CMOS implementation. Yet a state-of-the-art circuit has been designed which accomplishes all four of these objectives at a modest current drain. The design relies on a uniquely CMOS approach, and exploits the simplifications implied by direct-conversion [6]. Furthermore, the two components of the front-end, the LNA and mixer, are designed simultaneously.

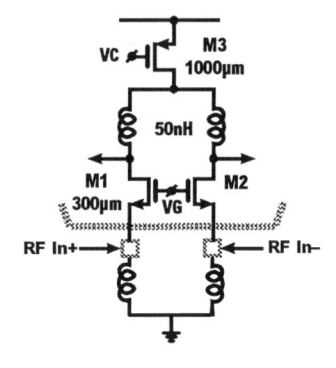

Figure 20: RF Low Noise Amplifier

When a 3 dB noise figure is acceptable in the front-end, it is much simpler to use a common-gate FET input stage whose transconductance is designed to be 1/50 S, than to use the more elaborate inductor matching techniques when the input is applied to the FET gate [1]. A broadband input match now requires the addition of only an off-chip shunt inductor to tune out the pad and FET capacitance at the source. This has been implemented in a quasi-differential circuit (Figure 20) [17]. On-chip inductor loads tune the drains of the FETs to 1 GHz, including the load capacitance of the directly-coupled downconversion mixer (Figure 21), and the Q is sufficiently high that a voltage gain larger than 22 dB is obtained. The input noise figure is set by the thermal noise at 1 GHz in the FET inversion layer, and is about 3 dB. The downconversion mixer (Figure 21) [17] comprises a linear FET voltage-to-current converter connected directly to the LNA output, whose output is commutated by the LO through four FET switches. Two common-source FETs with balanced input and output constitute a very linear transconductor. There is a small conversion loss of 67% when the switches commutate the RF current fully, and balanced FET triode-region loads set the overall mixer gain.

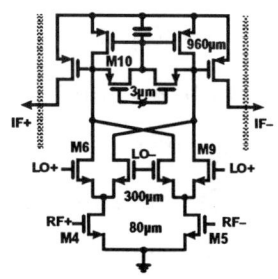

Figure 21: Downconversion Mixer

The measured data (Figure 22) verifies a satisfactory design. In a direct-conversion receiver, the appropriate noise figure is measured with a double-sideband input [18], because both sidebands around the LO carry useful signal energy. The LNA and *one* mixer take 8 mA from a 3V supply when driving an on-chip load. The overall mixer voltage gain is about −3 dB, and it makes a negligible contribution to the overall noise figure. A 0 dBm LO drive is necessary for complete commutation of the mixer switches, and lowest overall noise figure.

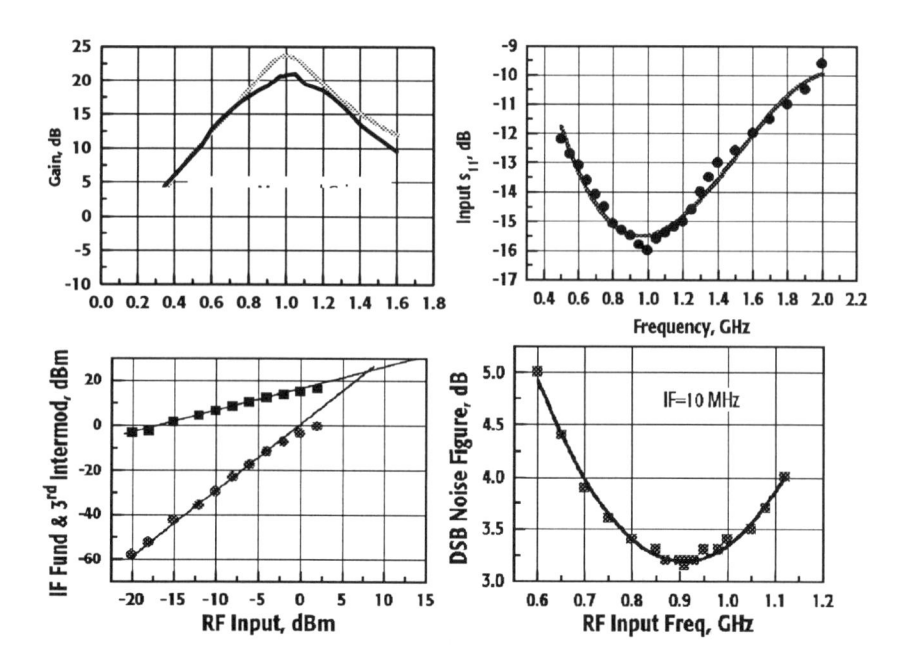

Figure 22: Measured data for LNA-mixer combination

Flicker noise in the baseband section at the mixer output will degrade the noise figure by another 3 dB at 160 kHz, where the peak of the downconverted spectrum lies. Although DC offset at the mixer output may be nulled out in an FSK receiver [6], if the offset varies substantially with LO hopping frequency it becomes difficult to distinguish it from the received symbol energy.

Channel-Select Lowpass Filter

An active on-chip lowpass filter with a sharp cutoff at 230 kHz selects the desired channel, and attenuates all other users within and nearby the ISM band. The filter must itself be low noise so as not to degrade the overall receiver noise figure, it must have wide dynamic range so that out-of-band interferers do not create in-band intermodulation, and it should preferably provide some baseband voltage gain to overcome the input-referred noise of the subsequent limiting amplifier. Receiver linearity remains important until the unwanted signals have been filtered, and thereafter the received signal is subject to the clipping nonlinearity of the limiting amplifier.

The filter is composed of a decimating switched-capacitor cascade, which progressively amplifies and filters to maintain the largest dynamic range [19]. The downconverted signal is first filtered by a 2nd-order Butterworth section sampling at 57 MHz, which guarantees a 50 dB attenuation at 14 MHz, then decimated by 4 (14.25 MHz clock) into a 6th-order elliptic filter with a 230 kHz

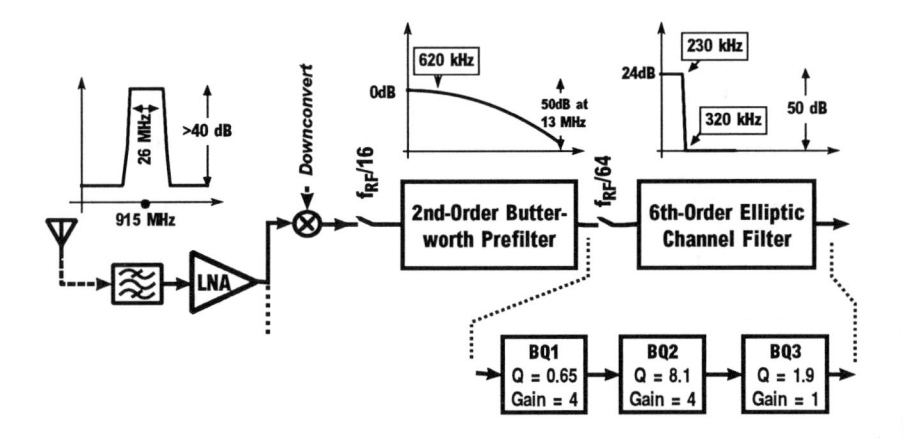

Figure 23: Channel-Select Filter Architecture

cutoff frequency, and the stopband edge at 320 kHz (Figure 23). This partition controls the capacitor spread, which nevertheless is 108:1, while maintaining a well-controlled stopband to 57 MHz. The unit capacitor size, and therefore the power dissipation, is to a large extent set by the specification on filter input noise.

The measured filter response is very close to what is expected. When consuming 4.6 mA from 3.3V, the input-referred noise in the passband is 70 nV/$\sqrt{Hz}$. This is dominated by the first stage sampling capacitor, and may be scaled down by redistributing the gain. The IP3 is measured for the case when two

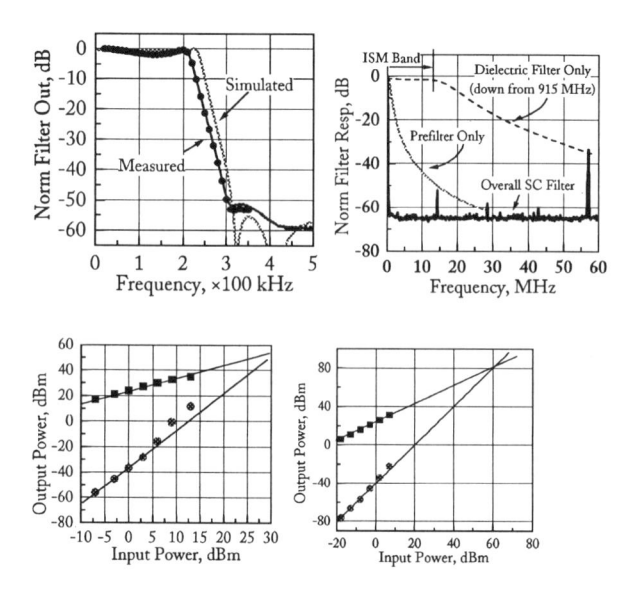

Figure 24: Measurements on Channel-Select Filter

large input signals in the stopband create intermodulation distortion in the passband. IP2 characterizes envelope-detection nonlinearity, when an AM signal in the stopband may create detected products lying in the passband. Both measurements bear out that the filter will not become a bottleneck to overall receiver linearity.

Limiting Amplifier and Detector

A limiting amplifier with 84 dB gain is implemented by a cascade of 7 identical clipping differential pairs. This operates on the selected baseband signal, and so need only have roughly a 1 MHz bandwidth. Rectified taps along this chain provide a successive-detection logarithmic measurement of the received signal strength (Figure 25) [20].

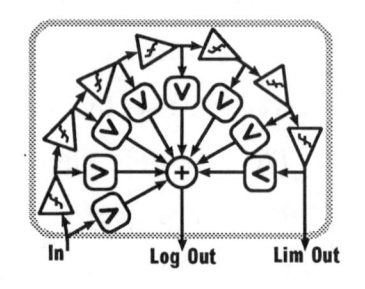

Figure 25: Limiting Amplifier with Log Signal-Strength Output

A capacitively-decoupled DC feedback loop is applied between the output and input of the 2nd stage, to suppress DC offset from dominating the limiting action. With this feedback, the circuit tolerates a ±100 mV offset on the input signal without loss of dynamic range. Typical of a sensitive limiting amplifier, in the absence of an input this circuit limits on its own noise, which is 50μV at the input in a 300 kHz bandwidth. The logarithmic output is 1 dB accurate over an 80 dB range of the input across the commercial temperature range. The circuit drains only 1 mA from 3.3 V.

A digital detector accompanies the limiting amplifier (Figure 26) [21]. This oversamples the limited output of the downconverted desired channel by 64×, then correlates with locally generated 1b *sin* and *cos* waveforms (square

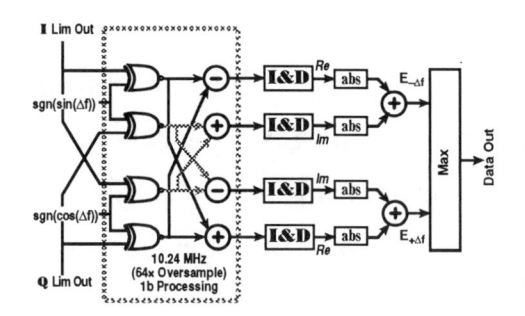

Figure 26: Correlating digital binary-FSK detector

waves) at the expected frequency offset. The correlations are integrated over 64 clock cycles, corresponding to one symbol, and a decision is made based on an estimate of the energy of the largest correlation. Various simplifications are made to retain 1b signal processing in the portions of the circuit clocking at high speed. When preceded by two limiting amplifier channels, and accompanied by a digital oscillator for the reference sinewaves, this circuit only dissipates 5 mW from 3.3 V. Yet with the limiting amplifier preceding it, the circuit operates at an acceptable error-rate for a 56µV rms input, and can tolerate inputs at least 82 dB larger than this.

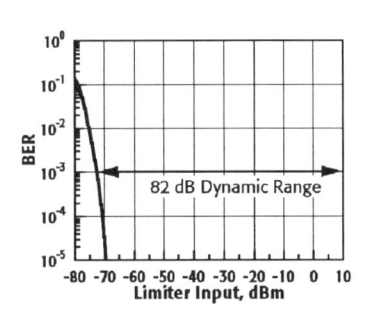

Figure 28: Measured FSK sensitivity of limiting amplifier and detector

The digital circuits for timing acquisition and frequency synchronization are described elsewhere [22].

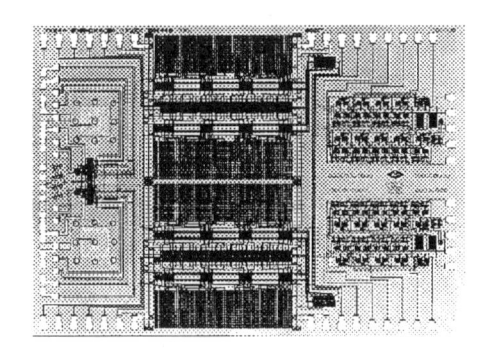

Figure 27: Integrated Receiver IC

A fully integrated receiver, including all functions from antenna to baseband data output, is now being evaluated (Figure 27).

REFERENCES

[1] A. A. Abidi, "Low-Power Radio-Frequency ICs for Portable Communications," *Proc. of the IEEE*, vol. 83, no. 4, pp. 544-569, 1995.

[2] I. A. W. Vance, "Fully Integrated Radio Paging Receiver," *IEE Proceedings*, vol. 129, Part F, no. 1, pp. 2-6, 1982.

[3] D. W. H. Calder, "Audio Frequency Gyrator Filters for an Integrated Radio Paging Receiver," presented at *Int'l Conf on Mobile Radio Systems & Techniques*, York, UK, pp. 21-26, 1984.

[4] P. E. Chadwick, "Analogue VLSI-applications of the technique," presented at *IEE Colloquium on 'Advances in Analogue VLSI'*, London, pp. 1/1-3, 1991.

[5] J. Min, A. Rofougaran, H. Samueli, and A. A. Abidi, "An All-CMOS Architecture for a Low-Power Frequency-Hopped 900 MHz Spread-Spectrum Transceiver," presented at *Custom IC Conf.*, San Diego, CA, pp. 379-382, 1994.

[6] A. A. Abidi, "Direct-Conversion Radio Transceivers for Digital Communication," *IEEE J. of Solid-State Circuits*, vol. 30, no. 12, pp. 1399-1410, 1995.

[7] H. T. Nicholas and H. Samueli, "A 150-MHz Direct Digital Frequency Synthesizer in 1.25µm CMOS with -90dBc Spurious Response," *IEEE J. of Solid State Circuits*, vol. 26, pp. 1959-1969, 1991.

[8] G. Chang, A. Rofougaran, M. K. Ku, A. A. Abidi, and H. Samueli, "A Low-Power CMOS Digitally Synthesized 0-13 MHz Agile Sinewave Generator," presented at *Int'l Solid State Circuits Conf.*, San Francisco, pp. 32-33, 1994.

[9] R. C. V. Macario and I. D. Mejallie, "The Phasing Method for Sideband Selection in Broadcast Receivers," *EBU Review (Technical Part)*, no. 181, pp. 119-125, 1980.

[10] M. Rofougaran, A. Rofougaran, C. Olgaard, and A. A. Abidi, "A 900 MHz CMOS RF Power Amplifier with Programmable Output," presented at *Symp. on VLSI Circuits*, Honolulu, pp. 133-134, 1994.

[11] D. B. Leeson, "A Simple Model of Feedback Oscillator Noise Spectrum," *Proc. of IEEE*, vol. 54, no. 2, pp. 329-330, 1966.

[12] A. A. Abidi, J. Chang, C. R. Viswanathan, J. Wikstrom, and J. Wu, "Uniformity in Flicker Noise Characteristics of CMOS IC Technologies," presented at *Ninth International Conference on Noise in Physical Systems*, Montreal, pp. 469-472, 1987.

[13] J. Chang, A. A. Abidi, and C. R. Viswanathan, "Flicker Noise in CMOS Transistors from Subthreshold to Strong Inversion at Various Temperatures," *IEEE Trans. on Electron Devices*, vol. 41, no. 11, pp. 1965-1971, 1994.

[14] A. Rofougaran, J. Rael, M. Rofougaran, and A. A. Abidi, "A 900 MHz CMOS LC Oscillator with Quadrature Outputs," presented at *Int'l Solid State Circuits Conf.*, San Francisco, CA, pp. 392-393, 1996.

[15] A. Rofougaran, G. Chang, J. J. Rael, M. Rofougaran, S. Khorram, M.-K. Ku, E. Roth, A. A. Abidi, and H. Samueli, "A 900 MHz CMOS Frequency-Hopped Spread-Spectrum RF Transmitter IC," presented at *Custom IC Conf.*, San Diego, CA, pp.209-212, 1996.

[16] J. Y.-C. Chang, A. A. Abidi, and M. Gaitan, "Large Suspended Inductors on Silicon and their use in a 2-μm CMOS RF Amplifier," *IEEE Electron Device Letters*, vol. 14, no. 5, pp. 246-248, 1993.

[17] A. Rofougaran, J. Y.-C. Chang, M. Rofougaran, S. Khorram, and A. A. Abidi, "A 1 GHz CMOS RF Front-End IC with Wide Dynamic Range," presented at *European Solid-State Circuits Conf.*, Lille, France, pp. 250-253, 1995.

[18] S. A. Maas, *Microwave Mixers*, 2nd ed. Boston: Artech House, 1993.

[19] P. J. Chang, A. Rofougaran, and A. A. Abidi, "A CMOS Channel-Select Filter for a Direct-Conversion Wireless Receiver," to be presented at *Symp. on VLSI Circuits*, Honolulu, 1996.

[20] S. Khorram, A. Rofougaran, and A. A. Abidi, "A CMOS Limiting Amplifier and Signal-Strength Indicator," presented at *Symp. on VLSI Circuits*, Kyoto, pp. 95-96, 1995.

[21] J. Min, H.-C. Liu, A. Rofougaran, S. Khorram, H. Samueli, and A. A. Abidi, "Low Power Correlation Detector for Binary FSK Direct-Conversion Receivers," *Electronics Letters*, vol. 31, no. 13, pp. 1030-1032, 1995.

[22] H.-C. Liu, J. Min, and H. Samueli, "A Low-Power Baseband Receiver IC for Frequency-Hopped Spread Spectrum Applications," presented at *Custom IC Conf.*, Santa Clara, CA, pp. 311-314, 1995.

BANDPASS DELTA-SIGMA AND OTHER DATA CONVERTERS

Rudy van de Plassche

Introduction

In telecom applications, hearing aids and audio systems low-power high performance analog-to-digital and digital-to-analog converters are needed. Examples of such converters will be presented. To conclude this section tools for automatic design of sigma-delta modulators will be introduced.

In the first paper by van der Eric Zwan a low-power CMOS sigma-delta modulator with a continuous time noise-shaping filter is presented. The paper describes the design of the analog part of the system.

In the second paper by Bosco Leung a passive filter implementation incorporating
a mixer function for applications at 10 MHz is described. The mixer function transforms the low-pass noise-shaping filter operation into a bandpass operation at 10 MHz.

In the third paper by Vincenzo Peluso continuous time gm-C filters are used to implement a bandpass noise-shaping characteristic at 10.7 MHz. Considerable attention is given to non-idealities such as noise, finite Q-factor and overall system linearity to reduce cross-modulation effects.

In the fourth paper by Armond Hairapetian a bandpass MASH structure is used to obtain a stable sixth order noise-shaping converter. The system includes a mixer stage to shift the signal frequency from about 81 MHz to 3.25 MHz with a 200 kHz bandwidth. The MASH structure uses a fourth order bandpass followed by a second order bandpass. As a result good cross-modulation performance is obtained.

The fifth paper by Tom Kwan describes a third-order switched-capacitor reconstruction filter for sigma-delta digital-to-analog converters. Coupled biquad structures using double-sampling switched capacitor techniques are used having the least component sensitivity to filter characteristic variations.

In the final paper by Angel Rodriguez-Vazques tools for automated design of sigma-delta modulators are described. Optimization procedures and behavioral simulation tools are included. The tools resulted in the design of practical 17 bit and 8 bit converter systems at different sampling frequencies.

Low-Power CMOS ΣΔ modulators for speech coding

Eric van der Zwan

Philips Research Laboratories, Eindhoven, The Netherlands

ABSTRACT

A ΣΔ modulator with continuous-time loopfilter has some important advantages compared to its discrete-time counterpart. Bandwidth requirements to the active elements of the loopfilter are relaxed, so that power consumption is reduced. Furthermore, aliasing is reduced, eliminating the need for an anti-aliasing filter at the modulator input. A 4th order, 64 times oversampling ΣΔ modulator with microphone input was designed and shows 80 dB dynamic range over the 300-3400 Hz voice bandwidth. THD is -72dB for a $40mV_{RMS}$ maximum input signal at 95 µA current consumption from a 2.2V supply voltage. The active die area of the modulator is 0.5 mm^2 in a standard 0.5µm CMOS process.

Introduction

In the past, a lot of effort was put into the realization of ΣΔ modulators as high resolution A/D and D/A converters. Nowadays these converters are widely used in communication and audio applications. The need arises for low power and low voltage converters for use in battery powered applications such as mobile telephones, hearing instruments or portable audio. The low power A/D converter described in this paper is suited for speech coding. Application examples are cellular telephones and hearing aids. Its principles can be used for other applications which need higher performance such as audio, at the cost of more chip area and power consumption.

In most ΣΔ A/D converters, the input signal is sampled and a discrete-time modulator is used. An anti-aliasing filter is needed in front of the sampler at the input in order to remove frequency components near multiples of the sampling frequency, which would otherwise be folded back into the signal band (see fig. 1). The demands on the slope of the anti-aliasing filter are relaxed by the oversampling ratio of the ΣΔ modulator. In a telephone or a hearing aid, the input signal comes from a microphone. Therefore a microphone preamplifier is included in order to get a reasonable input level for the ΣΔ modulator. The decimation filter at the back-end of the system will not be discussed in this paper.

Instead of a discrete-time modulator, a continuous-time configuration may be used in a ΣΔ modulator. In fact, Δ and ΣΔ modulators started that way [1],[2],[3]. Recently, several continuous-time ΣΔ modulators have been reported [4],[5],[6],[7],[8]. The power consumption of such a modulator can be significantly lower than in the discrete-time case. Furthermore, the requirements of the anti-aliasing filter are relaxed [3],[6],[9], and it can often be discarded.

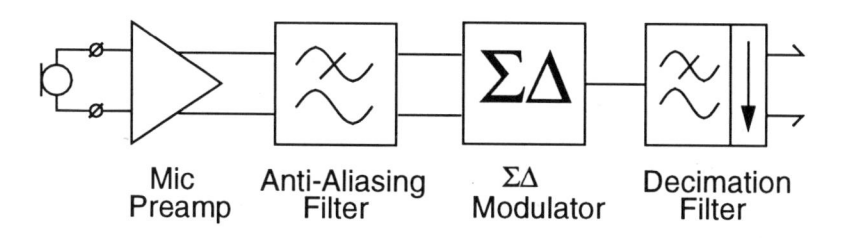

Figure 1 ΣΔ A/D converter system for speech coding

Discrete-Time ΣΔ modulators

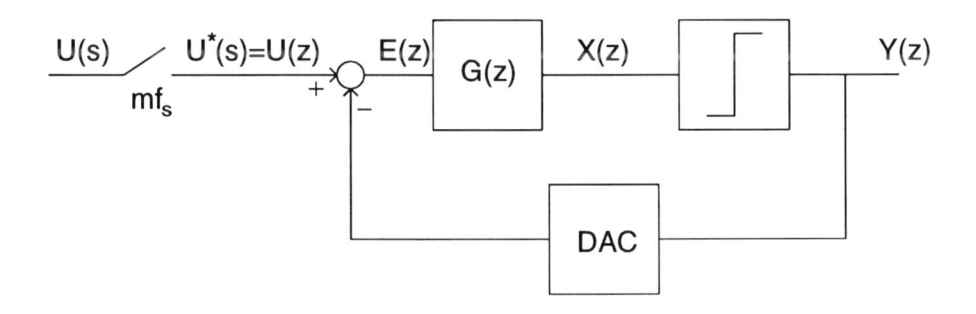

Figure 2   Block schematic of discrete-time ΣΔ A/D converter

The input of ΣΔ A/D converters usually is a sampler operating at a high oversampling ratio, which is followed by a discrete-time ΣΔ modulator (see fig. 2). The ΣΔ loop consists of a loopfilter G, a quantizer and a D/A converter in the feedback path. The loopfilter commonly is a switched capacitor filter or a switched current filter. The input signal is continuous-time and is indicated by U(s) in the Laplace domain. After sampling U(s) with frequency mf_s, where m is the oversampling factor and f_s the (Nyquist) sampling frequency, U*(s) is the sampled version of the input signal and can also be represented in the z-domain, with

$$z = e^{\frac{s}{mf_s}} \tag{1}$$

Since the quantizer is a nonlinear device, general linear system theory can not be applied to the ΣΔ modulator. Therefore, the quantizer is usually modelled as a gain stage c and an additive noise source (fig. 3). The D/A converter is modelled as a gain d. Although this model is not correct and does not explain certain high frequency effects, it can be used to get a feeling of the operation of a ΣΔ modulator and to predict its performance. From fig. 3 one can easily derive:

$$Y(z) = \frac{cG(z)}{1+cdG(z)} \cdot U(z) + \frac{1}{1+cdG(z)} \cdot N(z) \qquad (2)$$

Since G(z) is a lowpass filter, for low frequencies its gain is very large, $cdG(z) \gg 1$, so that

$$Y(z) \approx \frac{U(z)}{d} + \frac{1}{1+cdG(z)} \cdot N(z) \qquad (3)$$

Equation (3) shows that the output signal of the $\Sigma\Delta$ modulator is the input signal added to the filtered additive noise N(z). If the loopfilter is a lowpass filter and the additive noise is assumed to be white, the output signal has a noise frequency characteristic that increases with frequency with a slope that equals the order of the loopfilter. As an example, fig. 4 shows the output spectrum of a 4th order $\Sigma\Delta$ modulator at -20 dB input signal and a clock frequency of 512kHz. A lot of noise is present in the output signal. However the noise is shaped to high frequencies, and most of the noise power is concentrated around $mf_s/2$. In the signal band the noise level is very low.

Stability of the $\Sigma\Delta$ loop will not be addressed in detail in this paper. It

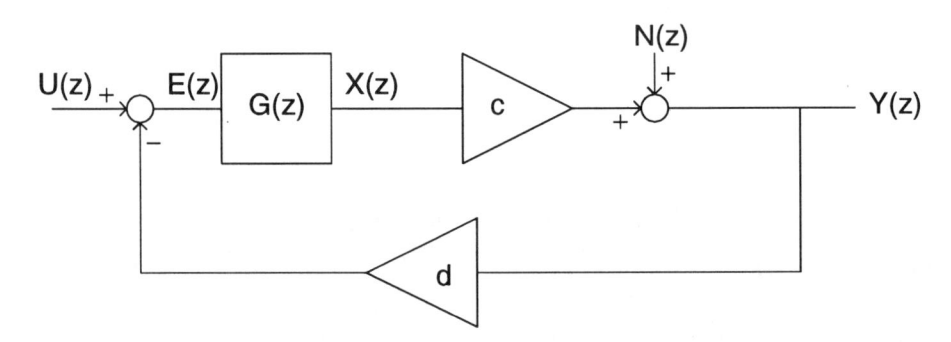

Figure 3   Modelling of the quantiser of a discrete-time $\Sigma\Delta$ modulator by a gain stage and an additive noise source

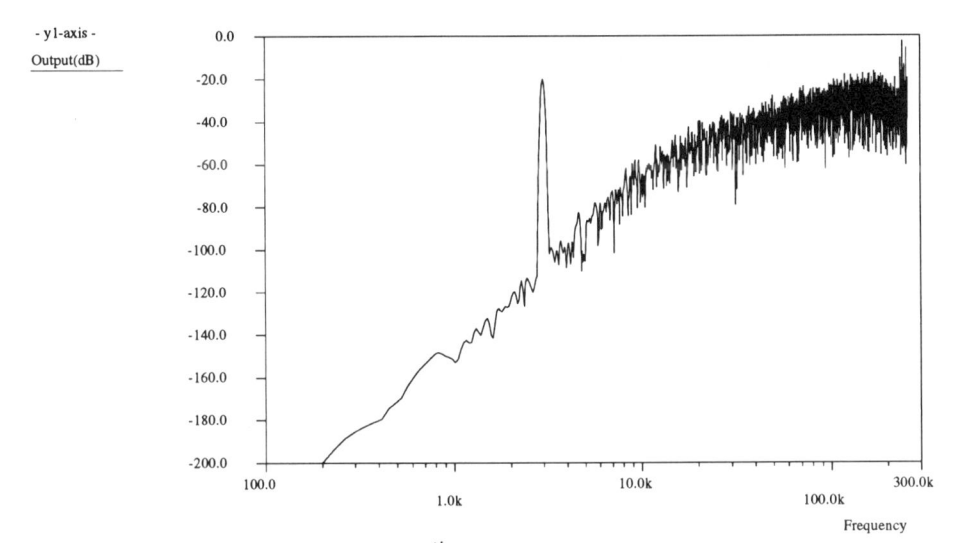

Figure 4   Output spectrum of a 4th order ΣΔ modulator with a -20dB 3kHz input signal and 512 kHz sampling frequency

is guaranteed by the coefficients of the loopfilter and internal clippers.

Several effects influence the power consumption of the ΣΔ modulator. Settling requirements in the discrete-time loopfilter impose bandwidth requirements on the active components inside this filter, thus effecting the current through these components. For example, in a switched capacitor filter the operational amplifier bandwidth must be at least five times the system clock frequency [10]. Noise requirements set a current requirement to the modulator input stage. Also distortion requirements generally play a role in the current consumption.

Since the input signal U is sampled directly, aliasing causes signals close to multiples of the sampling frequency to be folded back into the signal band, with no attenuation:

$$U^*(n \cdot m\omega_s - \omega) = U^*(\omega) \tag{4}$$

where $m\omega_s$ represents the oversampling frequency and n is an integer. The high oversampling ratio m relaxes the demands on the slope of an anti-aliasing filter compared to a system that samples at the Nyquist sampling frequency. Still such a filter is needed in order to decrease the sensitivity of the sampler to interference signals in the MHz range.

ΣΔ modulators with continuous-time loopfilter

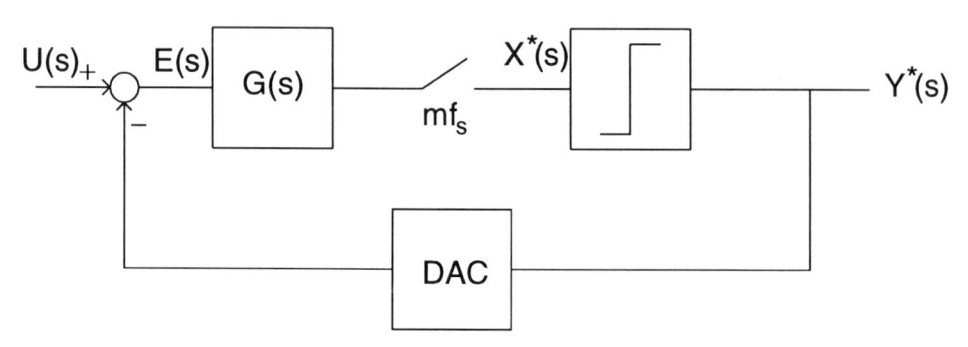

Figure 5   Block schematic of ΣΔ A/D converter with continuous-time loopfilter

If a continuous-time loopfilter is used in the ΣΔ modulator, the sampling can be performed *after* the loopfilter (fig. 5). Again, the quantizer can be modelled as a gain c and the D/A converter as a gain d (fig. 6). Note that the additive white noise and the output signal are

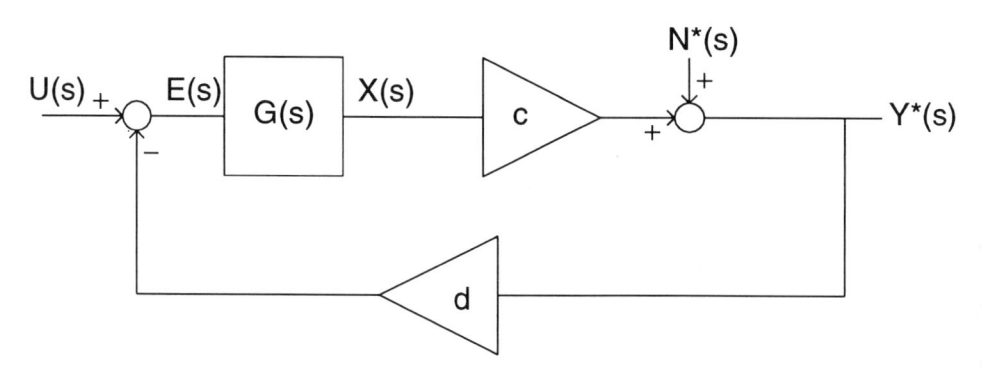

Figure 6   Linear model of ΣΔ A/D converter with continuous-time loopfilter

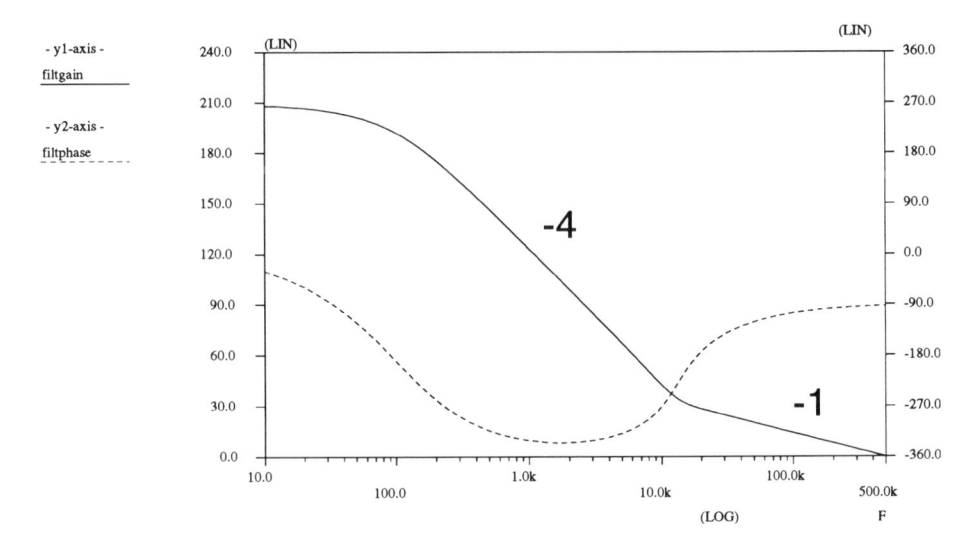

Figure 7   Frequency characteristics of a 4th order loopfilter

discrete-time signals, which is indicated by the asterisks. This means that gain c represents the sampler and the quantizer, and that gain d includes a hold stage.

A similar analysis as in the discrete-time case reveals the formula for the output spectrum of the modulator:

$$Y^*(s) \approx \frac{U(s)}{d} + \frac{1}{1 + cdG(s)} \cdot N^*(s) \qquad (5)$$

No settling requirements are imposed to the loopfilters. The unity gain frequencies of the integrators inside the loopfilter are only in the order of magnitude of 100 kHz. The power consumption is mainly determined by the noise and distortion, and not by speed requirements. It can therefore be lower than in the discrete-time case. In fig. 7 the frequency characteristics of a 4th order loopfilter are shown. For high frequencies, the loopfilter is 1st order for stability reasons.

A possible disadvantage of continuous-time $\Sigma\Delta$ modulators is their sensitivity to clock jitter. The output of the feedback D/A converter are rectangular voltage or current pulses. Clock jitter causes variation in the pulse widths, resulting in noise in the D/A converter output signal. However, the dynamic range requirements for speech coding are fairly relaxed (about 80dB), so that the noise caused by clock jitter is still of minor importance.

Less aliasing occurs than in the discrete-time case. Consider a full scale input signal $U(\omega)$ with a frequency close to the sampling frequency $m\omega_s$. In a discrete-time modulator, this signal would be sampled at the input and a frequency component $m\omega_s-\omega$ would be aliased back into the baseband with 0dB gain. However, if the loopfilter is continuous-time, and the sampling is performed after the loopfilter, one can easily show that the component aliased back into the baseband is reduced:

$$Y(m\omega_s - \omega) = \frac{G(\omega)}{dG(m\omega_s - \omega)} U(\omega) \tag{6}$$

where d is the DAC gain and $G(\omega)$ is the loopfilter transfer function.

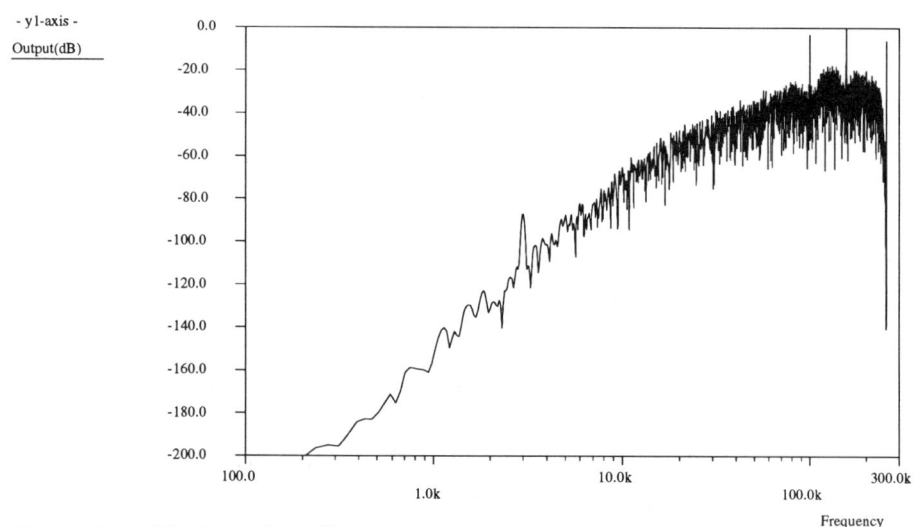

Figure 8   Aliasing of a full scale input signal 3kHz next to the sampling frequency

This is confirmed by the simulation result of fig. 8, where a full scale 509 kHz input signal was applied to a ΣΔ modulator with 512 kHz sampling frequency. It shows that the aliased frequency component at 3 kHz is at -87dB with respect to full scale. Usually additional anti-aliasing filtering is not necessary.

Low power ΣΔ modulator implementation

Apart from the choice between a discrete-time or a continuous-time loopfilter, there are several other considerations to be made when a ΣΔ modulator must be designed with low power consumption.

- The circuit noise should dominate the quantization noise of the modulator. The latter can be decreased by increasing the modulator order, without adding to much power consumption, whereas the circuit noise is much stronger related to the power consumption.
- The maximum possible output signal, or the modulator overload level, should be as large as possible, which means as close as possible to the bitstream output levels, so that maximum dynamic range with respect to the circuit noise is achieved.
- Design for low power does not necessarily mean that the supply voltage must be as low as possible. At very low supply voltages, the signal swings may become so small that it becomes difficult to achieve the desired dynamic range at reasonable current levels, so that the power consumption actually increases!
- No overkill in dynamic range or distortion should be designed, because both are strongly related to the power consumption.

The input signal of the A/D converter comes from a microphone and was designed at maximum 40 mV$_{RMS}$. Dynamic range for the voice band ΣΔ modulator must be 80dB, so that the total input referred noise should not exceed 4 µV. The harmonic distortion demands are usually quite relaxed for speech coding, for example typically -50 dB for telephony. Other applications may need better linearity.

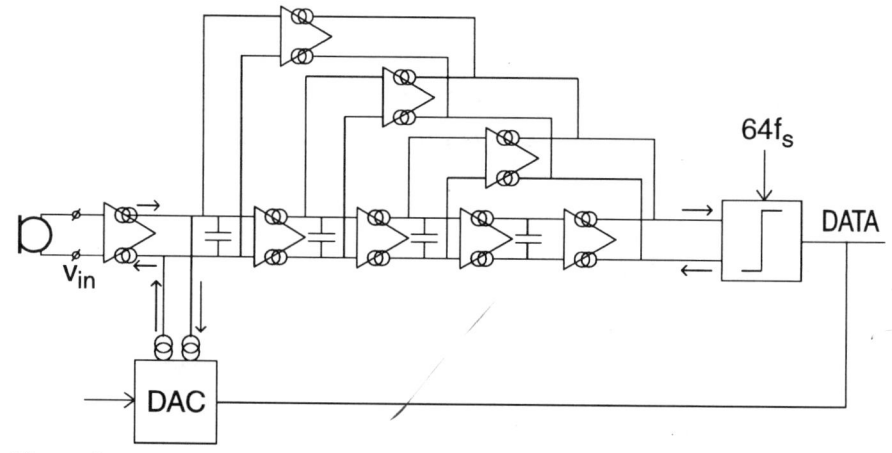

Figure 9   ΣΔ modulator block diagram

In fig. 9 the block diagram of a low power ΣΔ modulator for speech coding is shown. It is a 4th order modulator operating at an oversampling ratio of 64, giving an in-band quantization noise level of -94dB. This is well below the specified ΣΔ modulator dynamic range of 80dB. The circuits can now be designed such that this desired noise level is obtained, and the quantization noise can be neglected.

The loopfilter was implemented using transconductance-C integrators (fig. 9). The feedforward coefficients are also transconductors, so that their outputs can easily be added by connecting them together. Stability of the ΣΔ modulator during overload is guaranteed by the limited input range of the transconductors, which effectively results in clipping the integrator outputs. The filter coefficients are scaled such that if the input signal exceeds the overload level, first the fourth integrator clips. Due to the feedforward structure of the filter still a third order ΣΔ modulator is left. If the input signal is even higher, also the third integrator clips, and a second order modulator is left, which is always stable. Thus the ΣΔ modulator reduces its order during overload (graceful degradation).

In fact the ΣΔ modulator of fig. 9 has a current input and one could speak of a "current domain ΣΔ modulator". However, the first transconductor, which is outside the feedback loop, is necessary to convert the microphone output voltage to an input current for the modulator. Note that the noise and distortion of this first transconductor is dominant and will determine the overall performance of the system! Most of the current is therefore spent in this first transconductor, which can be seen as the "microphone preamplifier" (fig. 1).

From fig. 9 it appears that the first transconductor can be designed in a current path together with the feedback D/A converter (see fig. 10). The transconductor is a differential pair degenerated by resistor R, so the transconductance is approximately $\frac{1}{R}$. The transistors of the differential pair are operating in weak inversion. The output currents of

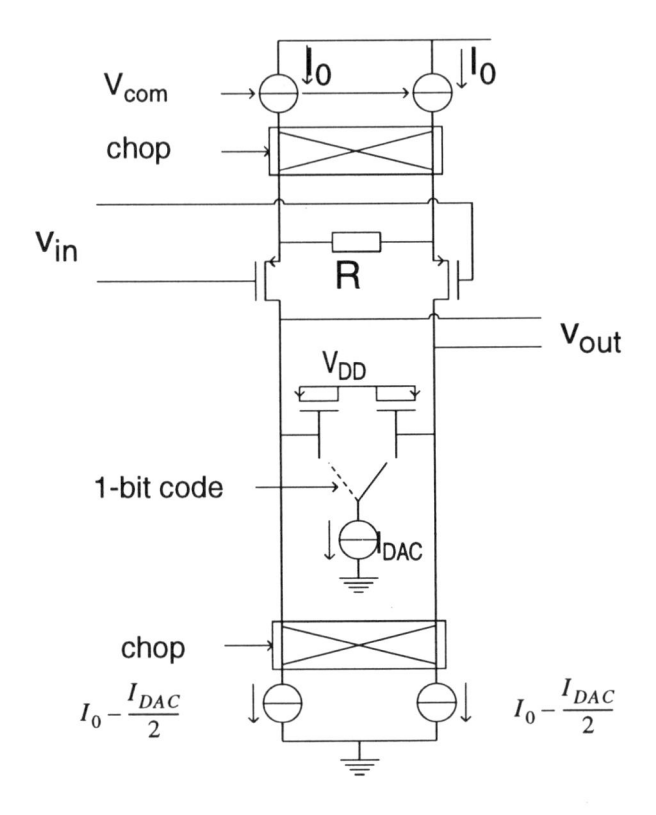

Figure 10 Circuit implementation of input transconductor, first integrator and feedback D/A converter

the differential pair are applied to the first integrator, which is the series connection of two gate oxide capacitors. The feedback D/A converter of the $\Sigma\Delta$ modulator is DC current source I_{DAC}, which is switched between the capacitor terminals by the bitstream output code. Thus, the current through the first integrator (capacitor) is

$$I_{C_1} = \frac{v_{in}}{R} \pm \frac{I_{DAC}}{2} \qquad (7)$$

The DC bias current sources are chopped in order to modulate their contribution to the 1/f noise and offset to a frequency outside the signal band. The remaining noise of the circuit consists of the thermal noise of the resistor and the DC bias current sources, and the 1/f noise of the input transistors. The input transistors are not chopped because switching at the modulator input may introduce aliasing problems again. Note that the noise is almost current independent: only the thermal noise contribution of the DC bias current sources varies with the current, but this is only a small contribution.

Once the transconductance based on the noise specifications has been chosen by choosing the appropriate degeneration resistor R, the current consumption is determined by the linearity that has to be achieved in the degenerated input pair. If the transconductance of the input transistors is g_m, the overall transconductance G_m of the degenerated pair is

$$G_m = \frac{g_m}{2 + g_m R} \qquad (8)$$

The larger g_m, the better the linearity, but also the larger the current.

The other transconductors in the $\Sigma\Delta$ modulator of fig. 9 have far more relaxed requirements with respect to noise and distortion. They are implemented as shown in fig. 11. Since at least some linearity is required in order to prevent quantization noise to be folded back into the signal band, degenerated differential pairs are used, but the degeneration is done by MOS transistors [11]. A supply current of only

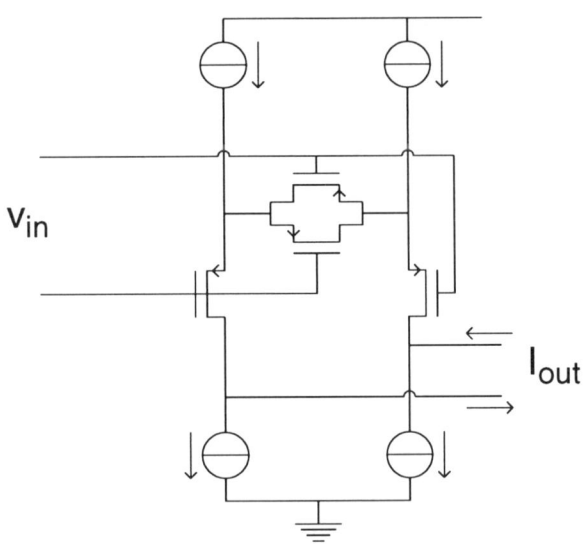

Figure 11 Implementation of the other loopfilter transconductors

a few µA per stage is sufficient to meet the relaxed noise and distortion requirements.

The last part of the ΣΔ modulator is the quantizer. It is implemented as a cross-coupled latch. Since the loopfilter output is a current, the signal can be applied to the latch by a folded cascode configuration (fig. 12).

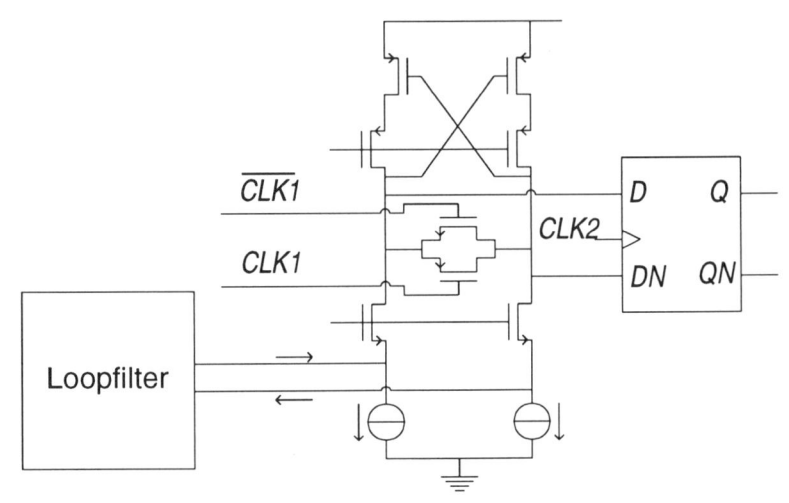

Figure 12 Sampler/quantizer of the ΣΔ modulator

Simulation of noise and distortion

The simulated noise performance of the $\Sigma\Delta$ modulator operating at a supply voltage of 2.2V and a current of 95 µA is shown in fig. 13. Clearly the circuit noise dominates the quantization noise in the signal band of 300-3400 Hz. The effect of the chopping of the DC bias current sources is also clearly shown: It reduces the total in-band noise from 8.6 to 3.9 µV$_{RMS}$. As expected, the majority of the noise is due to the first transconductor, which therefore also consumes the majority of the current (65 µA).

The noise and distortion of the loopfilter as a function of the supply current is shown in fig. 14. The noise is fairly constant, as expected; a slight increase with the current can be observed because of the increase of the noise contribution of the DC bias current sources. From this picture it is also clear that linearity increases with the supply current. THD = -50dB is obtained at a current of only about 30 µA. However, the actual circuit design was made with 95 µA supply current, resulting in a THD of better than -70 dB.

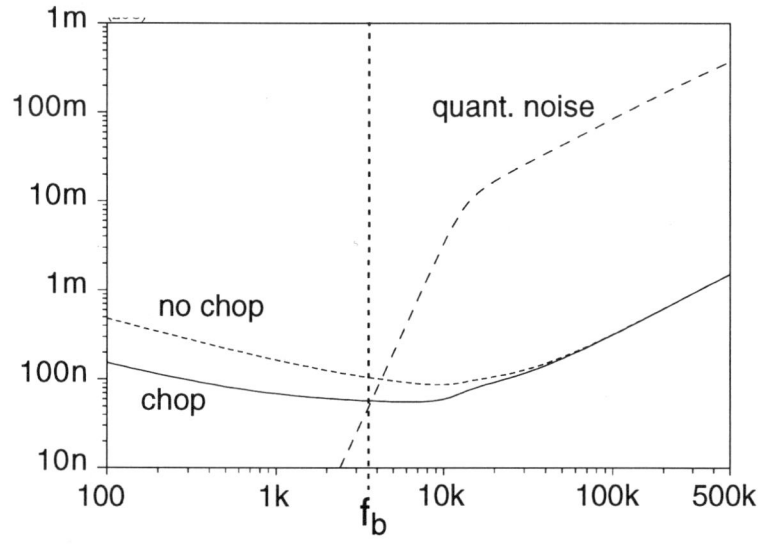

Figure 13   Simulated modulator input referred noise

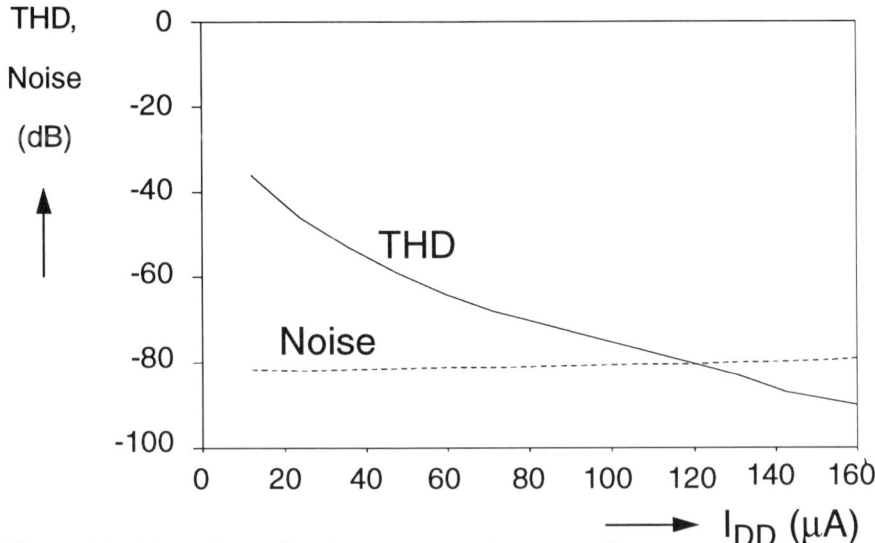

Figure 14 Linearity and noise vs. current consumption

Measured results

The ΣΔ modulator was processed in a standard 0.5 μm digital CMOS process. The active area is 0.5 mm^2. The measured noise and distortion of the ΣΔ modulator operating at a 2.2V supply voltage and at 95 μA

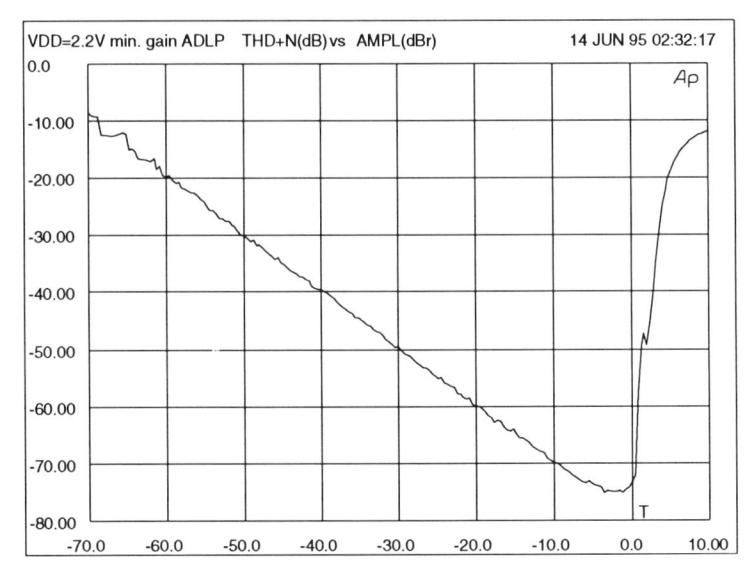

Figure 15 Measured noise and distortion of the ΣΔ modulator

current consumption are shown in fig. 15. A dynamic range of 80dB achieved and harmonic distortion at full signal is below -72dB. Aliasing was measured applying a full-scale signal with a frequency close to the sampling frequency to the input. The aliased frequency components in the signal band were below -63dB. The discrepancy with the theoretical result is probably due to the fact that the signal is mixed back to the signal band by distortion rather than by aliasing.

Conclusions

This paper shows that a ΣΔ modulator with a continuous-time loopfilter has several advantages compared to its discrete-time counterpart. Due to more relaxed bandwidth requirements to the active elements the power consumption is lower. Aliasing is reduced, so that an external anti-aliasing filter can be discarded. A possible disadvantage of this type of ΣΔ modulator, especially for higher end applications, may be its sensitivity to clock jitter.

A ΣΔ modulator for speech coding was designed. It has a microphone input and the maximum input level is 40mV$_{RMS}$. Its input referred equivalent noise level is 4μV$_{RMS}$ for 80 dB dynamic range. It was optimized for low power on system level by choosing a 4th order modulator with a high overload level and by using a continuous-time loopfilter. The circuit was optimized by minimizing the number of current paths and by scaling of the current in circuit parts that have a minor impact on noise and distortion. Operating at a 2.2V supply voltage and with 95 μA supply current (0.2mW), THD is below -72dB. It was shown that if more distortion is allowed, the current may be further reduced. For example, a THD of -50dB can be obtained at a current consumption of 30μA (0.07mW).

References

[1] F. de Jager, "Deltamodulation, a method of P.C.M. transmission using the 1-unit code," *Philips Res. Rep*, vol. 7, pp. 442-466, Nov. 1952.

[2] H. Inose, Y. Yasuda, J. Murakami, "A telemetring system by code modulation - $\Delta\Sigma$ modulation," *IRE Trans. Space Electronics and Telemetry*, vol. 30, pp. 204-209, Sep. 1962.

[3] J.C. Candy, "A use of double integration in sigma-delta modulation," *IEEE Trans. Commun.*, vol COM-33, no. 3, pp. 249-258, Mar. 1985.

[4] P.J.A. Naus, E.C. Dijkmans, "Multibit oversampled $\Sigma\Delta$ A/D converters as front end for CD players," *IEEE J. Solid-State Circuits*, vol. 26, no. 7 pp. 905-909, July 1991.

[5] V. Comino, M.S.J. Steyaert, G.C. Temes, "A first-order current-steering sigma-delta modulator," *IEEE J. Solid-State Circuits*, vol. 26, no. 3, pp. 176-183, Mar. 1991.

[6] R.G. Lerch, M.H. Lankemeyer, H.L. Fiedler, W. Bradinal, P. Becker, "A monolithic sigma-delta A/D and D/A converter with filter for broad-band speech coding," *IEEE J. Solid-State Circuits*, vol. 26, no. 12, pp. 1920-1927, Dec. 1991.

[7] Y. Matsuya, J. Yamada, "1V power supply, low-power consumption A/D conversion technique with swing-suppression noise shaping," *IEEE J. Solid-State Circuits*, vol. 29, pp. 1524-1530, Dec. 1994.

[8] E.J. van der Zwan, E.C. Dijkmans, "A 0.2mW CMOS $\Sigma\Delta$ modulator for speech coding with 80 dB dynamic range," *ISSCC Dig. Tech. Papers*, pp. 232-233, Feb. 1996.

[9] S.H. Ardalan, J.J. Paulos, "An analysis of nonlinear behavior in delta-sigma modulators, "*IEEE Trans. Circuits Syst*, vol. CAS-34, no. 6, pp. 593-603, June 1987.

[10] R. Gregorian, G.C. Temes, "Analog MOS Integrated Circuits for Signal Processing," New York, Wiley, 1986.

[11] F. Krummenacher, N. Joehl, "A 4-MHz CMOS continuous-time filter with on-chip automatic tuning," *IEEE J. Solid-State Circuits*, vol. 23, no. 3, pp. 750-758, June 1988.

Passive Sigma-Delta Modulators with Built-in Passive Mixers for Mobile Communications

Bosco Leung and Feng Chen

Electrical and Computer Engineering Department
University of Waterloo, Canada, N2L 3G1

1. Introduction

The push towards implementing radio receivers in a power efficient and cost-effective manner has led to the development of new circuits and architectures, especially for IF digitization. Conventionally baseband digitization is performed only after the second IF stage because of speed and power constraints imposed by the ADC. Pushing the analog/digital boundary to ever higher frequency allows more filtering functions (e.g. channel selection) to be performed digitally, which leads to further simplification of system, reduction of power consumption, higher degree of integration and lower cost.

2. Approaches of IF Signal Digitization

2.1. Digitization of IF Signal with Bandpass ΣΔ Modulator

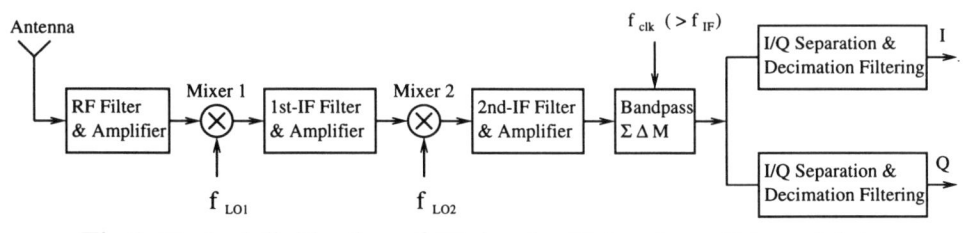

Fig.1: Typical digitization of IF signal with bandpass ΣΔ modulator

For narrow-band receiver, bandpass ΣΔ ADC can be used to digitize an IF signal directly [1], as shown in Fig. 1. Because the IF signal is not mixed down to baseband, the approach does not suffer from DC offset and low frequency noise problem. In addition, there is no I/Q mismatch problem for quadrature modulation since the two paths can be matched perfectly in digital domain. As with lowpass oversampled ADC, the high sampling rate relaxes the requirement on pre-filtering. However, one of the major problems with the approach is that it has to sample the IF signal at a frequency higher than IF carrier frequency, typically by a factor of 4. This makes the approach unattractive for high IF digitization in terms of power consumption. The other drawback of the approach is that any center frequency shift (e.g. due to pole location movement as a result of circuit imperfections) can translate into significant SNR degradation as shown in Fig. 2. In contrast, if digitization is done by first mixing the IF signal down to baseband and then quantizing using a lowpass sigma-delta modulator, (our proposed scheme as discussed in the next section) the center frequency is always at DC. Hence any potential pole movements will not shift the quantization noise null position, which stays at DC.

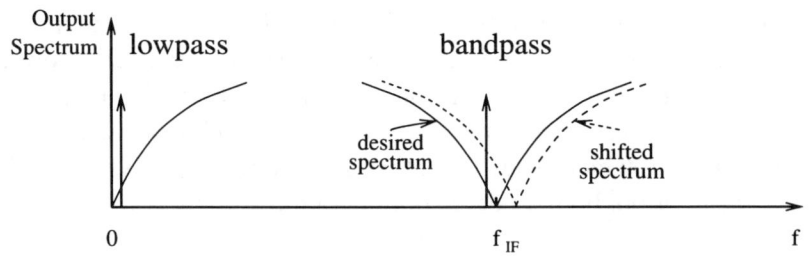

Fig.2: Typical output spectra for bandpass and lowpass ΣΔ modulators

2.2. IF Sampling/Digitization with Lowpass ΣΔ Modulator

Our proposed approach is shown in Fig. 3 and digitizes the output of the 1st-IF stage. However, since the I and Q components are separated in the analog domain, there will be path mismatch. Here the dashed box in Fig. 3 is redrawn and is shown in Fig. 4(a). It consists of a mixer for mixing the IF signal down to the DC by setting $f_{LO} = f_{IF}$, a lowpass filter for filtering out the frequency component at the sum frequency of the LO and IF carrier frequencies as well as an ADC for digitizing the baseband signal.

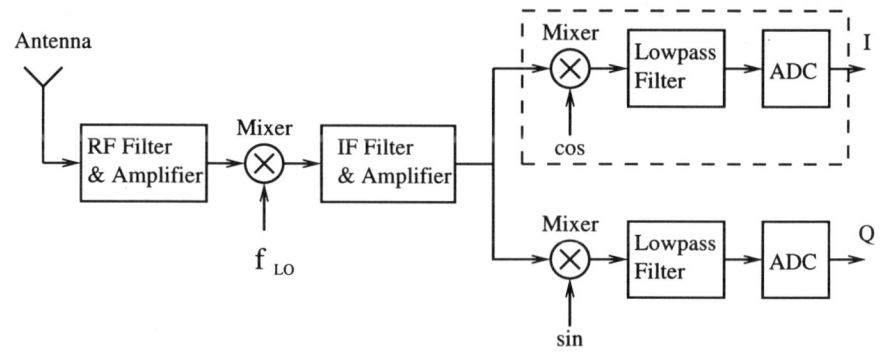

Fig.3: Typical digitization by mixing IF signal down to baseband

Assuming the mixer drives a capacitive load and the mixer is implemented passively, e.g. a simple MOS switch, the resulting circuit is shown in Fig. 4(b) in which the low-pass filter is removed because it is not necessary to filter out the replica at multiples of the sampling frequency in the sampled data domain. By examing the circuit in Fig. 4(b), it can be seen that the part formed by the switch mixer and its loading capacitor is identical to a S/H (sampled and hold) circuit, which is inherent in any ADC. Therefore the mixing function can be merged with the sampling function in the ADC to simplify the system design.

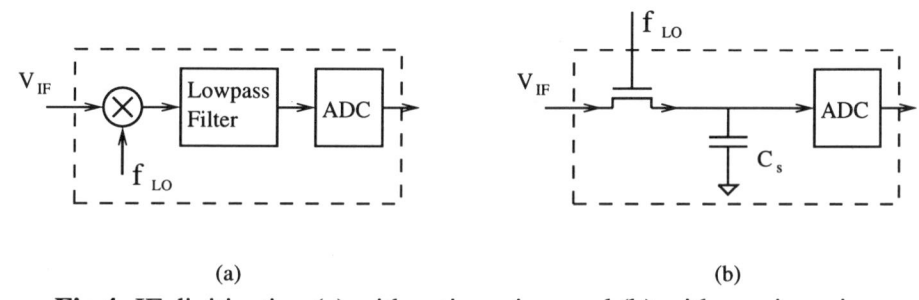

Fig.4: IF digitization (a) with active mixer and (b) with passive mixer

3. System Level Considerations of Passive ΣΔ Modulator with Built-in Passive Mixer

3.1. Introduction

Among all the active components in an active sampled-data time based ΣΔ modulator, the operational amplifiers (OpAmp) in the integrators consume most of the power. It can be expected that considerable amount of power would be saved if the active loop filter is replaced with a passive network. Past implementation of 1-bit ΣΔ modulator with a passive RC loop filter (using discrete R and C) [2] consumes significant amount of power. In the following sections critical design issues which affect the SNR performance and power consumption of a passive ΣΔ modulator will be identified.

Since passive loop filter does not have any gain, the comparator is required to have high enough resolution so that the required SNR is achieved. High resolution of the comparator, however, can translate into high power consumption. To alleviate the requirement for power saving, a switch only (hence passive) gain-boost network will be proposed.

To reduce circuit complexity and power consumption even further, the sampler in the A/D converter is redesigned and used for mixing as well, as discussed in the last section. Even though the sampler can be used as mixer, one difference to note though is this sampler (mixer) has an input (actually more than one if one assumes there are strong adjacent channel interfering signals in addition to the small desired signals) whose frequency is higher than the sampling (mixing) frequency f_{sample} (f_{LO}). In addition in a mixer performance measures such as IM3 (third order intermodulation product) that is not normally important in a S/H circuit can become an issue. This poses quite a different set of design requirement on the design of a switch mixer from that of a baseband sampling switch [3].

3.2. System Level Design of the Modulator

In this section we concentrate on the discussion of the modulator. The block diagram of the 1-bit passive $\Sigma\Delta$ modulator is shown in Fig. 5. The loop filter is replaced with a passive lowpass filter. Its operation is basically the same as in the active case. The digital output contains a highpass quantization noise which can be filtered out with a digital lowpass filter. But the large loop gain needed to suppress the quantization noise is provided by the comparator.

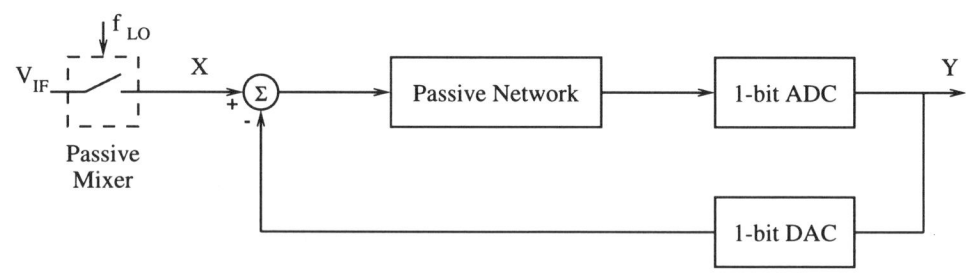

Fig.5: The block diagram of a passive $\Sigma\Delta$ modulator with 1-bit quantizer

The modulator with a linear model for the 1-bit quantizer is shown in Fig. 6. The H here is the transfer function for the loop filter. The comparator input referred noise is modeled as an additive noise source E_{com}, and the quantization noise as E_Q. The G here is the equivalent gain factor in the comparator. The G is defined as the ratio of the output rms value of comparator to its input rms value. It is assumed to be constant and is usually on the order of thousands. Since the $\Sigma\Delta$ modulator is a nonlinear system, G can only be determined from simulation. We may, however, get an estimate of G as follows: suppose a zero input is applied to a first-order modulator (e.g. the loop filter consists of one RC section); ideally the output code of the quantizer will oscillate at $f_s/2$. Therefore the output of the feedback DAC (or the input of the loop filter) will consist of a square wave oscillating at $f_s/2$. The output of the loop filter (or input to the comparator) will consist approximately of a square wave oscillating at $f_s/2$, but with its amplitude attenuated. The attenuation is dependent on the transfer function H of the loop filter at $f_s/2$. Assuming the reference voltage to be $\pm 1V$, this amplitude is roughly $1/|H(f_s/2)|$ and serves to give an estimate of the value G defined above.

With the linear model, the transfer function of the modulator is given by the following expression.

$$Y = \frac{GHz^{-1}}{1+GHz^{-1}} X + \frac{E_Q z^{-1}}{1+GHz^{-1}} + \frac{Gz^{-1}}{1+GHz^{-1}} E_{com} \approx X + \frac{E_Q}{GH} + \frac{E_{com}}{H} \quad \text{(in baseband)} \quad (1)$$

Examining the above baseband expression, it is found out that the first two terms here are the same as in the active case: the input signal and the attenuated quantization noise. The large loop gain GH needed in the baseband is provided by the comparator. However, there is a third term here which is related to the comparator input referred noise E_{com}.

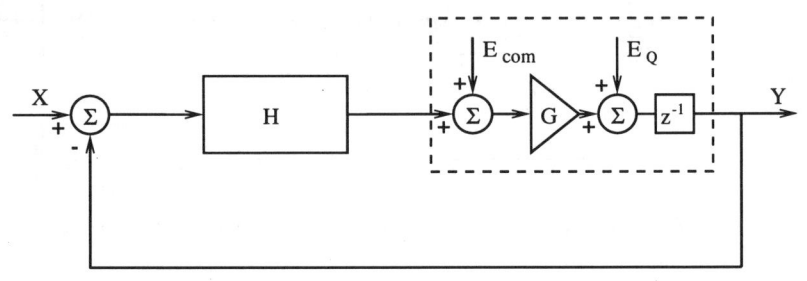

Fig.6: A passive ΣΔ modulator modeled with a linear quantizer model

The transfer function for quantization noise E_Q is defined as:

$$T_{E_Q} = \frac{1}{GH} \quad (2)$$

and for the equivalent comparator input noise E_{com}, its transfer function is defined as

$$T_{E_{com}} = \frac{1}{H} \quad (3)$$

It can be seen that both T_{E_Q} and $T_{E_{com}}$ have a highpass transfer function. One difference is that in the baseband E_{com} in the $T_{E_{com}}$ expression is not experiencing any attenuation because the only term in the denominator, H of the loop filter, does not have any DC gain. On the other hand E_Q in the T_{E_Q} expression is still attenuated by the large G factor (provided by the comparator) in the denominator. Therefore, for a given amount of E_{com}, one can always design a modulator with a large loop gain (e.g. using a higher order loop filter) and suppress the quantization noise to a small enough level, so that the E_{com} becomes dominant. In this sense it is seen that the resolution of comparator plays a key role in determining the overall SNR of the system. This is one of the essential differences between a passive and an active modulator.

Once the order of the loop filter is fixed, the pole locations have to be determined. Let us take a look at a first order modulator again. From the above estimate, it seems that G can be increased arbitrarily by pushing the poles' frequencies of the loop filter(H) to as low a frequency as possible (hence ensuring as low a $|H(f_s/2)|$ as possible). This increases the G factor in the denominator of equation (2). Fig. 7 shows the frequency response H of a first order (real pole only) lowpass filter with two different pole locations. As an example it can be seen that as f_{pole} moves from 8KHz to 1KHz, the increase of high frequency attenuation of H at $f_s/2$ (i.e. the increase of the estimate of G) is the same as the decrease of H at the baseband edge (i.e. at 20KHz). Therefore, the product GH at $f_{band-edge}$ in equation (2) remains relatively constant with respect to the pole location. Accordingly, the quantization noise term E_Q/GH remains relatively constant with respect to the pole location.

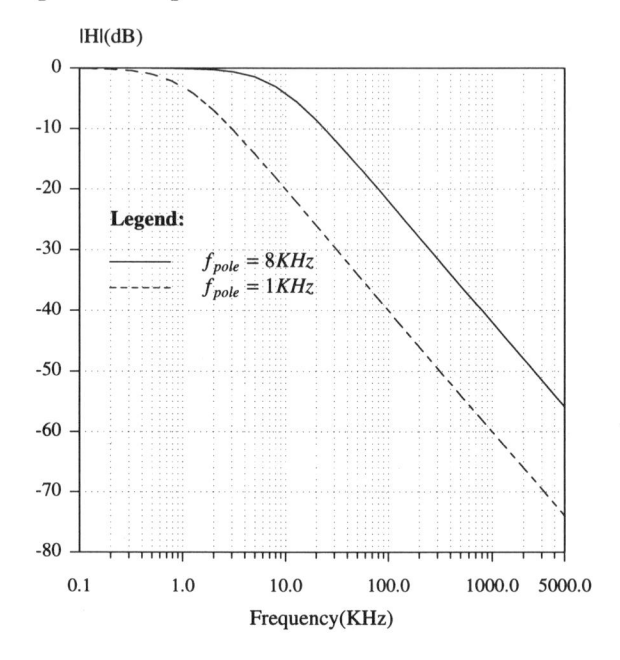

Fig.7: The frequency response of a 1st-order loop filter (RC assumed)

On the other hand the impact of poles placement on E_{com} is such that the factor $1/|H|$ in equation (3) increases in the baseband as the poles are moved to lower frequencies and hence the input referred E_{com} increases according to eqn (3). Eventually E_{com} will dominate the overall noise contribution and determine the modulator SNR for a given filter order. Therefore it seems that once the filter order is fixed, one design strategy is to fix the pole location such that E_{com} dominates. Then the resolution (SNR) of the modulator is determined by E_{com} which depends on the comparator power dissipation. This in turns determines the overall modulator power consumption and gives a rough correlation between the achievable SNR and the modulator power dissipation for the given filter order. As an illustration the above analysis is applied to the design of a modulator for a 10MHz IF input with a 13-bit resolution and 20kHz bandwidth. It is shown that a 2nd-order modulator can achieve the desired SNR (assuming quantization noise is dominant) without resulting in an unacceptably low comparator input level.

4. Circuit Design of the Prototype Modulator

In this section the circuit design aspects of the prototype second order modulator with built-in mixer are discussed.

4.1. Switch Mixer Design for a 10MHz IF Input

The switch mixer is designed for a 10MHz IF input. Its major difference from that of a mixer for RF input is that it does not have to drive a 50Ω load. The mixer here only drives capacitive load, instead, which usually displays a relatively high impedance. For example, in a passive $\Sigma\Delta$ modulator a sampling capacitor has a typical value of 0.2pF which results in an impedance of $500K\Omega$ at a 10MHz sampling rate. Therefore the switch mixer can be designed with a smaller W/L ratio, and the LO does not have to drive as large a load. The LO thus can be designed to have a rather sharp edge and this helps to improve the intermodulation distortion performance. In addition, since the strategy here is to adapt the sampler to be the switch mixer, the difference in the set of design requirement between a switch mixer and a baseband sampling switch should be considered. As an example in order to reduce the error caused by the nonlinear channel resistance which can cause severe IM3 the switch has to have a larger W/L [3] (it is increased from 3μm/1.2μm to 20μm/1.2μm). To further reduce the effect of this nonlinear channel resistance (due to changing V_{gs} during the falling edge of f_{LO}) a sharp falling edge is used. Finally to eliminate the charge injection error from the switch (a problem for both a mixer and a sampler), bottom plate sampling can be used.

4.2. Switched Capacitor Based Loop Filter Implementation

If the loop filter is implemented using cascaded RC sections, this may result in a few potential problems. First of all, the subtraction of the IF input and the feedback v_{ref} is most easily done through two resistors and the inherent voltage divider in this arrangement will attenuate both signals by 6dB. Secondly, the resistors are not easily integrable. Finally since the modulator is now operating in a continuous time domain, the mixer and the sampling function cannot be merged easily.

Fig. 8 shows the circuit diagram of a complete passive $\Sigma\Delta$ modulator with a built-in switch mixer. This is one half of a differential switched capacitor implementation. All switches are CMOS switches. During ϕ_{m2}, the 10MHz IF signal is mixed down to the baseband by the switch mixer that is composed of two switches, S_{m1} and S_{m2}, and is sampled and held on capacitor C_{R1}. Each mixing switch is designed to have a W/L of $20\mu m/1.2\mu m$. Their control clock, ϕ_{m2}, is equivalent to the LO and is set at 10MHz. It is derived from the modulator clock ϕ_2 via buffering and thus achieves a sharp falling edge of 0.3ns. Bottom plate sampling is used to eliminate the charge injection error. The loop filter is designed to have two poles at 8kHz and 34kHz. To compensate for the phase shift introduced by the poles, a zero at 750kHz is introduced in the loop filter by the switched capacitor C_{R0}. C_{com} is the input capacitance of the comparator. Because of the passive loop filter's attenuation, the comparator has to resolve a small input voltage of 150μVrms, thus posing a high requirement on the comparator resolution.

4.3. Comparator Design

The comparator architecture is selected to contain a pre-amplifier followed by a regenerative latch. One major consideration in the comparator design is its input referred noise coming from the preamplifier. Assuming the two MOS input transistors of the preamp are operating in the saturation region, the preamp thermal noise power is given by

Mixing&Sampling

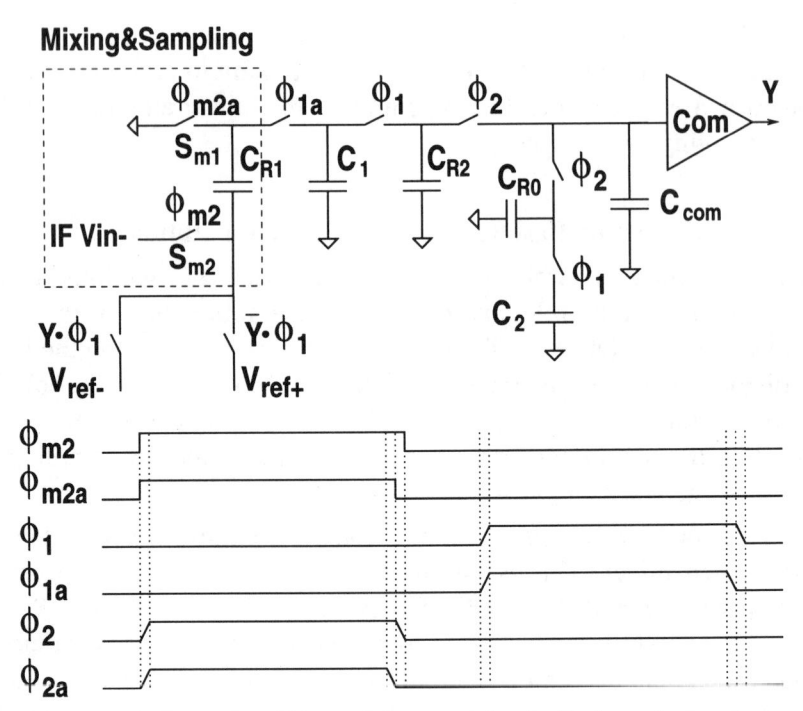

Fig.8: A SC passive ΣΔ modulator with a built-in switch mixer

$$E^2_{preamp} = \frac{8kT}{3}\left\{\frac{1}{g_{m\_in}}\right\} \times 2 \qquad (4)$$

where g_{m_in} is the transconductance of the input transistors in the preamplifier. In the present design the total comparator input referred noise (here assume to be dominated by the thermal noise of the preamplifier input pairs) is designed to be lower than the smallest input level applied to the comparator (150μVrms) as derived from the last section. System level simulations verify that with this level of comparator noise, the overall SNR is still primarily limited by the quantization noise. With this design E_{com} is estimated to be $13nV/\sqrt{Hz}$, according to SPICE simulations and serves to determine the value of g_{m_in} in equation (4). This is achieved with a bias current of 6.25uA in each input transistor (PMOS) whose W/L ratio is 200μm/1.2μm. With this value of W and L the loading due to C_{gs} on the loop filter remains acceptable.

5. A Switch-only Gain-boost Network

5.1. Mechanism

Even though the passive loop filter is very power efficient, the noise immunity is sacrificed because there is no gain in it. Fig. 10(a) shows the conventional switched capacitor lowpass filter implementation. To achieve voltage gain without using any operational amplifier, a passive gain-boost network with a lowpass feature is proposed and shown in Fig. 10(b).

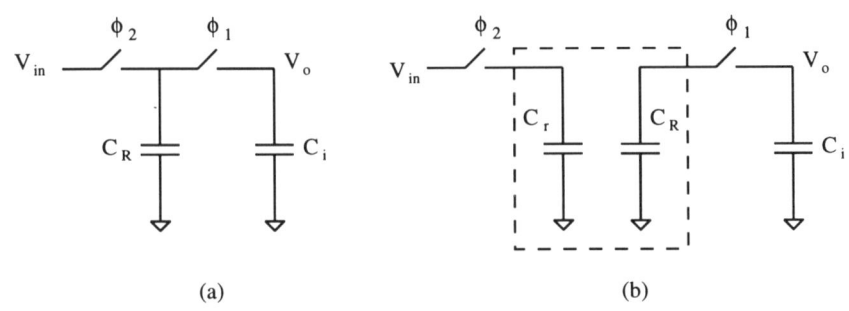

Fig.10: Switched-capacitor lowpass filter, (a) with no gain and (b) switch-only gain-boost implementation

The operation of the circuit is based on the principle that if the total sampling capacitance of the network is reduced from c_r to C_R during ϕ_1 while keeping the amount of signal charge on the sampling capacitor unchanged, a net voltage gain can be achieved. The transfer function for the gain-boost network was derived and shown as follows:

$$H_{SC} = V_o(z)/V_{in}(z) = \frac{z^{-1}C_r/(C_R+C_i)}{1-C_i/(C_R+C_i)z^{-1}} \qquad (5)$$

C_r is the sampling capacitance during ϕ_2 and C_R is the total sampling capacitance during ϕ_1. Therefore, a desired gain can be obtained by choosing a proper ratio for C_r/C_R. The size of C_R is determined once the pole position of the lowpass filter is fixed. Together with the desired gain boosting ratio the value for c_r is next determined. Notice this network does not provide power gain and therefore is used only at the input of the $\Sigma\Delta$ modulator.

5.2. Circuit Implementation

Fig. 11 shows the circuit schematic of the network used in our passive ΣΔ modulator. The operation of the circuit is now explained. Assume that we have N (designed to be 6 in the present design) identical sampling capacitor c_r. During ϕ_2, all N sampling capacitor c_r are in parallel and their top plates are charged to either V_{ref+} or V_{ref-} and their bottom plates are charged to V_{in}. Then during ϕ_1, they are reconfigured in a series connection.

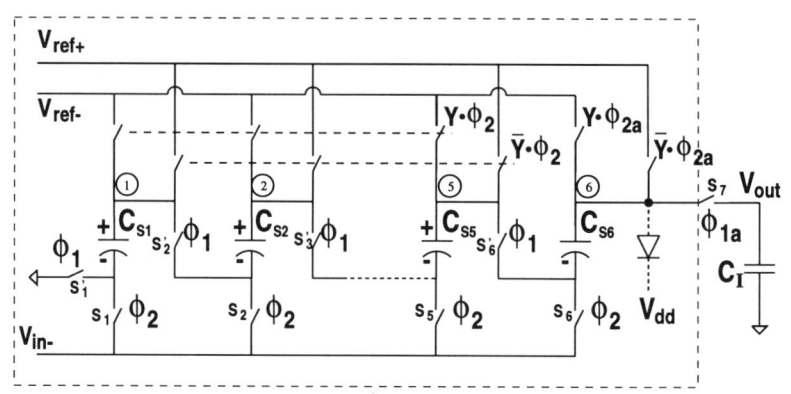

Fig.11: The circuit implementation of the switch-only gain-boost network in passive ΣΔ modulator

5.3. Charge Injection Error Cancellation

As in other switched capacitor circuits, charge injection error is also present in the network. Since virtual ground is not available, a straightforward bottom plate sampling is not effective here and a special timing scheme is proposed here to achieve first order error cancellation. First of all, bottom plate sampling is only applied to C_{S6} and no bottom plate sampling is applied to the rest of capacitors. Therefore, we only have one node (node 6) which is error free because its DC path has been cut off by using an advanced version of ϕ_2. Error charges are still sampled on all other nodes. Ideally if all switches (S1-S6) controlled by ϕ_2 are matched and all ϕ_2's falling edges are sharp enough to ensure an equal split of the switches' channel charges, equal amount of error charges will be sampled on all other nodes except node 6. For example, at node 5, ideally the opening of S5 and S6 will result in equal but opposite charges being dumped onto node 5. Complete cancellation of error is similarly achieved at node 1 to 4 because there is no net error charge associated these nodes, as evidenced by the opposite polarities assumed by the charges at these nodes as shown in the diagram. In addition, the switches s_1' to s_6' are turned off by ϕ_1, after the switch s_7 is turned off by ϕ_{1a} so that errors from switches s_1' to s_6' will not get dumped onto c_i.

6. Experimental Results

The design has been implemented in a 1.2μm CMOS technology and its micro-photograph is shown in Fig. 13.

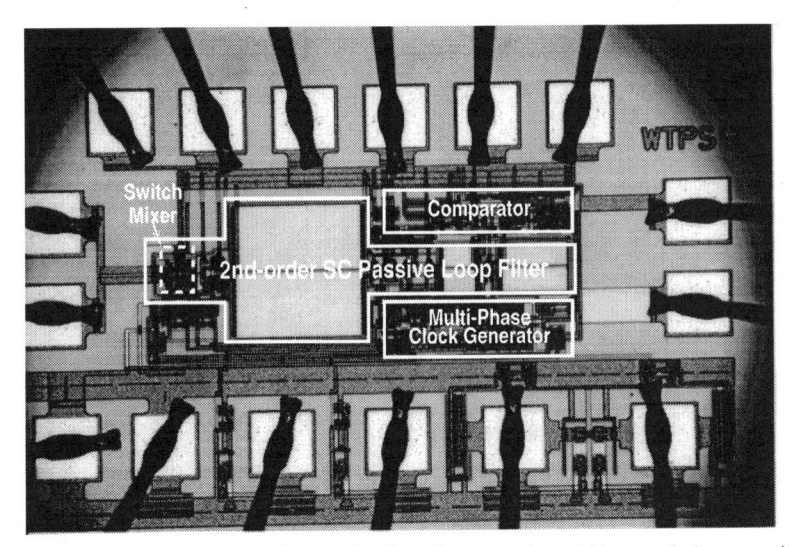

Fig.13: Micro-photograph of the 2nd-order passive $\Sigma\Delta$ modulator with built-in mixer

Its active area is about $0.4mm^2$. The capacitors in the passive loop filter can be clearly seen. The differential IF input is routed from the left hand side and applied to the loop filter. The comparator is put on the top and the logic block for the multi-phase clock as well as LO generation is placed at the bottom. Fully differential clocks are applied and outputs are taken in a differential manner to reduce substrate coupling effect. The chip runs from a 3.3V power supply at a 10MHz sampling rate for a 10MHz IF input.

Testing results show that the designs with and without gain-boost network achieve similar noise performance at a 10MHz sampling rate. But improved performance was observed when the sampling rate is reduced to 2MHz, as shown in Fig. 14. It can be seen that the idle channel noise for the design with the gain-boost network is 8dB lower than that for the design without.

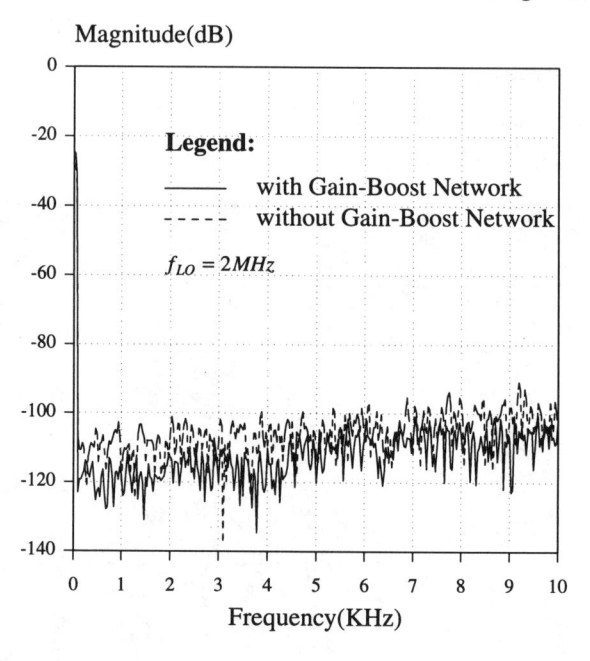

Fig.14: Measured output spectrum of 2nd-order passive $\Sigma\Delta$ modulator with gain-boost network

Since the overall architecture is designed for a 10MHz sampling rate, the following testing results are obtained from the design without the gain-boost network. Fig. 15 shows the measured signal-to-(noise+distortion) ratio (SNDR) versus the input level for the modulator with a 10.005MHz IF single tone input for a clock frequency (LO has the same frequency) of 10MHz. A peak SNDR of 67dB at a -11dB full-scale input and a dynamic range of 78dB have been achieved. The 0dB level corresponds to a 1.3Vpp input sine wave. Fig. 16 shows the output spectrum of the modulator with a 10.005MHz IF input. With an input level of -11dB, the 2nd-order and third-order harmonic distortion components are below the noise floor. To characterize the design in

SNDR (dB)

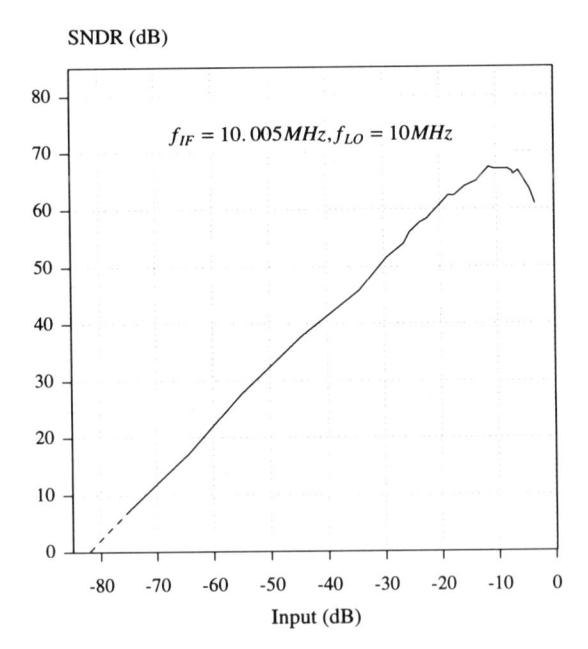

$f_{IF} = 10.005 MHz, f_{LO} = 10 MHz$

Input (dB)

Fig.15: Measured signal-to-(noise+distortion) ratio versus the IF input level

terms of its intermodulation distortion performance, a two-tone input at 10.005MHz and 10.02MHz respectively is applied. Fig. 17 shows that the IM3 component is 82dB down from the fundamentals with an input level of -13dB for both tones.

Testing conditions and measured performance are summarized in Table 1. The total bias current for the comparator is 25μA and the total power consumption is about 0.25mW, (this does not account for dynamic power from the logic gates). It can be seen that the IF digitizer achieves a 13bit resolution for a 10MHz IF input at a very low level of power consumption. A similar design for baseband application where the input signal lies within the 20KHz bandwidth has also been fabricated and tested. The design uses smaller sampling switches (W/L=3/1.2), as no mixing is required. With a clock rate of 10MHz, it dissipates 0.23mW of power, has a measured peak SNDR of 77dB and a dynamic range of 87dB over a bandwidth of 20KHz for a single tone input of 5KHz (summarized in column3 of Table 1). This demonstrates that the mixer is limiting the performance for a 10MHz IF input.

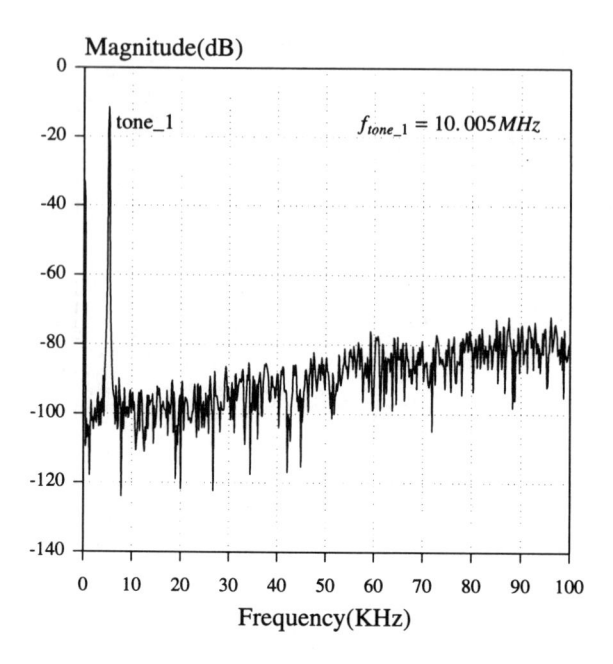

Fig.16: Measured output spectrum of 2nd-order passive $\Sigma\Delta$ modulator with single tone input

References

[1] Bang-Sup Song, "A 4-th-Order Bandpass $\Sigma\Delta$ Modulator with Reduced Number of Opamps," Digest of Technical paper of ISSCC'95, pp.204-205, Feb. 1995, San Francisco

[2] Ulrich Roettcher, Horst L. Fielder and Guenter Zimmer, "A Compatible CMOS-JFET Pulse Density Modulator for Interpolative High-Resolution A/D Conversion," IEEE J. Solid-State Circuits, pp.446-452, Vol.SC-21, No.3, June 1986

[3] Cynthia Diane Keys, Ph.D thesis, UC Berkeley, 1994

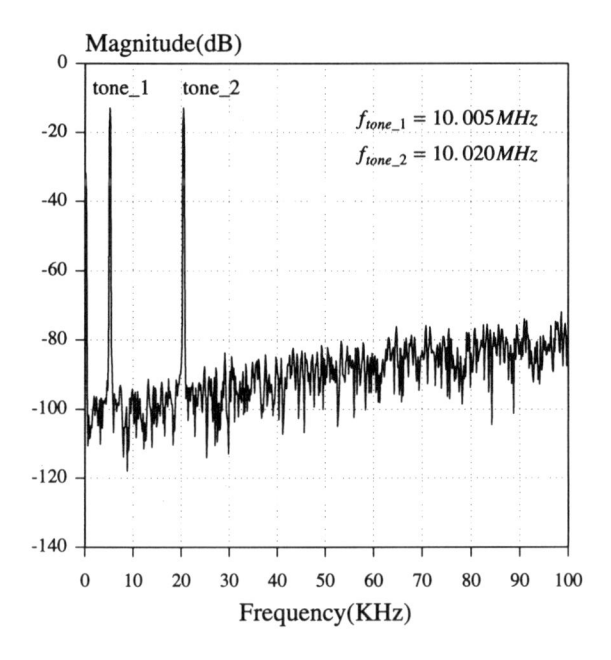

Fig.17: Measured output spectrum of 2nd-order passive $\Sigma\Delta$ modulator with two-tone input

Table1: Testing conditions and measured performance

IC Technology	1.2μm CMOS	
Active Area	0.4mm^2	
Power Supply	3.3V	
Full Scale	1.3V	2.0V(baseband)
Sampling Rate(LO Frequency)	10MHz	
Analog Input Freq.	10.005MHz	5KHz(baseband)
Passband Bandwidth	20KHz	
Total Bias Current	25μA	
Power Consumption	0.25mW	0.23mW(baseband)
Peak SNDR	67dB	77dB(baseband)
Dynamic Range	78dB	87dB(baseband)
IM3 at -13dB input	-82dB	
IM3 at -10dB input	-70dB	

Design of Continuous Time Bandpass ΔΣ modulators in CMOS

Vincenzo Peluso, Michiel Steyaert and Willy Sansen
KULeuven, ESAT-MICAS, Heverlee, Belgium

Abstract - An implementation procedure for continuous time bandpass ΔΣ modulators as a fully integrated circuit is discussed. An overview is given of the synthesis theory. A methodology for the filter design yielding practical design equations is explained. A topology for the loop filters is proposed: they are Gm-C filters. The specifications of the various building block are discussed. Considerable attention is given to non-idealities such as circuit noise, finite Q-factor and non-linearity of the transconductance amplifiers, which affect the performance of the modulator. Circuit realizations are discussed.

I. Introduction

The high quality wireless communication links we know today are to a great extent a consequence of the advent of digital signal processing. It has enabled complex modulation algorithms and accurate digital filtering. In this respect it is advantageous in digital receivers to perform the demodulation and the filtering of the *IF* signal in the digital domain as well. The building block needed for this operation is an appropriate ADC. In recent years bandpass delta sigma modulators have been used for this [5,6]. They allow to carry out an analog to digital to conversion of a narrowband signal at an *IF* frequency. The advantage is that the overhead of two mixers, followed by low pass filters and separate low pass AD converters is avoided.

All of the reported bandpass ΔΣ modulators were realised with the switched capacitor technique [5,1,2]. The switched capacitor technique has a rather low upper limit in frequency. However, it is also possible to implement bandpass ΔΣ modulators with continuous time loop filters. These are therefore called continuous time ΔΣ modulators. They enable operation at higher frequencies than possible with switched capacitors. Many receiver systems use two or three *IF* frequencies. Bandpass ΔΣ modulators with switched capacitors have been applied only to the lower one of these [6]. The use of continuous time bandpass ΔΣ modulators at the higher *IF* frequencies would allow to eliminate yet another analog downconversion.

There is no need for a continuous time implementation specific stability theory. The synthesis of a continuous time bandpass ΔΣ modulator relies on two transformations. The first one transforms the discrete time baseband modulator to a bandpass one. The second one is a discrete time to continuous time transformation. Once having understood that the continuous time ΔΣ modulator actually behaves as a normal discrete time one, it is clear that its properties are kept under both transformations. I.e. a stable modulator is obtained with the same *SNR*.

The implementation of a continuous time bandpass ΔΣ modulator with discrete components is not problematic. However, in integrated receiver systems it is more beneficial to eliminate the expensive use of discrete components. What is needed is an integrated continuous time bandpass ΔΣ modulator.

There are several goals intended in this paper. The first one is to give a good understanding of the synthesis and operation of a continuous time bandpass ΔΣ modulator. This is done by means of an overview of existing theory. Secondly, by taking this as a starting point the synthesis of a fully integratable modulator is explained. For this matter practical design equations are deduced, design considerations are made and some transistor circuitry is discussed.

II. Some Definitions and Performance

The Oversampling Ratio is the ratio of the actual sampling rate f_s and the Nyquist rate for the signal band $2B_s$:

$$OR = \frac{f_s}{2 \cdot B_s} \qquad (1)$$

This definition applies for both Low Pass modulators - i.e. operating on baseband - and bandpass modulators. In a bandpass ΔΣ the signal band is around the band center frequency, in this text referred to as *IF* (Intermediate Frequency). Figure 1 illustrates the definitions.

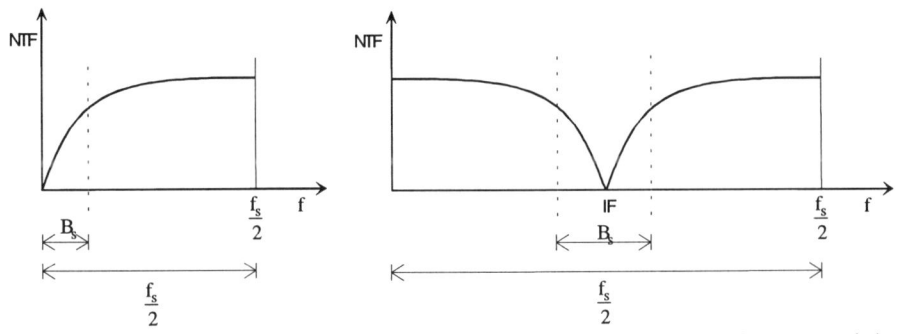

Figure 1 : Typical NTF of lowpass and bandpass case. (linear scale)

For the low pass modulator it holds that the sampling frequency is much higher than the inband frequencies. For the bandpass case however this does not hold. In fact it is convenient to choose the ratio between *IF* and f_s as:

$$IF = \frac{2 \cdot m + 1}{4} \cdot f_s , \quad m = 0,1,2... \quad (2)$$

Bandpass sampling allows digital quadrature modulation. Choosing the sampling frequency according to this formula leads to very simple digital mixers. The samples of the orthogonal carriers only consist of ones, zeroes, and minus ones. The case $m=0$ yields $f_s=4IF$: *IF* is placed at a quarter of the sampling frequency. This is called the sampling case. For $m>0$ the analogue signal is subsampled.

The *SNR* of the lowpass and bandpass modulator are exactly the same in terms of oversampling ratio and number of filter sections:

$$SNR_{1\,bit} = a^2 \cdot \frac{3 \cdot \pi}{2} \cdot (2n+1) \cdot \left(\frac{OR}{\pi}\right)^{2n+1} \quad (3)$$

It has to be pointed out though that in the bandpass case *n* corresponds to the number of bandpass "stages" and not to the order. For the lowpass case the order and the number of stages are equal. The order for the bandpass is twice the number of stages or $2n$.

IV. The baseband to bandpass transformation

An overview of the synthesis of the continuous time bandpass ΔΣ modulator [7,8] is given here concisely for the reader to understand the starting point for the derivation of the design equations. The synthesis of the continuous time bandpass ΔΣ modulator can be done by a couple of transformations of the discrete time lowpass ΔΣ modulator.

In this section the bandpass loop filter structure will be derived from the single loop lowpass structure. The family of single loop second order baseband modulators is described by the block diagram of Figure 2 [9].

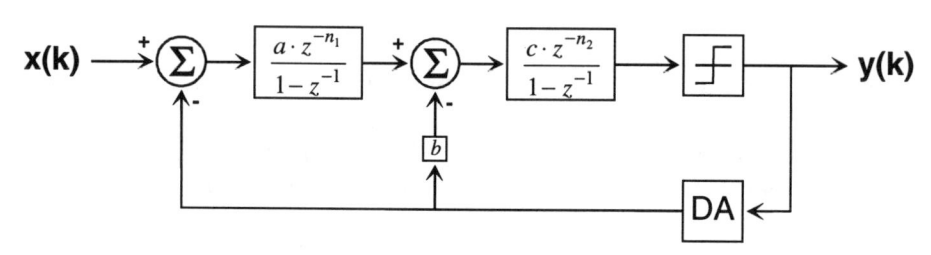

Figure 2 : Block diagram of second order baseband modulator

The output can be written as:

$$Y(z) = \frac{K\,a\,c\,z^{-(n_1+n_2)}}{D(z)} \cdot X(z) + \frac{1}{ac} \cdot \frac{(1-z^{-1})^2}{D(z)} \cdot N(z) \qquad (4)$$

with: $\quad D(z) = 1 - 2z^{-1} + z^{-2} + K\,b\,c\,z^{-n_2} - K\,b\,c\,z^{-(n_2+1)} + K\,a\,c\,z^{-(n_1+n_2)} \qquad (5)$

To eliminate the poles and the zeros in the signal and noise transfer functions, we should make $D(z)=1$. Experimentally it has been found that the operation of the loop is such that Kac is made unity [10]. So two solutions exist: 1) $n_1=0$, $n_2=1$, $b=a$ and 2) $n_1=1$, $n_2=1$, $b=2a$.

The lowpass to bandpass conversion can be made by substituting z by $-z^2$. The integrators in the block diagram are transformed into resonators. The zero's of the noise transfer function will be moved from DC to $\frac{f_s}{4}$. The block diagram of the corresponding bandpass modulators are shown in Figure 3.

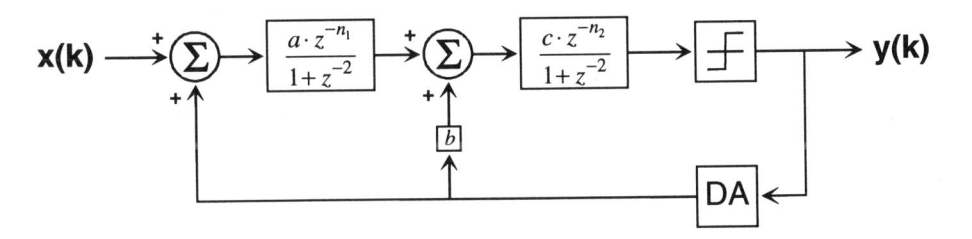

Figure 3 : Block diagram of two stage bandpass modulator

Now the output can be written as:

$$Y(z) = \frac{K\,a\,c\,z^{-(n_1+n_2)}}{D(z)} \cdot X(z) + \frac{1}{ac} \cdot \frac{(1+z^{-2})^2}{D(z)} \cdot N(z) \quad (6)$$

with: $\quad D(z) = 1 + 2z^{-2} + z^{-4} + K\,b\,c\,z^{-n_2} + K\,b\,c\,z^{-(n_2+2)} + K\,a\,c\,z^{-(n_1+n_2)} \quad (7)$

To make $D(z)=1$, again two solutions exist: 1) $n_1=0$, $n_2=2$, $b=a<0$ and 2) $n_1=2$, $n_2=2$, $b=2a<0$. They correspond respectively to the two solutions for the baseband case.

The structure corresponding to the first solution is chosen for the detailed elaboration of the continuous time bandpass modulator design equations. It is shown in Figure 4.

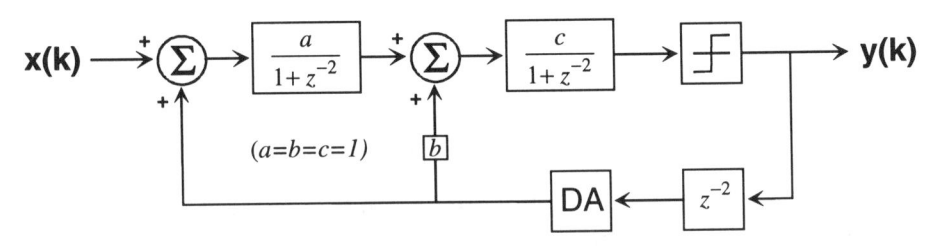

Figure 4 : Chosen topology for continuous time implementation

The two clock delays are shifted to the feedback. [5,8] The fact that a and c are negative actually causes the feedback signals to be summed rather than subtracted. But the signal is delayed two clock cycles. For a sinusoidal signal at *IF* this implies a phase shift of 180 degrees which is a sign inversion. So there is indeed negative feedback for inband signals.

The next step is to derive the continuous time modulator from the discrete time structure. This is done in the following sections.

V. The discrete time to continuous time transformation

Intuitively it seems obvious that a continuous time ΔΣ modulator can be obtained by replacing the discrete time integrators (for the baseband case) and oscillators (for the bandpass case) by continuous time ones. The question that remains is how the loop filter gain factors then should be chosen. The key to designing the continuous time loop filters is to realize that a so called continuous time ΔΣ still behaves as a discrete time ΔΣ. The reason is that the comparator is sampling. Only the value of the signal seen at the sample times is relevant. So, the response of the continuous time loop filter to both the analogue input signal and the pulsed wave form from the DAC must match that of the discrete time loop filter at the sample times. The simplest approach is to perform a pulse invariant design. This means a time domain analysis is used to match the pulse response of the continuous time filter at the sampling instants to the impulse response of the discrete time filter. For the input signal some considerations need to be made concerning stability.

A. The impulse response of discrete time loop filter

From Figure 4 the impulse response of the loop filter of the feedback signal can be seen to be:

$$H(z) = \left(b + \frac{a}{\left(1+z^{-2}\right)}\right) \cdot \frac{c}{\left(1+z^{-2}\right)} \qquad (8)$$

Although $a=b=1$ they are still explicited in the formula for a reason that will become clear later. The inverse z-transform is :

$$h(t) = c \cdot (a+b) \cdot \left(1 + \frac{(1-\lambda)}{2T} \cdot t\right) \cdot \cos\left(\frac{\pi}{2} \cdot (2m+1) \cdot \frac{t}{T}\right) \quad , \quad \lambda = \frac{b}{a+b} \qquad (9)$$

This is the time response of the discrete time loop filter. The pulse response of the continuous time loop filter will have to equal this.

B. The pulse response of the continuous time loop filter

Here the response of the continuous time loop filter to the feedback pulse is derived. Firstly the continuous time filter structure needs to be defined. The discrete time oscillator can be replaced by a continuous time oscillator. A possibility is the LC oscillator shown in Figure 5a.

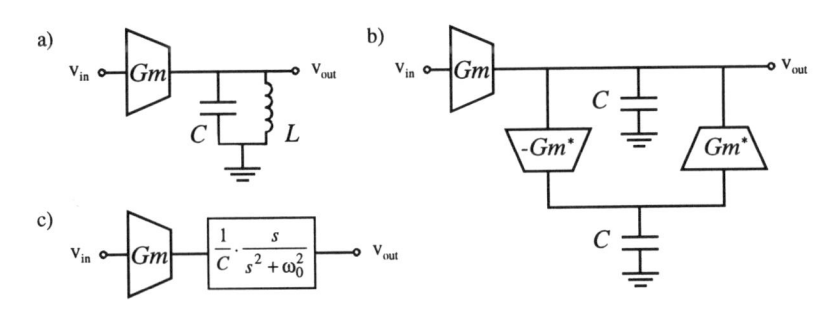

Figure 5 : Continuous time filter sections and block representation

Since a full chip implementation is desired the Gm-C biquad is chosen (Figure 5b). It can also be regarded as the LC oscillator with the inductor realized by a gyrator. Both sections have the same Laplace transform, shown in Figure 5c.

The loop filter's center frequency is $\omega_0 = \frac{1}{\sqrt{LC}}$ for the LC implementation or $\omega_0 = \frac{Gm^*}{C}$ for the Gm-C implementation.

The continuous time filters are modeled lossless. Nevertheless the performance of continuous time implementations will be determined by their quality factor Q. The main effect is that finite Q factors will limit the noise shaping. It will make the noise transfer function (NTF) flatten out in the center of the passband, and will reduce the inband noise suppression

The complete architecture of the continuous time $\Delta\Sigma$ is shown in Figure 6. The block R is a resistance. It converts the currents from G_0 and F_0 to a voltage again. The comparator is a voltage comparator. It would be also possible to leave out R and apply a current comparator. That does not change anything to the results of this analysis.

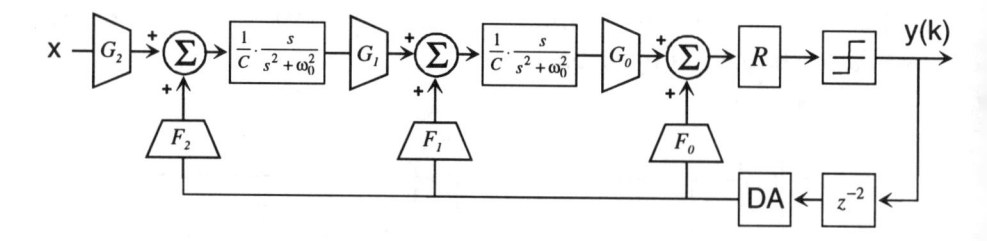

Figure 6: The complete continuous time bandpass ΔΣ modulator architecture

The complete pulse response of the continuous time loop filter can now be given:

$$G(s) \cdot DAC(s) = \left[F_2 G_1 G_0 \left(\frac{1}{C} \cdot \frac{s}{s^2 + \omega_0^2} \right)^2 + F_1 G_0 \left(\frac{1}{C} \cdot \frac{s}{s^2 + \omega_0^2} \right) + F_0 \right] \cdot R \cdot V_{ref} \cdot \left[\frac{e^{\left(\frac{s\tau}{2}\right)} - e^{\left(-\frac{s\tau}{2}\right)}}{s} \right]$$

In this equation the last part represents the DAC output pulse of width τ and centered on t=0. The pulse amplitude is Vref.

The pulse response in the time domain can be found by taking the inverse Laplace transform. The result is normalized to the reference voltage and given by:

$$PR(t) = \begin{cases} a) \; 0 & t \leq -\frac{\tau}{2} \\[6pt] b) \; \dfrac{F_2 G_1 G_0 R}{2\omega_0 C^2}\left(t+\dfrac{\tau}{2}\right)\cdot \sin\!\left(\omega_0\!\left(t+\dfrac{\tau}{2}\right)\right) \\ \quad + \dfrac{F_1 G_0}{\omega_0 C} R \cdot \sin\!\left(\omega_0\!\left(t+\dfrac{\tau}{2}\right)\right) & -\dfrac{\tau}{2} < t \leq -\\ \quad + F_0 R \\[6pt] c) \; \dfrac{F_2 G_1 G_0 R}{\omega_0 C^2}\left[\dfrac{\tau}{2}\cos\!\left(\omega_0\dfrac{\tau}{2}\right)\cdot \sin(\omega_0 t)+t\cdot \sin\!\left(\omega_0 \dfrac{\tau}{2}\right)\cdot \cos(\omega_0 t)\right] \\ \quad + \dfrac{2 F_1 G_0 R}{\omega_0 C} \cdot \sin\!\left(\omega_0 \dfrac{\tau}{2}\right)\cdot \cos(\omega_0 t) & t > \dfrac{\tau}{2} \end{cases}$$

(11)

This now needs to be equated to (9) *for t=kT*. In this way constraints for the transconductances F_i, G_i and R can be found. It should be pointed out that the continuous time pulse is centered on t=0 and has a width of half a period. Therefore the pulse should start a quarter period earlier than the two full periods delay of the z^{-2} block. The delay follows to be 1.75 T.

C. Matching the pulse responses

Matching the pulse response of the continuous time filter to the impulse response of the discrete time filter is obtained by equating terms in $\cos(\omega_0 t)$ and $t\cos(\omega_0 t)$ in expression (11) to those in (9). The reader is reminded that $\omega_0 = \frac{\pi}{2} \cdot (2m+1) \cdot \frac{1}{T}$

The equations found are:

$$\frac{2F_1 G_0 R}{\omega_0 C} \cdot \sin\left(\omega_0 \frac{\tau}{2}\right) = (a+b) \cdot c \tag{12}$$

$$\frac{F_2 G_1 G_0 R}{\omega_0 C^2} \cdot \sin\left(\omega_0 \frac{\tau}{2}\right) = \frac{(1-\lambda)}{2T} \cdot (a+b) \cdot c \tag{13}$$

$$\frac{F_2 G_1 G_0 R}{2\omega_0 C^2} \cdot \frac{\tau}{2} \cdot \sin\left(\omega_0 \frac{\tau}{2}\right) + \frac{F_1 G_0 R}{\omega_0 C} \cdot \sin\left(\omega_0 \frac{\tau}{2}\right) + F_0 R = (a+b) \cdot c \tag{14}$$

The variables are:. F_0, F_1, F_2, G_0, G_1, R. The parameters are: $\tau = T/2$, T, a, b, c, ω_0, C. There are six variables and only three equations. This means there are three degrees of freedom.

These equations can be simplified. Equations (12) and (13) are substituted into equation (14). The division of equations (12) and (13) yields a new equation (16). Equation (15) is taken unaltered from (12):

$$\frac{2F_1 G_0}{\omega_0 C} \cdot \sin\left(\omega_0 \frac{\tau}{2}\right) = \frac{(a+b) \cdot c}{R} \tag{15}$$

$$\frac{F_1 C}{F_2 G_1} = \frac{T}{(1-\lambda)} \tag{16}$$

$$\frac{(1-\lambda)}{8} \cdot \frac{\tau}{T} + \frac{F_0 R}{(a+b) \cdot c} + \frac{1}{2} = 1 \tag{17}$$

F_0 is determined by (17). The feedback through F_0 will appear to be a small correction term to equate the pulse responses at $t=0$. F_1 and F_2 could be determined by (15) and (16) respectively if extra constraints on G_0 and G_1 and R were found. Notice that no equation is obtained yet for G_2. This is logical since only the feedback has been considered. G_2 is in the input signal path. So before looking for extra design constraints, an equation for G_2 will be derived.

D. The response to the input signal

The response of the continuous time loop filter to the input signal is similar to the response of the discrete time loop filter to the sampled input signal. They both have very high gain in the passband. But these gains do not have to be matched. The operation of the loop is such that the value of the gain operating on the input signal (*a.c* in Figure 4 e.g.) is irrelevant. As already explained it is such that the loop gain is made one. The gain must only be large. In that case the error signal at the output of the first summer the input signal encounters, tends to zero. This is similar to an amplifier in negative feedback configuration in linear system analysis. A ΔΣ modulator is not a linear system, but also it has been found empirically to show this behavior. From these considerations a constraint can be put on the relationship between G_2 and F_2.

The error signal is the difference between the input and feedback signal quantity (SQ) $|SQ_{in} - SQ_{fb}|$ and on the average it tends to zero. That means that on the average $|SQ_{in}| = |SQ_{fb}|$. Since the continuous time filter is inputting current in the first summer node and the feedback current has a duty cycle of 50%, the signal quantity to be considered is integrated current or charge. So: $|V_{in}G_2T| = |V_{out}F_2T/2|$ (on the average). Since $V_{in} = V_{out}$ the constraint found is:

$$\frac{F_2}{G_2} = 2 \qquad (18)$$

E. Extra design equations

The extra equations are based on design choices. It is chosen to make the two filter sections identical. This implies:

$$G_1 = G_2 \qquad (19)$$

and $\qquad F_1 = F_2 \quad \text{or} \quad \dfrac{F_1}{G_1} = 2 \qquad (20)$

So one variable still remains. It is the resistor R. It will be implemented with a transconductance. Its resistive value will be given by $R = \dfrac{1}{G_r}$. In fact all G and F blocks will translate into a transconductance value Gm. The design equation chosen is:

$$G_r = G_0 \qquad (21)$$

This means amplification equal to unity for the signal on the second oscillator output node. It makes it even more obvious that the last summer, realized by G_r and G_0, is only needed to allow for the correction feedback term F_0 at $t=0$.

All the degrees of freedom have been used now. Nevertheless it would be an advantage to set:
$$G_0 = G_1 \tag{22}$$
in which case all the G_i would be equal. One and the same transconductance could be used, which shortens design time and is beneficial to matching. An extra degree of freedom can be obtained by freeing one of the parameters that were first assumed constant. R and c always appear as a ratio (equations (15) and (17)). Since the exact value of c is irrelevant for the operation of the discrete time modulator, c can be freed.

Other constraints can be imposed. Instead of making the two filter stages equal, the signal swing on the filter nodes can be normalized. Doing so the system can be made to take full advantage of the maximum filter dynamic range. This will be done further on by scaling of several loop coefficients starting from the found modulator.

VI. Practical Design Equations

In this section practical design equations in terms of transconductances are derived. The equations will allow to calculate the design parameters. The four equations can be manipulated keeping in mind the extra constraints. These were: $G_2 = G_1 = G_0 = G_f = Gm_{in}$ and $F_2 = F_1 = Gm_{fb} = 2Gm_{in}$. Remember also that $Gm^* = \omega_0 C$. Substituting G and F, equations (15) and (16) can be manipulated as follows:

$$2\left(\frac{Gm_{fb}}{Gm_{in}}\right)\left(\frac{Gm_{in}}{Gm^*}\right)\left(\frac{Gm^*}{C}\right)\frac{1}{\omega_0} \cdot \sin\left(\omega_0 \frac{\tau}{2}\right) = (a+b) \cdot c \tag{23}$$

$$\left(\frac{Gm_{fb}}{Gm_{fb}}\right)\left(\frac{C}{Gm^*}\right)\left(\frac{Gm^*}{Gm_{in}}\right) = \frac{T}{(1-\lambda)} \tag{24}$$

The three new design equations are:

$$4\left(\frac{Gm_{in}}{Gm^*}\right)\sin\left(\omega_0 \frac{\tau}{2}\right) = (a+b) \cdot c \tag{25}$$

$$\left(\frac{Gm^*}{Gm_{in}}\right) = \frac{\pi}{2(1-\lambda)} \tag{26}$$

$$\frac{(1-\lambda)}{8} \cdot \frac{\tau}{T} + \frac{1}{(a+b)c} \cdot \frac{F_0}{Gm_{in}} + \frac{1}{2} = 1 \tag{27}$$

The design then proceeds as follows. Firstly the needed capacitance value C is calculated based on noise considerations. This will be done in a later section. Then from $Gm^*=\omega_0 C$ the Gm^* value is obtained. The ratio of Gm^* to Gm_{in} can be calculated from equation (26). An example is given:. for $C=5pF$ and $\omega_0=2\pi 10.7 MHz$, $Gm^*=336\mu S$. It follows then that $\left(\dfrac{Gm^*}{Gm_{in}}\right)=\pi$. This is a difficult ratio to realize. Therefore it is rounded to 3. The implication of this is that λ (meaning a and b) is altered. The ratio b/a affects the stability of the loop and can be varied in a wide range without hampering the stability. By rounding the ratio of π to 3, λ acquires the value of 0.476 and $b=0.910a$, which yields a performance not significantly different from the optimum. So $Gm_{in}=112\mu S$ and therefore $Gm_{fb}=224\mu S$. Now from equation (25) (a+b)c can be determined and filled in into equation (27). Now F_0 can be calculated to be $27\mu S$.

All system parameters in the Figure 5b and Figure 6 have now been explicited. However there is still room for optimization. In CMOS circuit technologies the dynamic range is restricted. Therefore it is important to control the signal swing on the internal nodes of the loop filter. For the filter dimensioned as proposed above, simulations reveal that the signal swing is about 1.5 times the reference voltage or maximum signal swing. This is a too large restriction on dynamic range. Equation (16) shows that coefficients F_2, G_2 and G_1 can easily be scaled to alter the signal ranges in the first node, keeping in mind that F_2 and G_2 have a ratio of two. Equation (15) shows that F_1 and G_0 (and G_1) can be scaled to alter the signal swing in the second stage. Figure 8 shows the results. The signal range is now slightly larger than 0.5 times the reference voltage, keeping about a factor of two safety margin.

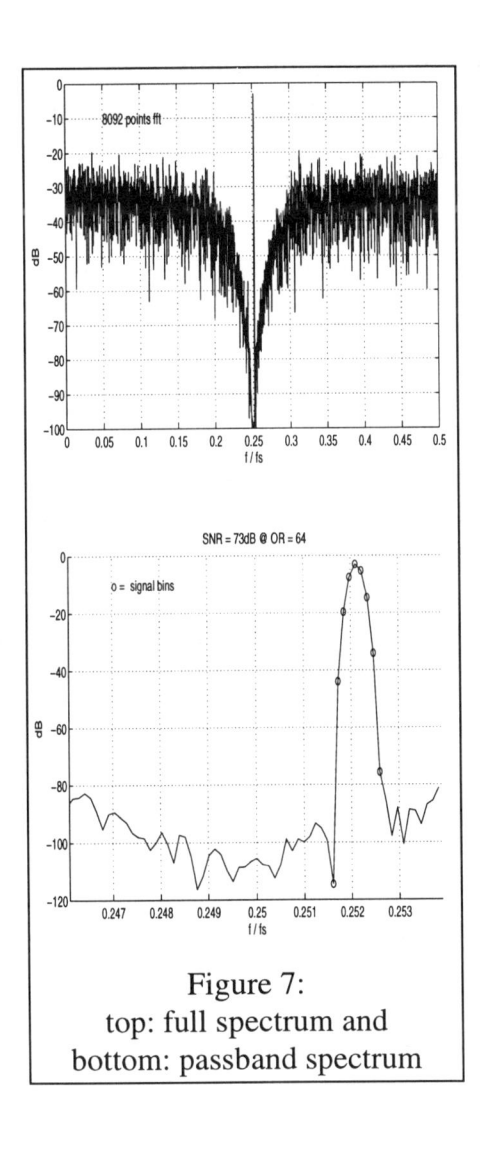

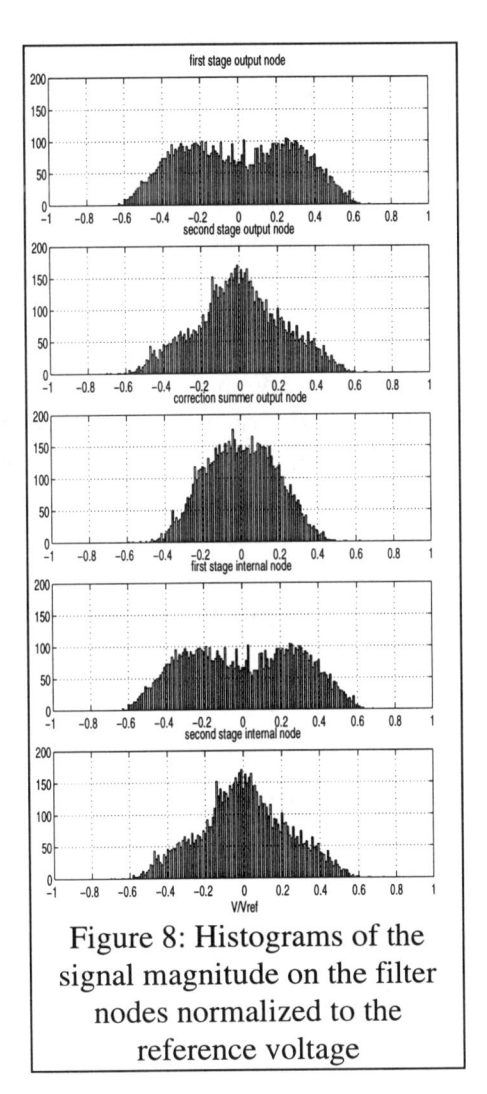

Figure 7:
top: full spectrum and
bottom: passband spectrum

Figure 8: Histograms of the signal magnitude on the filter nodes normalized to the reference voltage

VII. Simulations

The continuous time bandpass $\Delta\Sigma$ modulator of Figure 6 has been macro modeled in SABER. using voltage controlled current sources as transconductance elements. The simulation has been performed using the derived parameters.

Figure 7 shows the full output spectrum at the top and the pass band at the bottom. An *OR* of 64 yields an *SNR* of 73 dB for a signal of $0.779V_{ref}$ which is in good agreement with the theoretically predicted performance of 77dB.

VIII. Implementation specific considerations

The above part theoretically fully describes the functionality of the continuous time bandpass $\Delta\Sigma$ modulator. However, for circuit realization the non-idealities of the circuit building blocks need to be taken into consideration. They are circuit noise, finite Q-factor, relative mismatches of the bandbass filter center frequency and non-linearity of the transconductance amplifiers.

A. Circuit Noise

The dynamic range of the theoretical modulator is only determined by the quantization noise floor. But in a circuit implementation circuit noise will be generated by the active elements. In order not to degenerate the dynamic range, the inband circuit noise spectral density should be made lower than the desired resolution.

The equivalent input noise for a two stage bandpass filter is [12]:

$$v_{eq}^2 = F\frac{kT}{A_1 C}\left(\frac{Q}{A_1}+1\right)+\frac{1}{A_1^2}\left(F\frac{kT}{A_2 C}\left(\frac{Q}{A_2}+1\right)\right) \quad (28)$$

F represents a noise figure and is dependent on the particular implementation of the transconductance amplifier.

The noise of the second stage will be negligible. In this design the gain is of the same order of magnitude as the Q-factor and can be varied. So the circuit noise can be reduced by making the gain of the first stage high.

B. Q-factor

The theory ideally requires an infinite Q-factor.

Figure 9 shows the effect of Q-factor on the performance. A finite Q-factor makes the noise transfer function flatten out in the center of the passband instead of continuing to fall off. So the largest effect will be on the performance of modulators with a high oversampling ratio: they require large Q-factors to yield a performance close to the theoretical value. The main conclusion to be drawn is that for a certain performance there is a minimum bound on the Q-factor. It can also be concluded that for low oversampling ratio modulators Q-factor requirements are much relaxed. The Q-factor has an effect on the internal node signal swing. Lower Q-factors reduce the swing.

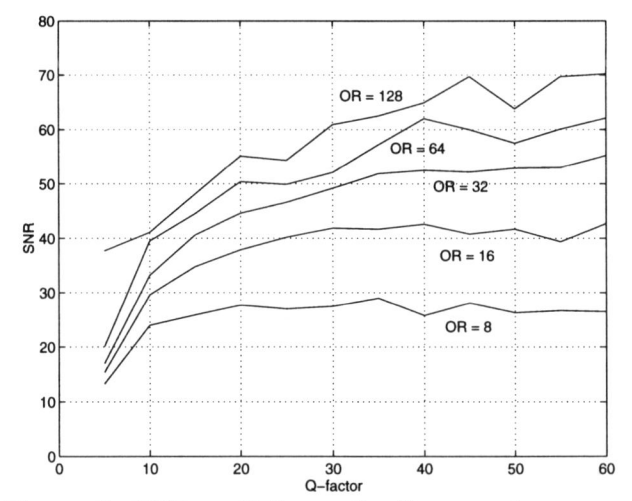

Figure 9: SNR vs Q-factor for first topology type

C. Passband center frequency mismatch

The passband center frequencies will not be exactly equal to each other but will show a relative deviation due to device mismatch. It is important to know how much relative frequency mismatch can be tolerated in a certain bandpass ΔΣ modulator. Simulations were done with the center frequency of the first stage a fraction lower and the center frequency of the second stage a fraction higher than the nominal value.

Figure 10 shows the results. For high oversampling ratios the performance of the modulator actually is slightly better when a slight mismatch is present. The reason for this is that the inband noise suppression is enhanced. It is also clear that the lower the oversampling ratio the less center frequency mismatch matters.

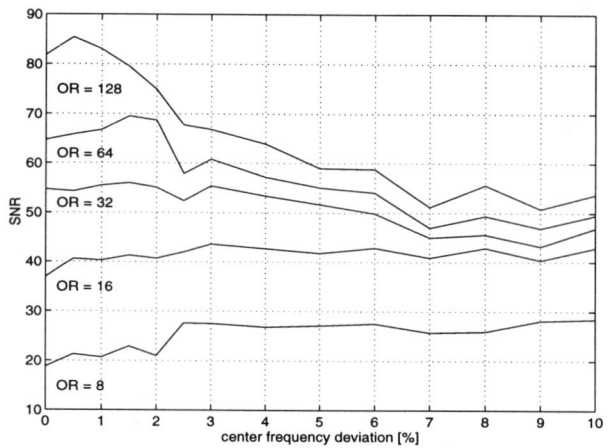

Figure 10: Effect of center frequency mismatch

D. Transconductance nonlinearity

Transconductance nonlinearity introduces harmonic distortion. Especially third order intermodulation (*IM3*) is important for a bandpass type operation. Two strong carriers can have a third intermodulation product in the signal band, masking the real inband signals. Also two inband tones can have an inband third intermodulation product degrading the SNR.

The analysis of transconductance nonlinearity on the performance of the modulator is cumbersome. In order to see the effect of it at different locations in the loop filter several simulations were done: each time a nonlinearity of 1% for an amplitude equal to the reference voltage was introduced in a different transconductance. Two inband signals with inband intermodulation product were applied at the modulator input. Although the system is nonlinear and the effects cannot be superimposed, this experiment nevertheless qualitatively shows the sensitive locations in the loop. The input transconductance G_2 is not in the feedback loop. Therefore it generates intermodulation products that cannot be distinguished from the input signal by the loop and are thus not suppressed. Nonlinearity in the feedback transconductances doesn't seriously degrade the performance of the loop. They are fed with a pulse. A nonlinearity only changes the pulse value somewhat. This is equivalent to a small change in the value of that feedback coefficient. And it is a known fact that a $\Delta\Sigma$ modulator is quite insensitive to a slight change in loop coefficients [10]. The most severe effect is in the transconductances in the biquad loops of the loop filter. The noise floor is raised a lot and thus the modulator performance is reduced drastically. The output spectrum for this case is shown in the right hand graph of Figure 11. On the left hand side the ideal case is shown. The *SNR* is indicated at the top of each figure and can be used for comparison.

It is clear that transconductances need to be highly linear in order to achieve high performance.

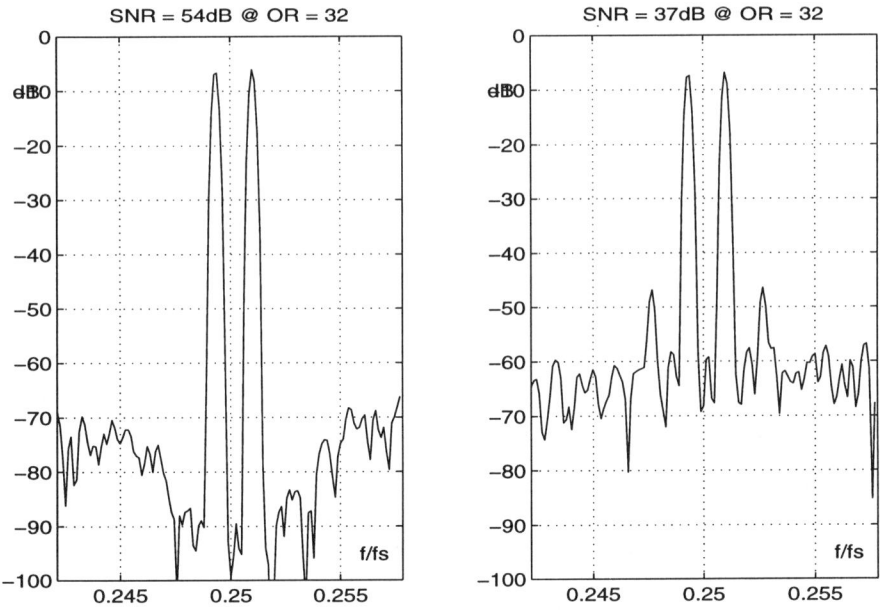

Figure 11: Effect of transconductance nonlinearity

IX. Implementation

The building blocks to be implemented are a comparator and some digital circuitry taking care of the feedback, a 1 bit DAC for the feedback and a transconductance amplifier for application in the Gm-C filter. The latter is the most critical for the performance and will be discussed in more detail here. The main specs for the transconductance are very high Q-factor, high linearity and tuneability.

A. Q-factor

Using a two pole model for a transconductor based integrator and assuming all integrators have identical dominant and non-dominant poles, the following approximative expressions can be found for the resonance frequency and the quality factor of the biquad [13]:

$$\frac{1}{Q} = \frac{2 p_d}{\omega_0} - \omega_0 \left(\frac{2}{p_{nd}} \right) \qquad (29)$$

It can be seen that Q can be made very large when the two terms compensate each other. The following constraint can be found:
$\omega_0 = A \cdot p_d$, $p_{nd} = A \cdot \omega_0$. This means that the dominant and non-dominant poles are spaced equally around the gainbandwidth product of the integrator. The reason is that the phase shift of the non-ideal integrator is then exactly 90 degrees at the resonance frequency of the biquad, as it is for the ideal integrator.

The way to make a high Q is to push the dominant pole down and the equivalent non-dominant pole/zero up in frequency as far as possible, and more or less equally spaced around ω_0. In the next section a circuit is presented with which Q-factors exceeding 100 are achievable.

B. Transconductance amplifier structure

In an earlier section it was concluded that for a certain performance the Q-factor needs to be larger than a minimal value. This is to the advantage of a high Q design: it is not necessary to add a tuneable load. Therefore the dominant pole is determined by the output impedance which can be made quite large by cascading the output stage. A safety margin can be used to guarantee a minimal Q value.

Pushing the non-dominant poles in frequency is done by employing the structure in fig. 12 that we call the high frequency loop [14]. It is actually a current amplifier in unity feedback, and acts as a current buffer. The input transistor M1 acts as a source follower on the input signal. The effective g_m of M1 is boosted by the loop gain and makes a very good source follower. The voltage drop over the R_{deg} generates the ac current. The feedback loop draws the ac current through transistor M4 and it is mirrored to M5. The advantage is that no p type transistors are in the signal path.

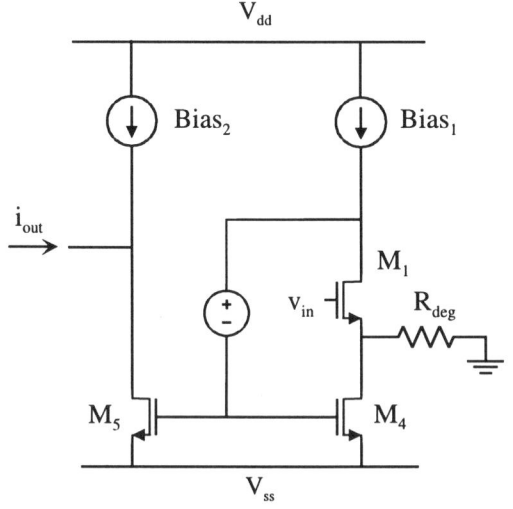

Figure 12: The high frequency loop

The *GBW* of the high frequency loop will be the non dominant pole of the transconductance amplifier:

$$GBW = \frac{1}{2\pi} \frac{g_{m4}}{C_{gs4} + C_{gs5} + C_{db\_bias}}. \qquad (30)$$

The non dominant pole of the high frequency loop is

$$p_{nd} = \frac{1}{2\pi} \frac{g_{m1}}{C_{gs1}} \qquad (31)$$

If all transistors are made of equal size the *GBW* is about $f_t/2$ and the non dominant pole is f_t, inherently ensuring stability.

The complete transconductance amplifier is given in fig. 13. R_s realises a level shift while C_s closes the loop in ac. Transistor M3 is added to cascode the mirroring transistor. The output stage is double cascoded on the n-MOS side to compensate for the lower boosting capability due to the high V_{GS}-V_T and small lengths that are necessary to achieve high f_t's. The degeneration resistance R_{deg} is realised by a series connections of a polysilicon resistance and a MOS transistor in the linear region for tuning the effective G_m. The latter is tuned via the node V_{tune}. The current biasing is via transistor M10. The common mode is measured at the V_{CM} node and the CMFB amplifier feeds back on the gate of M9.

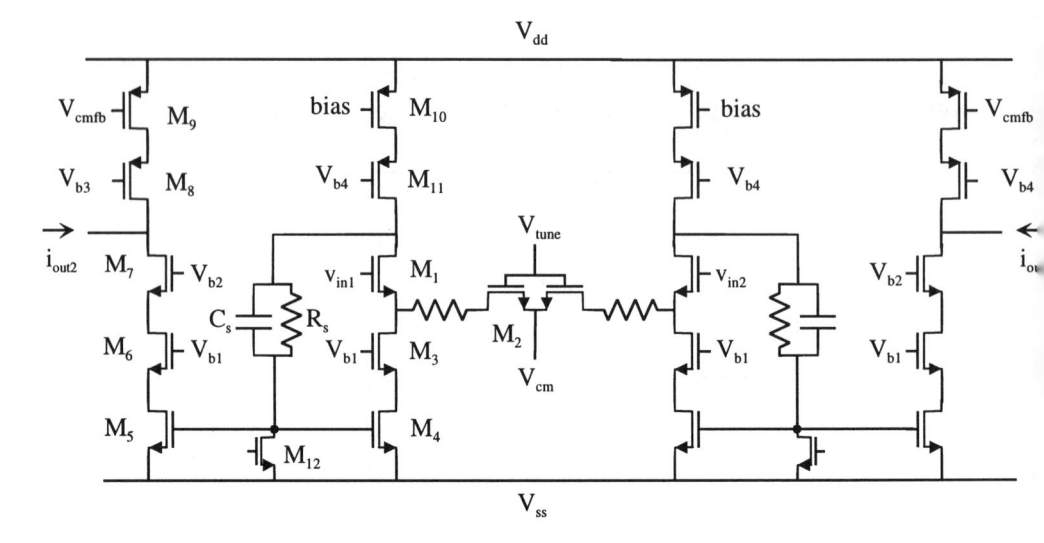

Figure 13: The full transconductance circuit

IX. Conclusions

A comprehensive overview of the synthesis of continuous time band pass $\Delta\Sigma$ modulator has been given. From the pulse response matching equations design equations have been derived for the proposed Gm-C topology for the continuous time loop filters. Simulations show that the technique yields stable band pass $\Delta\Sigma$ modulators with the performance predicted by the linear model. The effect of the relevant circuit non-idealities on the performance of the continuous time band pass $\Delta\Sigma$ modulator have been analyzed. Circuit noise can be reduced. Critical are Q factor, ω_0 matching and non-linearity of the filter transconductances. A tunable transconductance amplifier circuit suitable for high Q and low distortion applications has been discussed.

REFERENCES

[1] S. Jantzi, W.M. Snelgrove, P.F. Ferguson, "A Fourth-Order Bandpass Sigma-Delta Modulator," *IEEE J. Solid-State circuits*, vol. 28, pp. 282-291, Mar. 1993.

[2] F.W. Singor, W.M. Snelgrove, "Switched-Capacitor Bandpass Delta-Sigma A/D Modulation at 10.7 Mhz," *IEEE J. Solid-State Circuits*, vol. 30, pp. 184-192, Mar. 1995

[5] L. Longo, B-R Horng, "A 15b 30kHz Bandpass Sigma-Delta Modulator", *proc. ISSCC 93*, pp. 226-227.

[6] L.Longo, R. Halim, B-R Horng, K. Hsu, D. Shamlou, "A cellular Analog Front End with a 98dB IF Reeciver", *proc. ISSCC 94*, pp. 36-37.

[7] A.M. Thurston, T.H. Pearce and M.J. Hawksford, "Bandpass Implementation of the Sigma-Delta AD conversion technique", *IEE conference on ADC and DAC*, Swansea, Sep. 1991.

[8] A.M. Thurston, T.H. Pearce, M.D. Higman, M.J. Hawksford, "Bandpass Sigma Delta A-D Conversion," *Analog Circuit Design*, Kluwer Academic publishers pp.259-281, 1993

[9] B.E. Boser, Bruce A. Wooley, "The design of Sigma-Delta Modulation Analog-to-Digital Converters," *IEEE J. Solid-State circuits*, Vol.23, pp 1298-1308, Dec. 1988

[10] James C. Candy, "A Use of Double Integration in Sigma Delta Modulation," *IEEE Trans. on Communications*, VOL. COM-33, pp. 249-258, Mar. 1985.

[11] F. Op 't Eynde, W.Sansen, *Analog interfaces for digital signal processing systems*, Kluwer Academic Publishers, 1993

[12] Y.-T. Wang, A.A. Abidi, "CMOS Active Filter Design at Very High Frequencies," *IEEE J. Solid-State Circuits*, vol. 25, pp. 1562-1574, Dec. 1990.

[13] J. Silva-Martinez,M. Steyaert, W. Sansen, *High performance CMOS continuous-time filters*, Kluwer Academic Publishers,1993

[14] P.Kinget, M. Steyaert, "A 1 GHz CMOS upconverter", *proc. IEEE - CICC*, May 1996

Bandpass Delta-Sigma Converters in IF Receivers

Armond Hairapetian

Newport Microsystems Inc.
111 Pacifica Suite 250
Irvine, CA 92718
(714) 450-1080
e-mail: armond@newportmicro.com

Abstract

Performing the analog to digital conversion of the IF signal in a heterodyne receiver has many advantages. These advantages which include removal of DC offset and flicker noise as well as eliminating the gain mismatch in the I and Q signal paths are presented.

By minimizing the quantization noise at an intermediate frequency, bandpass Delta-Sigma converters offer the most efficient method of analog-to-digital conversion of the IF signal. Moreover, due to their high resolution and linearity, bandpass Delta-Sigma converters allow a lower gain in the RF section. An IF sampling receiver architecture utilizing a sixth-order bandpass Delta-Sigma converter is presented and the circuit implementation of the receiver is discussed.

1. Introduction

In most digital radio systems, the use of CMOS technology is limited to baseband processing functions. However, as device geometries are reduced, CMOS becomes a viable technology to perform the IF and RF functions. Moreover, in order to take full advantage of the smaller geometries in reducing the die size, it is desirable to perform more of the radio functions in the digital domain. Replacing the dual baseband ADCs in the receiver with an IF sampling ADC reduces the analog content of the receiver and takes advantage of the high frequency capabilities of modern submicron CMOS process. In addition to increased robustness, IF sampling receivers do not suffer from problems such as DC offset, flicker noise, phase error in the final LO path, and I/Q gain mismatch.

Traditional IF sampling systems make use of a high speed Nyquist-rate ADC to digitize the entire frequency band from DC to Fs/2, where Fs is the sampling frequency of the converter. Because the bandwidth of the IF signal is typically a small fraction of the carrier frequency, the use of a wide-band Nyquist-rate converter does not result in the optimum solution for digitizing the IF signal. An optimum solution for digitizing a narrowband IF signal is a converter which provides high resolution in the narrow band of interest and is capable of handling large out-of-band signals. Due to their oversampling and noise shaping nature, bandpass Delta-Sigma converters provide the most optimum solution for performing analog to digital conversion on narrow band IF signals. By digitizing only the band of interest and not the entire Nyquist band, bandpass Delta-Sigma converters provide high dynamic range with relatively low power consumption.

In section 2, an IF sampling heterodyne receiver architecture is presented and compared with a baseband sampling receiver architecture. In section 3 the circuit implementation of an 81MHz IF receiver which consists of a continuous-time IF amplifier, a subsampling gain stage and a sixth-order bandpass Delta-Sigma converter is presented and the measured results are discussed in section 4.

2. IF Sampling Receiver Architecture

As shown in Figure 1, a typical superheterodyne receiver consists of 3 main sections: 1. the RF section, 2. the IF section and 3. the baseband processor.

In the RF section, the RF signal is filtered, amplified and downconverted to an intermediate frequency. In the IF section similarly, the IF signal is filtered, amplified and downconverted to baseband, producing in-phase and quadrature components. And finally, in the baseband section the I and Q components are digitized and passed on to the DSP where the original signal is recovered.

One disadvantage of this architecture is that it requires peripheral circuitry to perform DC offset cancellation and gain calibration between I and Q signal paths.

Another disadvantage of most of these type of receivers is that they require bipolar technology to perform the RF and IF functions, which prevents their integration with the CMOS baseband section.

Because of higher noise figure and excessive power consumption, integrating the RF section in a 0.8micron CMOS process is not practical for most applications. However, the IF section can be realized in CMOS with no significant noise or power penalty.

In an IF sampling receiver of Figure 2, by combining the IF and the baseband sections and performing the analog to digital conversion at an IF frequency using a bandpass Delta-Sigma ADC, higher level of integration is achieved and the need for DC offset cancellation and I/Q gain calibration is eliminated.

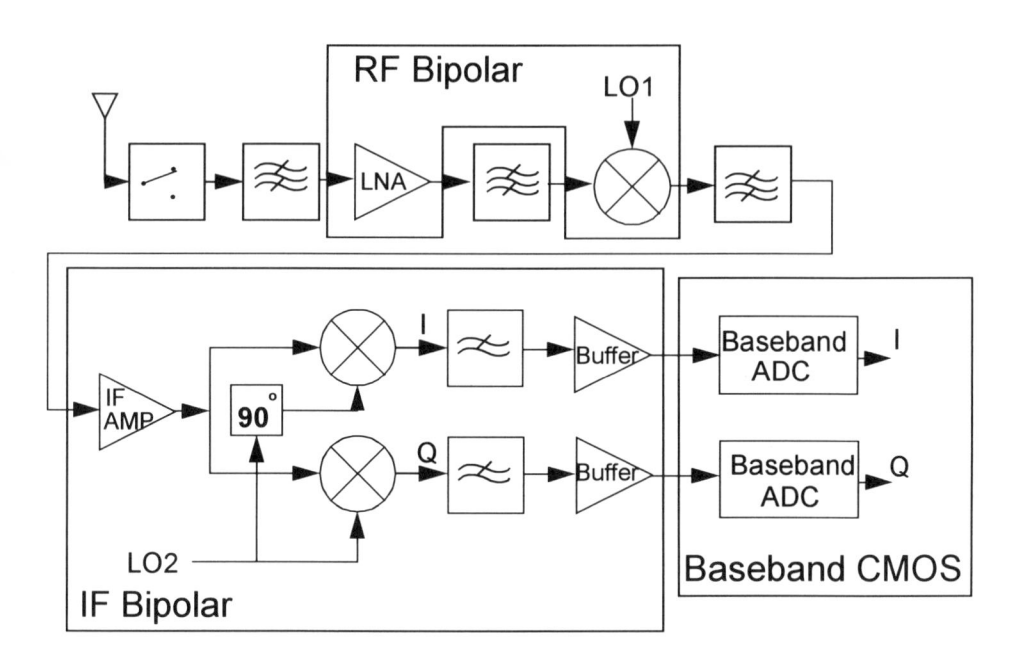

Figure 1. A typical superheterodyne receiver.

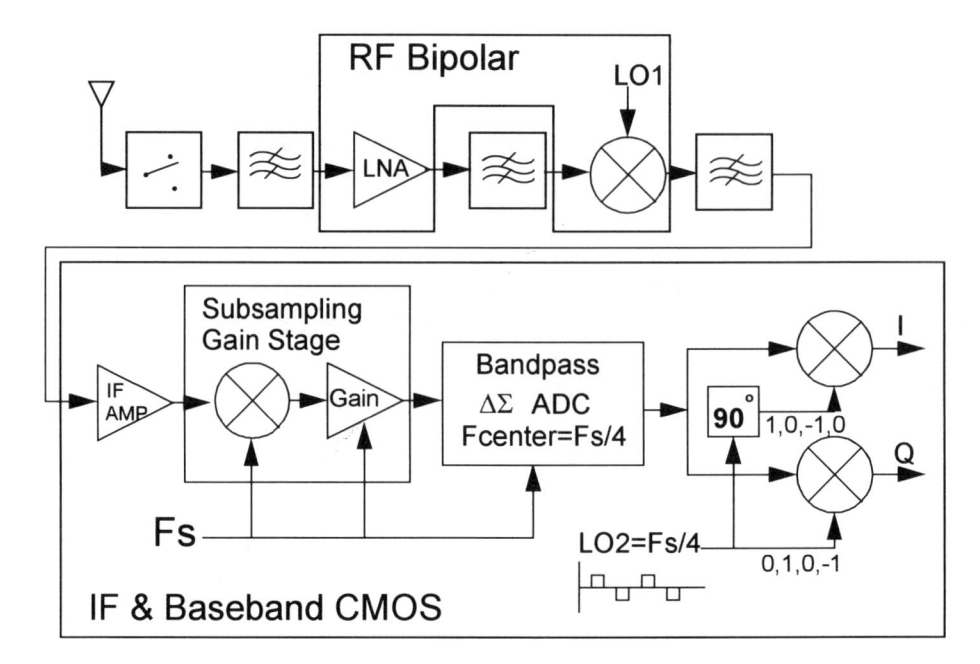

Figure 2. An IF sampling receiver architecture.

The second mixer in this architecture is a sampling stage which downconverts the signal from the first IF to the second IF by means of aliasing. The second IF signal is then digitized by a bandpass Delta-Sigma ADC. The output of the bandpass ADC is passed on to a digital mixer which performs the final downconversion and generates the baseband I and Q components.

The center frequency of the bandpass ADC is designed to be at the second IF or Fs/4. This greatly simplifies the design of the digital mixer. The sampling frequency Fs, which is normally a multiple of the output sampling rate, dictates the location of the first IF. With this frequency plan the first IF should be an odd multiple of Fs/4.

To summarize, by using an IF sampling receiver we can remove the flicker noise, eliminate the need for DC offset cancellation and I/Q gain calibration as well as eliminating two analog mixers. The system is more robust, since I and Q mix to

3. Implementation of the IF Receiver

An 81MHz IF receiver which performs all the IF functions of an IF sampling receiver is implemented. As shown Figure 3, the receiver consists of a continuous-time IF amplifier, a subsampling switched-capacitor gain stage and a sixth-order bandpass Delta-Sigma ADC. The input signal is at the first IF or 81.25MHz. The second IF is at Fs/4 or 3.25MHz. As mentioned in the previous section, the first IF should be an odd multiple of the second IF. In this case the first IF is 25 times the second IF. 24dB of programmable gain is distributed among the three blocks. The IF amplifier provides 0 and 6dB, the subsampling gain stage 0, 6 and 12 dB and the bandpass ADC provides 0 and 6 dB of switchable gain. This receiver is designed for a channel bandwidth of 200kHz.

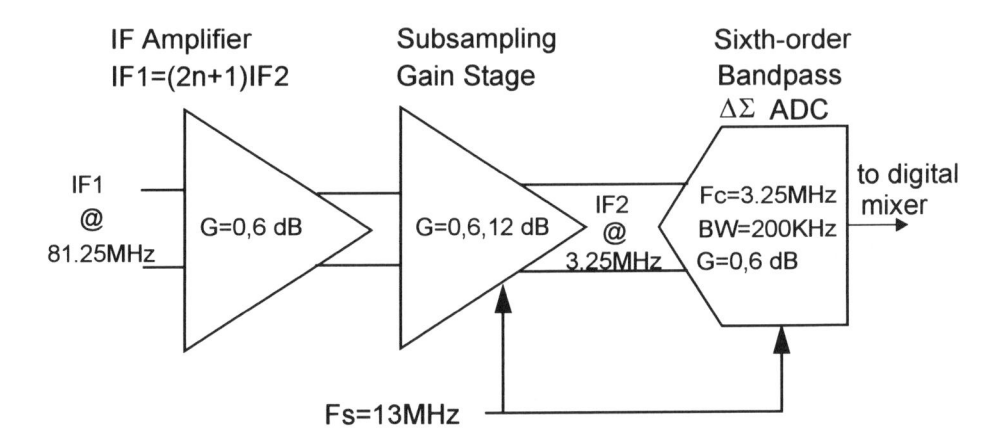

Figure 3. Block diagram of the IF receiver.

3.1. IF Amplifier

The first stage of the receiver is the IF amplifier of Figure 4. The main function of this block is to isolate the external LC filter from the switched-capacitor gain-stage. This isolation is necessary to prevent ringing on the external LC filter which can be triggered by coupling from the switched-capacitor clock signals. This amplifier is required to have a gain accuracy of +/- 0.5 dB. In order to meet this requirement with an open loop amplifier a replica biasing scheme is used. This scheme adjusts the bias current to make the transconductance of the input transistors inversely proportional to an on-chip resistor. This makes the gain of the amplifier proportional to the ratio of two resistors.

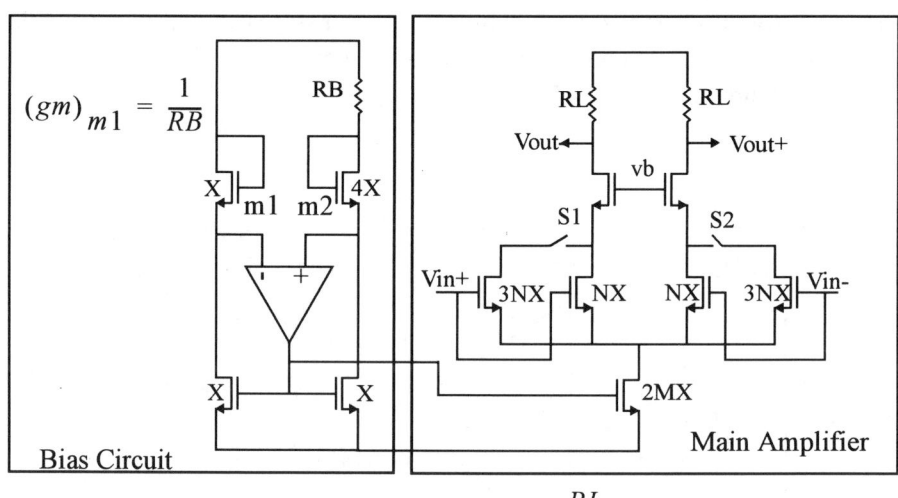

S1 and S2 open: $\quad Gain = \dfrac{RL}{RB} \times \sqrt{M \times N}$

S1 and S2 closed: $\quad Gain = \dfrac{2RL}{RB} \times \sqrt{M \times N}$

Figure 4. IF amplifier.

In the 0dB mode the gain is given by the ratio of RL and RB multiplied by a factor which is determined by ratio of device sizes. In the 6dB mode when S1 and S2 are closed the sizes of the input devices are quadrupled, which causes the gain to increase by 6dB.

In order to achieve good linearity the input devices are designed to have a sufficiently large vdsats. This however reduces their transconductance and increases the input referred noise. Therefore, sufficiently high bias current must be chosen to meet the noise requirement. Since the input to the IF amplifier is ac coupled, the value of the input common mode voltage is designed to optimize the linearity and gain accuracy of the amplifier.

Another precaution that must be taken in designing this circuit is related to it's inherently poor power supply rejection. Separate power and ground pins were dedicated to this circuit to eliminate any noise coupling from other blocks through the power and ground lines.

3.2. Subsampling Gain Stage

The next block in the IF receiver is the subsampling gain stage [1]. Subsampling a signal means sampling it at a frequency which is lower than the Nyquist rate. It is a well known fact that such sampling will cause aliasing and as shown in Figure 5 will move the signal from its original frequency IF1 to a frequency between 0 and Fs/2 which is defined as IF2. Assuming that the sampling frequency is higher than twice the channel bandwidth, no information is lost in this process. However, along with the desired channel, the wideband noise as well as all the unwanted components of the input signal will be aliased and appear between 0 and Fs/2. This increases the amount of noise that appears at IF2 and degrades the signal to noise ratio. Therefore, in order to minimize the amount of unwanted noise that appears at IF2, the input signal should be filtered as much as possible before sampling takes place.

Another important issue associated with subsampling downconverters is the jitter of the sampling clock. Although the frequency of the sampling clock is lower than that of a local oscillator in a conventional mixer, the phase noise requirement remains the same. Therefore, the sampling clock can not be treated like other digital clocks in the circuit.

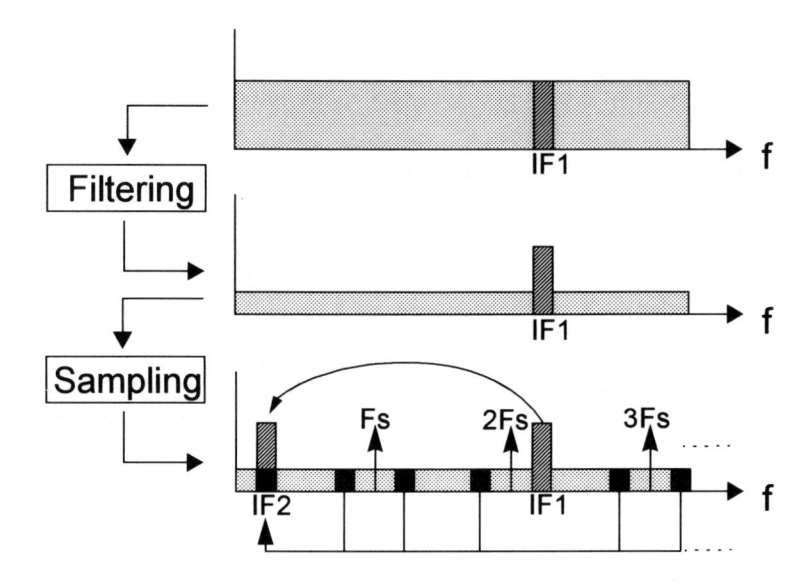

Figure5. Downconversion using subsampling.

Shown in Figure 6, is the subsampling switched-capacitor gain stage which performs down-conversion and provides 0, 6, and 12 dB of programmable gain. By sampling the input signal at 13MHz, the IF of 81.25MHz is down-converted to an IF of 3.25MHz.

As shown in Figure 6, the input signal is sampled differentially by two capacitors and three switches. The common sampling switch is driven by a low-jitter clock to minimize the aperture uncertainty.

The value of the sampling capacitors are chosen to minimize the kT/C noise without limiting the bandwidth of the preceding IF amplifier.

In order to keep the bandwidth of the IF amplifier constant and minimize the complexity of the critical input stage, fixed sampling capacitors and programmable feedback capacitors are used. The DC feedback to the input of the amplifier is provided by the feedback switches. The opamp used in this stage is a single stage folded cascode opamp with switched-capacitor common mode feedback.

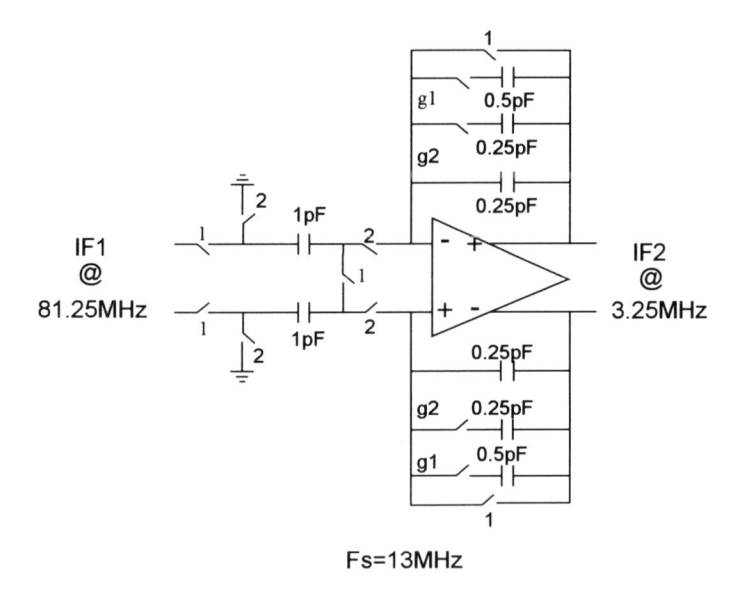

Figure 6. Subsampling gain stage.

3.3. Sixth-Order Bandpass ΔΣ ADC

The subsampling gain stage is followed by a sixth-order bandpass Delta-Sigma analog to digital converter [2]. Similar to lowpass Delta-Sigma ADCs, increasing the order of the modulator reduces both the inband quantization noise and the idle channel tones. As shown in Figure 7, the bandpass ADC implemented in this receiver consists of a fourth-order first stage and a second-order second stage bandpass modulators. The interstage gains 1/g, h, and 1/k are optimized to yield the maximum dynamic range. The output of the bandpass ADC includes the input signal and the quantization noise shaped by a notch filter.

The architecture of the sixth order bandpass ADC is derived by performing a z^{-1} to $-z^{-2}$ transformation on a 2-1 lowpass prototype [3]. As a result of this transformation zeros of the noise transfer function move from DC to Fs/4, where Fs is the sampling frequency of the converter.

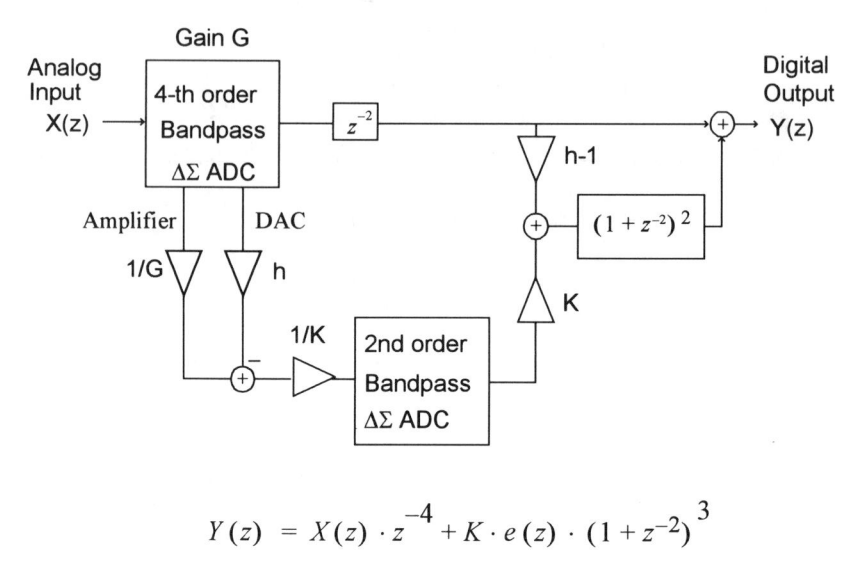

$$Y(z) = X(z) \cdot z^{-4} + K \cdot e(z) \cdot (1 + z^{-2})^3$$

Figure 7. Sixth-order Bandpass ΔΣ ADC

As shown in Figure 8, the first stage of the ADC is derived by performing lowpass to bandpass transformation on a second-order Delta-Sigma modulator. The result is a fourth-order modulator which consists of two second order resonators a one-bit ADC and a one bit DAC. The delays in the loop can be distributed in several ways. In this design, the delays are distributed in such a way to avoid longer settling time due to cascaded amplifiers. The second stage of the ADC is derived by performing lowpass to bandpass transformation on a first order Delta-Sigma modulator. The major impact of the lowpass to bandpass transformation on the circuit implementation is that the integrators in the lowpass modulators are replaced by second order resonators.

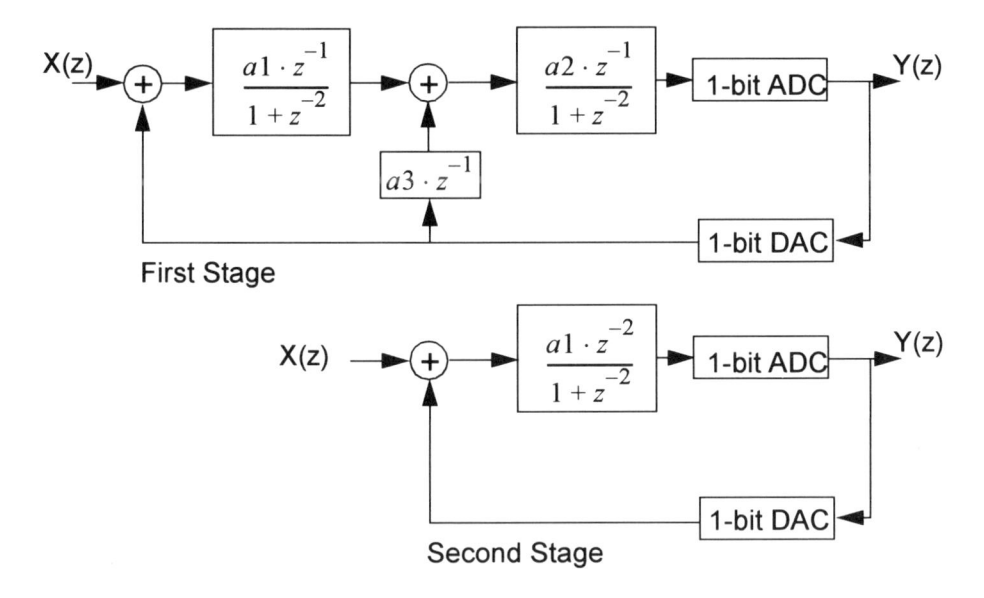

Figure 8. First and second stages of bandpass ADC.

While integrators require a single opamp, resonators are often implemented with two opamps. However, in order to maintain low power consumption, it is essential not to increase the number of active circuits in the receiver.

In this implementation, using a pseudo two path architecture of Figure 9, the resonator is realized with a single opamp [4]. Two time-interleaved channels, each operating at half the sampling frequency are used to achieve the two clock delays that is necessary to perform the resonator function. While one channel is in the idle mode, the other is being charged or discharged by the opamp.

Unlike most switched-capacitor filters, where the location of the pole is dependent on the capacitor ratio, in this structure the location of the poles are independent of capacitor ratios, therefore the notch frequency is immune to capacitor mismatch. This is an important property of this resonator since a predictable and stable center frequency is imperative for a high performance bandpass ADC.

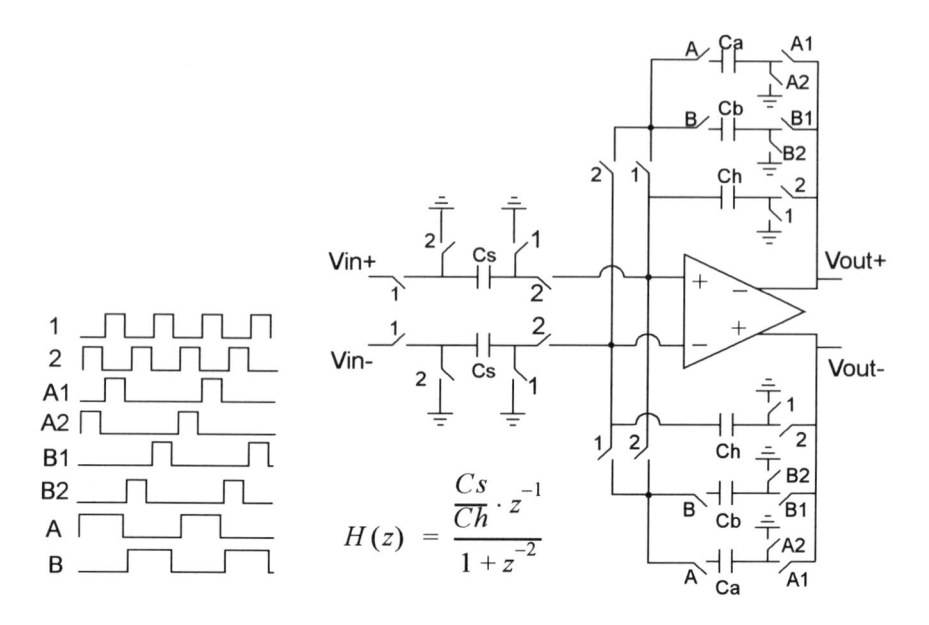

Figure 9. Pseudo two-path resonator.

$$H(z) = \frac{\frac{Cs}{Ch} \cdot z^{-1}}{1 + z^{-2}}$$

In addition to the usual non-overlapping clocks, this structure requires 6 extra clocks at half the sampling frequency. The proper timing of these phases is extremely important in order to avoid signal dependent clock feedthrough. Any signal dependent clock feedthrough which is caused by the slower Fs/2 clock will appear inband as an image of the input frequency with respect to Fs/4.

Providing sufficient isolation between the two channels is essential for proper operation. Any leakage from one to the other channel will move the zeros of the noise transfer function away from Fs/4, thus increasing the inband quantization noise.

The operation of this resonator can be explained in detail by looking at its configuration during different phases. As shown in Figure 10a, during phase A2, both the input capacitors and the channel A capacitors transfer their respective charges to Ch. The charge on Ch therefore, represents the difference of the input signal and the output signal of the previous A cycle. Sampling the output voltage at this phase will result in the desired function.

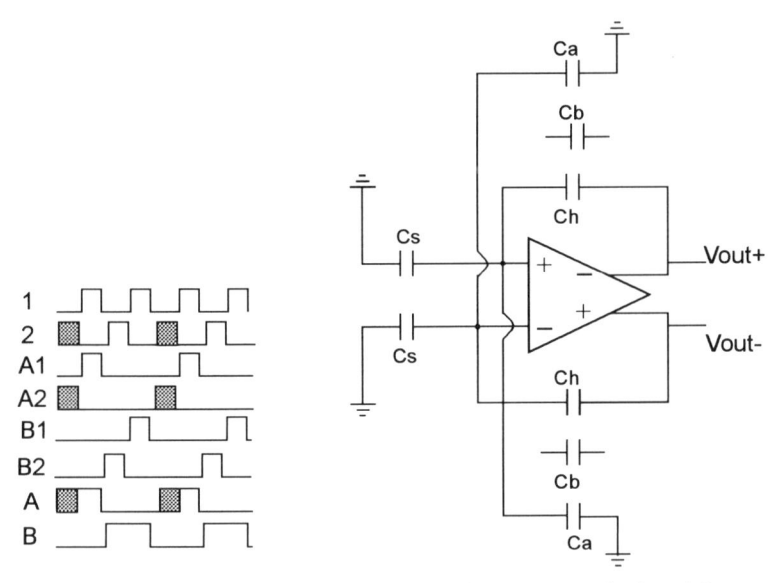

Figure 10a. Pseudo two-path resonator during A2.

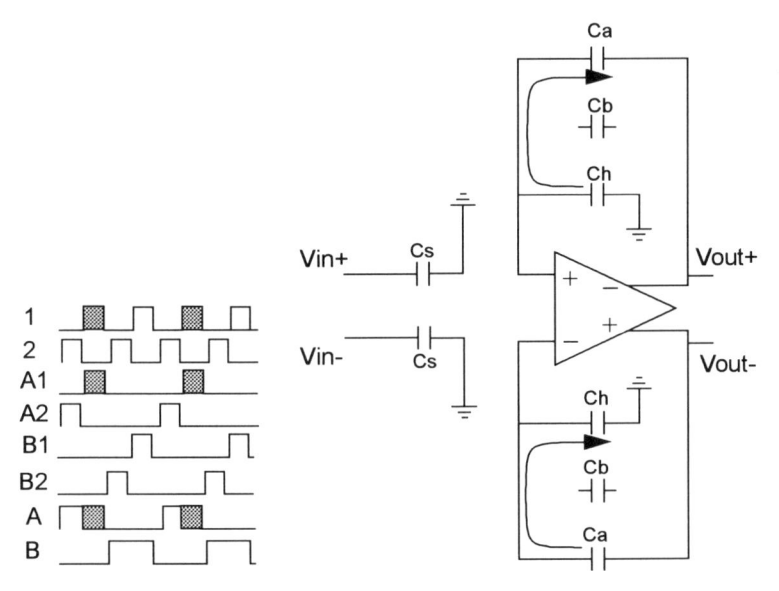

Figure 10b. Pseudo two-path resonator during A1.

Shown in Figure 10b is the configuration of the resonator at phase A1. During this phase, when the new input is being sampled on the input capacitors the main integrating capacitor Ch is transferring it's charge back to channel A capacitors preparing it for the next A cycle. The same sequence is repeated during the B cycle using the B channel capacitors. Once again fully differential, folded cascode opamps are used to implement the resonators the resonators

4. Measured Results

To measure the performance of the bandpass ADC the receiver gain is set to 0 dB and a single tone is applied to the input of the receiver. Comparing the power spectral density of the first stage fourth-order and the complete sixth-order bandpass ADC, as shown in Figure 11, a 7.5dB of improvement in the dynamic range was measured.

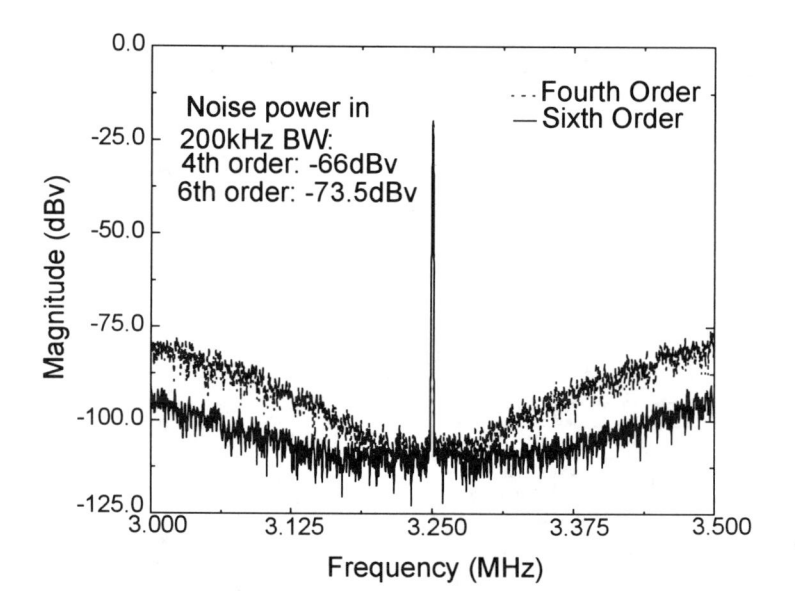

Figure 11. Measured PSD of 4th and 6th order $\Delta\Sigma$ ADC.

With a sampling frequency of 13 MHz, a center frequency of 3.25 MHz and a channel bandwidth of 200kHz a dynamic range of 72dB is achieved in the ADC. This dynamic range however is limited by the kT/C noise. Since 72dB of dynamic range in the ADC was more than sufficient for the receiver, increasing the dynamic range at the expense of increasing the power was not necessary.

The linearity of the receiver, which is limited by the linearity of the IF amplifier is measured by applying two tones at 81.25 and 81.26 MHz.

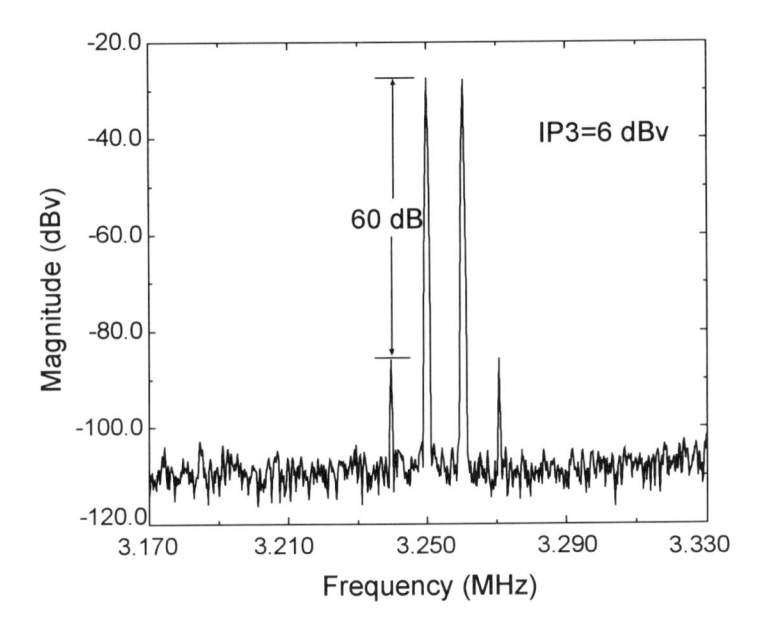

Figure 12. Measured two-tone performance.

As shown in Figure 12, with the input signals at -24dBv, a third order intermod product of -84dBv is measured. This results in an IP3 of 6dBv.

The receiver performance is evaluated by applying an 81.26MHz sinewave for different gain settings and performing a 64K point FFT on the output of the band-pass ADC. The noise and distortion are integrated over a 200kHz bandwidth centered at 3.25MHz. Measured performance curves for the five gain settings are shown in Figure 13. The curves indicate a dynamic range of 92dB. In 0dB mode the peak SNR is limited by the IF amplifier, whereas in other gain settings the peak SNR is limited by the ADC. The modulator 0dB input level corresponds to 1.2V peak differential signal or -1.4 dBv.

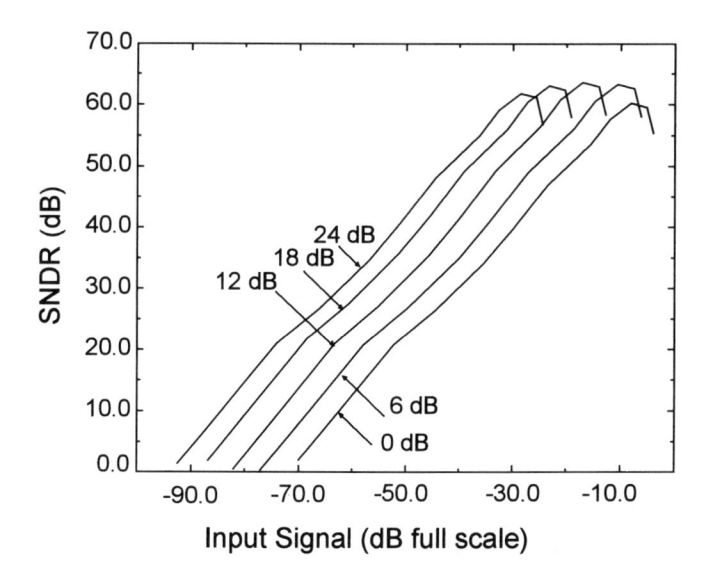

Figure 13. Measured SNDR vs. input amplitude.

To summarize, with an input frequency of 81.25 MHz, and a signal bandwidth of 200kHz the receiver achieves a dynamic range of 92 dB as well as an input IP3 of 6dBv. The sampling frequency of the subsampling gain stage and the bandpass ADC is 13 MHz. The total power consumption of the three blocks with a 3V power supply is 14.4 mW. The IF amplifier consumes 4.2mW, the subsampling gain stage 3.2 and the bandpass ADC 7 mW. This chip was fabricated in a 0.8micron double poly CMOS process.

5. Conclusions

The advantages of the IF sampling receiver are discussed. It has been argued that Delta-Sigma bandpass ADCs provide the most efficient means of digitizing an IF signal. It is shown that an IF sampling receiver, based on a sixth-order bandpass Delta-Sigma converter and a subsampling gain stage provides high bandwidth, low power and a wide dynamic range.

References

[1] Chan, P.Y., Rofougaran, A., Ahmed, K.A., Abidi, A.A., "A Highly Linear 1-GHz CMOS Downconversion Mixer," European Solid-State Circuits Conf., pp.210-213, Sevilla, Spain, 1993.

[2] Longo, L., Horng, B.R., "A 15b 30kHz bandpass Sigma-Delta modulator," ISSCC Digest of Technical Papers, pp. 226-227, Feb., 1993.

[3] Longo, L., Copeland, M., "A 13 bit ISDN-band oversampled ADC using two-stage third order noise shaping," IEEE Proc. CICC., pp. 21.2.1-21.2.4, May. 1988.

[4] Schreier, R., et al. "Multibit bandpass delta-sigma modulators using N-path structures," Proc. of ISCAS, pp. 596-598, 1992.

Design and Optimization of a Third-Order Switched-capacitor Reconstruction Filters for Sigma-Delta DAC's

Tom Kwan
Analog Devices, Santa Clara, CA. USA

Abstract

Several popular filter structures are compared in the design of a third-order switched-capacitor reconstruction filter for a one-bit 64x oversampling 4th-order digital modulator. A coupled-biquad structure implemented using double-sampling switched-capacitors is found to exhibit the least component sensitivity and require the lowest total capacitance for a given KT/C noise budget when compared to a cascade, inverse-follow-the-leader and a similar coupled-biquad structure all using single-sampling switched-capacitors.

Introduction

Early audio DAC's that operated at the Nyquist-rate required reconstruction filters with steep rolloffs. Filters with orders exceeding 10 were not uncommon. Digital interpolation filters alleviated the image rejection problem by removing the images closest to baseband and widening the required transition band of the analog reconstruction filter. This lowered the order and the difficulty of the analog filter considerably (eg. 3rd order for 8x oversampling). With sigma-delta DAC's, the oversampling ratio is increased further but the employment of single-bit quantizers can generate large amounts of out-of-band high frequency noise which in combination with DAC clock jitter can limit the DAC's overall signal-to-noise ratio (SNR)[1]. Clock jitter is often present when the source of audio is remote and a local DAC clock must be extracted from the data stream. In addition, out-of-band noise can excite high frequency non-linearities of the DAC's follow on circuitry (possibly slew limited) to cause intermodulation distortion products to appear in the audio band. To attenuate this out-of-band noise, switched-capacitor filters are often used because as a discrete time filter, it's operation is insensitive to clock jitter given sufficient time to settle. As long as the out-of-band noise is much reduced at the output of the switched-capacitor where the discrete-time to continuous-time conversion takes place, the deleterious effects of clock jitter will be reduced as well. Fig. 1 shows a block diagram of a sigma-delta DAC along with typical signal spectrums at the input and output of the lowpass filter. Among the blocks in Fig. 1, the digital interpolation filter and noise shaper are relatively straightforward to design and implement. The theoretical inband SNR of the noise-shaper can be designed to be well in excess of the expected analog performance which makes the analog circuitry (1-bit D/A and lowpass filter) the limiting component in the performance of the overall DAC.

This note describes the design and optimization of a switched-capacitor reconstruction filter for a one-bit DAC driven by a 64x oversampling 4th-order noise-shaper. Behavioral simulations show that a third order filter is required to meet an out-of-band noise specification of less than -45dBr. Other filter specifications are a 0.1dB passband ripple, with a bandwidth of 25kHz and a sampling frequency of 3072kHz. Several filter structures are considered including a cascade (CAS), inverse follow the leader feedback (IFL), coupled-biquad (LAD1) and coupled-biquad with double-sampling switched-capacitors (LAD2).

These filter structures are used to implement an identical transfer function to determine the one that requires the least total capacitance for a given KT/C noise budget. The main benefit to minimizing the total capacitance is decreased capacitor area and cost, but the load on the amplifiers is reduced as well which imply additional power and area savings from the amplifiers. The capacitance minimization problem for integrator-based switched-capacitor filters given a KT/C noise budget is formulated as a linear programming problem with a non-linear constraint. A simple geometric interpretation is described which give rise to a closed-form solution. This solution is applied to the minimization of total capacitance for each filter structure given a 90dB SNR target. In addition, a magnitude sensitivity formula is derived which enables the calculation of sensitivity curves for each parameter in a filter structure. These curves are computed for each filter structure to compare their passband sensitivity to parameter variations.

Target Transfer Function

The target frequency response is a third order Chebychev filter with the following transfer function.

$$H_{target}(z) = \frac{g(1+z^{-1})^3}{(1+cz^{-1})(1+d_1 z^{-1}+d_2 z^{-2})} \quad (1)$$

where $c = -0.95162, d_1 = -1.947364, d_2 = 0.9516723$ and g determines the passband gain. The transmission zeros at $z = -1$ are not easily implemented. Fortunately for a highly-oversampled filter, these zeros can be eliminated by approximating $(1+z^{-1})^3$ with 8 while incurring a passband error of only 8.5mdB.

The signal flow graphs of the four filter structure candidates are shown in Fig. 2. The first one is a standard cascade implementation using a biquad and a 1st-order filter. The second structure is sometimes known as an "inverse follow the leader" structure and have been applied in sigma-delta modulators and DAC's[2]. The third structure is known as a coupled-biquad, or leap frog structure[10]. The fourth structure is the same as the third except double-sampling switched-capacitors are used which extends the frequency of operation [4, 5] and increases the filter's signal-to-noise ratio [3]. This technique can only be implemented differentially. As an example, Fig. 3 shows a third order implementation of a coupled-biquad structure with double-sampling switched-capacitors.

Transfer Functions: Applying Mason's gain formula on the signal flow graphs in Fig. 2 yields the following transfer functions for the "CAS", "IFL", "LAD1" and "LAD2" filter structures respectively:

$$H_{cas}(z) = \frac{a_1 a_2 a_3 z^{-3}}{(1-z^{-1})^2 + b_2 z^{-1}(1-z^{-1}) + b_3 a_2 z^{-2}(1-z^{-1} + b_1 z^{-1})} \quad (2)$$

$$H_{ifl}(z) = \frac{a_1 a_2 a_3 z^{-3}}{(1-z^{-1})^3 + b_1 z^{-1}(1-z^{-1})^2 + b_2 a_3 z^{-2}(1-z^{-1}) + b_3 a_2 a_3 z^{-3}} \quad (3)$$

$$H_{lad1}(z) = \frac{a_1 a_2 a_3 z^{-3}}{(1-z^{-1})^3 + u_2 z^{-1}(1-z^{-1})^2 + u_1 z^{-2}(1-z^{-1}) + u_0 z^{-3}} \quad (4)$$

where,

$$u_2 = (b_0 + b_3) \quad (5)$$
$$u_1 = (a_2 b_1 + a_3 b_2 + b_0 b_3) \quad (6)$$
$$u_0 = (a_3 b_0 b_2 + a_2 b_1 b_3) \quad (7)$$

$$H_{lad2}(z) = \frac{8 a_1 a_2 a_3 z^{-1}}{(1-z^{-1})^3 + u_2(1-z^{-1})^2 + u_1(1-z^{-1}) + u_0} \quad (8)$$

where,

$$u_2 = (b_0 + b_3)(1 + z^{-1}) \quad (9)$$
$$u_1 = 4z^{-1}(b_1 a_2 + b_2 a_3) + b_0 b_3 (1 + z^{-1})^2 \quad (10)$$
$$u_0 = 4z^{-1}(1 + z^{-1})(b_0 b_2 a_3 + b_3 b_1 a_2) \quad (11)$$

To solve for the filter parameters in the above transfer functions, each denominator is expanded and like powers in z are collected and matched against that of the target transfer function (Eqn. 1). There are two degrees of freedom in the structures "CAS" and

	CAS	IFL	LAD1	LAD2
b_0	-	-	0.0775	0.0165
b_1	0.0484	0.1018	0.0455	0.0216
b_2	0.0526	0.0618	0.0591	0.0165
b_3	0.0551	0.0304	0.0750	0.0330
a_1	0.0381	0.0304	0.0715	0.0294
a_2	0.0782	0.0618	0.0279	0.0224
a_3	0.0700	0.1109	0.0260	0.0415

Table 1: Capacitor ratios after scaling for maximum dynamic range.

"IFL" and these are used to scale the outputs of the first two integrators for maximum dynamic range[1]. For "LAD1" and "LAD2", there is an additional degree of freedom which is used to minimize the maximum capacitor spread. The resulting capacitor ratios are shown in Table 1.

Capacitance Minimization

The switched-capacitor filter structures above are synthesized from building blocks made up of integrators. The transfer function of these filters depend only on the ratios between the input sampling and integrating capacitors and not on their absolute values. This leads to one degree of freedom in sizing the capacitors for each integrator stage in the filter, which can be used to minimize the total capacitance while meeting a fixed KT/C noise budget. The converse problem was consider in [7, 8] in which the total capacitance is fixed and the problem is to allocate the fixed capacitance among the integrator stages to give the minimum KT/C noise. An alternative formulation is given here where the KT/C noise target

---

[1] In a reconstruction filter for sigma-delta DAC's, some of the dynamic range is devoted to the out-of-band noise. The exact proportion depends on the particular noise-shaper and the DAC filter. The out-of-band noise is not included in the above dynamic-range scaling.

is fixed and the total capacitance is minimized. This formulation has a simple closed-form solution which can be derived via a geometric view of the problem.

Problem Definition: A second-order filter shown in Fig. 4 is used as a vehicle to illustrate the capacitance minimization problem and its solution. The results are then generalized to higher order filters. Let the transfer function from the voltage across each sampling capacitor C_{ai} to the filter output be defined as $H_i(f)$. The output noise power v_{ni}^2 due to the sampling capacitor C_{ai} is given by[6]:

$$v_{ni}^2 = \frac{4KT}{C_{ai}f_s} \int_0^{f_b} |H_i(f)|^2 df \qquad (12)$$

where f_b is the bandwidth of interest, f_s is the sampling frequency, K is Boltzmann's constant, and T is the absolute temperature. The total noise power at the output of the filter is the sum of v_{ni}^2 over all sampling capacitors. The capacitance minimization problem can be defined as finding C_{I1} and C_{I2} such that C_t is minimized. Formally,

Minimize: $\quad C_t = g_1 C_{I1} + g_2 C_{I2}$
Subject to: $\quad N_o = v_{n1}^2 + v_{n2}^2 + v_{n3}^2 + v_{n4}^2$

where,

$$g_1 = a1 + a2 + 1$$
$$g_2 = a3 + a4 + 1$$

and N_o is the desired KT/C noise target. The noise constraint can be written explicitly as:

$$N_o = \frac{k_1}{C_{I1}} + \frac{k_2}{C_{I2}} \qquad (13)$$

where k_1 and k_2 are constants, characteristic of a particular filter structure.

$$k_1 = \frac{4KT}{f_s} \int_0^{f_b} (|H_1(f)|^2/a1 + |H_2(f)|^2/a2) df \qquad (14)$$

$$k_2 = \frac{4KT}{f_s} \int_0^{f_b} (|H_3(f)|^2/a3 + |H_4(f)|^2/a4) df \quad (15)$$

The above formulation is a linear programming problem with a non-linear constraint. As shown in Fig. 5, the surface of the constraint forms a curve in the positive quadrant. Since the objective function C_t is a line in two dimensional space, a necessary (and sufficient in this case) condition for optimality is to move the line by changing C_t until it is tangent to the constraint curve (where they intersect at only one point). This is equivalent to the gradient of the constraint curve having the same direction vector as the objective function[9]. This gives:

$$g_1 = U \frac{\partial N_o}{\partial C_{I1}} \quad (16)$$

$$g_2 = U \frac{\partial N_o}{\partial C_{I2}} \quad (17)$$

where U is a constant. These two equations along with the constraint provide a total of 3 equations in 3 unknowns (U, C_{I1}, C_{I2}). The solution for a biquad is:

$$C_{I1} = \frac{1}{N_o} \sqrt{\frac{k_1}{g_1}} \left(\sqrt{k_1 g_1} + \sqrt{k_2 g_2} \right) \quad (18)$$

$$C_{I2} = \frac{1}{N_o} \sqrt{\frac{k_2}{g_2}} \left(\sqrt{k_1 g_1} + \sqrt{k_2 g_2} \right) \quad (19)$$

Fig. 6 shows the results of applying Eqns. [18,19] to a biquad with a -3dB bandwidth of 100kHz, a sample-rate of 6144kHz and a SNR of 100dB relative to a 1V-RMS signal over 0 to 22kHz at 25degC. Note that the noise of the second stage is reduced considerably by the large gain in the first stage of the filter over the audioband. This is reflected in the significantly smaller capacitance required in the second stage to maintain the desired SNR. In constrast, Fig. 7 shows a biquad with damping in the first stage implementing an identical transfer function. Since the gain of the first stage is approximately one over the audioband, it is expected that the KT/C noise of the sampling capacitors in the second stage will be as equally dominating as the first stage. When Eqns. [18,19] are applied to the second biquad, the resultant capacitor values in Fig. 7 show that about twice as much capacitance is needed for the same SNR.

In general, equating the gradient function of the constraint tangent surface plane with that of the objective function gives N equations but $N+1$ unknowns, where N is the order of the filter. The capacitor values can always be explicitly solved for by adding the stated constraint (Eqn. 13) to give $N+1$ equations. The general solution for a filter with N stages is given by:

$$C_{Ii} = \frac{1}{N_o}\sqrt{\frac{k_i}{g_i}} \sum_{n=1}^{N} \sqrt{k_n g_n} \qquad (20)$$

The results above are applied to each of the four third-order filter structures shown in Fig. 2. The desired signal to KT/C noise ratio at the filter output relative to a 1V-RMS signal is set at 90dB. k_i for each filter structure can be calculated by plotting and integrating the appropriate transfer functions. Eqn. 20 is used to compute the various capacitor sizes and the results are shown in Table 2.

Note that the filter structure LAD2 requires less than half the total capacitance of other filters structures. Much of the improvement is due to the 6dB signal to noise ratio improvement of double-sampling vs single-sampling switched-capacitors (see Fig. 8). The expected 4x reduction in capacitance value is not realized because double-sampling switched-capacitors doubles the capacitance ratio of the integrators for the same integrator gain. Although sampling capacitors have reduced by 4 for the same SNR, the integrating capacitor is only reduced in half. In filters with a high sampling to passband ratio, the total capacitance is dominated by the integrating capacitors and the savings are more modest.

	CAS (pF)	IFL (pF)	LAD1 (pF)	LAD2 (pF)
$Cb0$	-	-	1.16	0.11
$Cb1$	0.83	0.91	1.10	0.15
$Cb2$	0.60	1.07	0.47	0.09
$Cb3$	1.25	0.69	0.33	0.11
$Ca1$	0.86	0.69	1.20	0.20
$Ca2$	0.89	1.07	0.77	0.12
$Ca3$	1.20	0.99	0.75	0.14
$CI1$	22.6	22.8	15.4	6.84
$CI2$	11.4	17.3	16.9	5.56
$CI3$	17.2	8.97	12.7	3.30
$Ctot$	56.9	54.5	50.9	16.6

Table 2: Capacitor values for 90dB SNR relative to 1V-RMS single-ended.

Filter Transfer Function Sensitivity

Component tolerances are inevitable in any practical filter implementation. It is desirable that a filter frequency response not deviate significantly from the design value in the presence of these errors. In switched-capacitor filters, the filter parameters are determined by capacitor ratios which are prone to matching and lithography quantization errors. To help define a sensitivity measure, three passbands are shown in Fig. 9. The target third order Chebychev passband has a gain of 0dB and a ripple of 0.1dB. In the presence of parameter errors, the passband may be distorted as shown in the figure. A shift in passband gain is not as critical as distortion over the passband. It is useful to estimate the passband distortion (relative to the DC gain the filter) given a fractional change in a filter parameter. Let the value ΔdB (as shown in Fig. 9) to be the change in dB of the distorted passband relative to the target passband (whose DC gain matches that of the distorted passband). Defining $H(f)$ to be the target transfer function and $\Delta|H(f)|^2$ the deviation from target (due to a parameter error), we have[10];

$$\Delta dB = 10 log \frac{|H(f)|^2 + \Delta|H(f)|^2}{|H(f)|^2 + \Delta|H(0)|^2} \quad (21)$$

Let x and Δx represent an ideal filter parameter value and its associated error, then Eqn. 21, under the condition that $\Delta x/x$ is much smaller than one, becomes:

$$\Delta dB = \frac{20}{\ln(10)} \frac{x}{|H(f)|^2} \Re \left\{ \frac{\partial H(f)}{\partial x} H^*(f) - \frac{\partial H(0)}{\partial x} H^*(0) \right\} \frac{\Delta x}{x} \quad (22)$$

where $\Re\{\cdot\}$ denotes the "real part of" and "*" is the complex conjugate operator.

A computation tool such as *mathematica*[11] can be used to calculate the partial derivatives and generate sensitivity curves for each parameter in the four filter structures described previously. An example of these curves for the LAD2 structure is shown in Fig. 10. These curves can be used to estimate passband deviation given a fractional change in a particular parameter. For example, given a change in the parameter b_3 of -10% in the filter structure "LAD2", the expected maximum change in the passband occurs at 20kHz with a value of $-0.1 \times 1.6 =$ -160mdB. To compare the sensitivities of the different filter structures, the root-mean-squared (RMS) value of all parameter sensitivities from each filter is plotted in Fig. 11 together. By this measure, the plot shows that the "LAD2" filter structure has the best sensitivity performance over the passband.

Conclusion

A coupled-biquad structure with double-sampling capacitors is found to exhibit the minimum passband sensitivity to parameter variations and require the minimum capacitance for a given KT/C noise budget among a cascade, an inverse-follow-the-leader and coupled-biquad structure, the latter three implemented using single-sampling switched-capacitors. Also, an alternative formulation of the capacitance minimization problem under KT/C noise constraints is presented along with a derivation of a closed-form solution which can be used to identify capacitance requirements of integrator-based switched-capacitor filters.

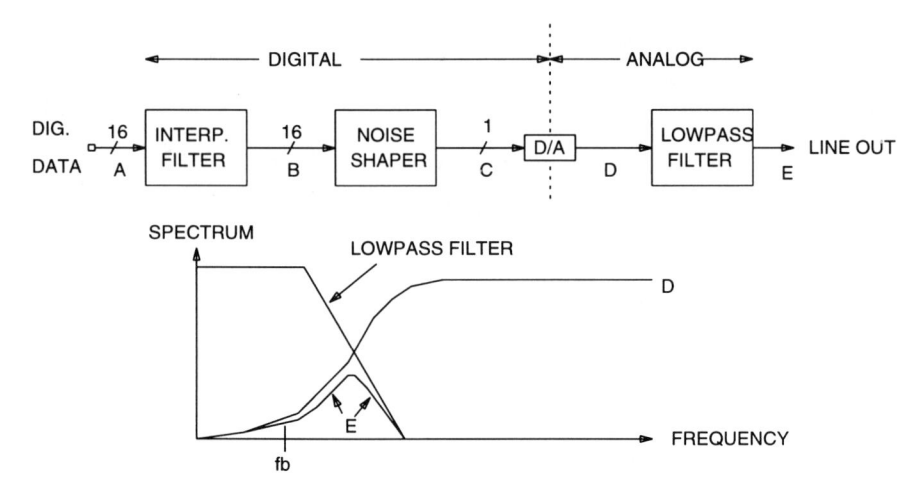

Figure 1: A block diagram of a sigma-delta D/A converter along with typical spectrums at the input and output of the lowpass filter.

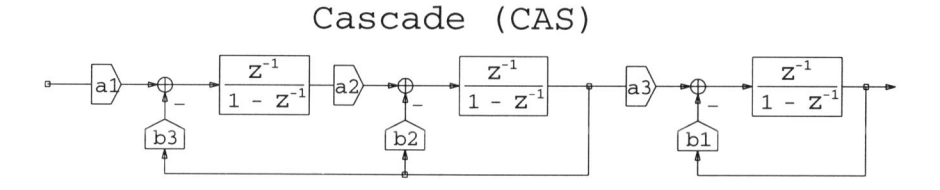

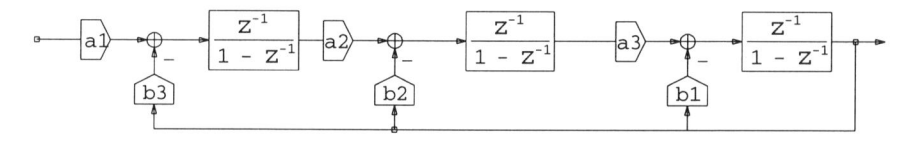

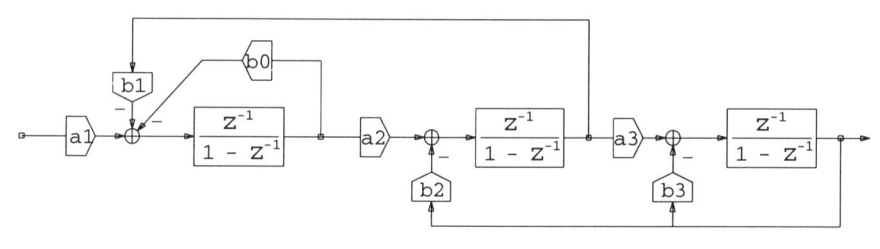

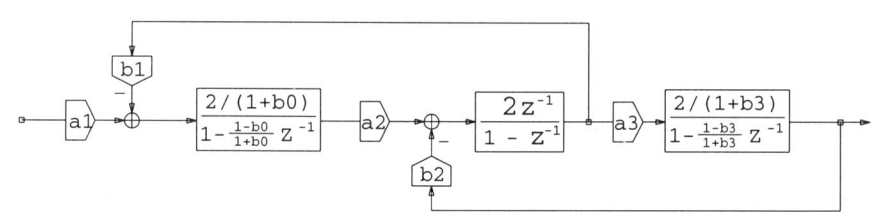

Figure 2: Z-domain signal flow graphs of "CAS", "IFL", "LAD1" and "LAD2" filter structures.

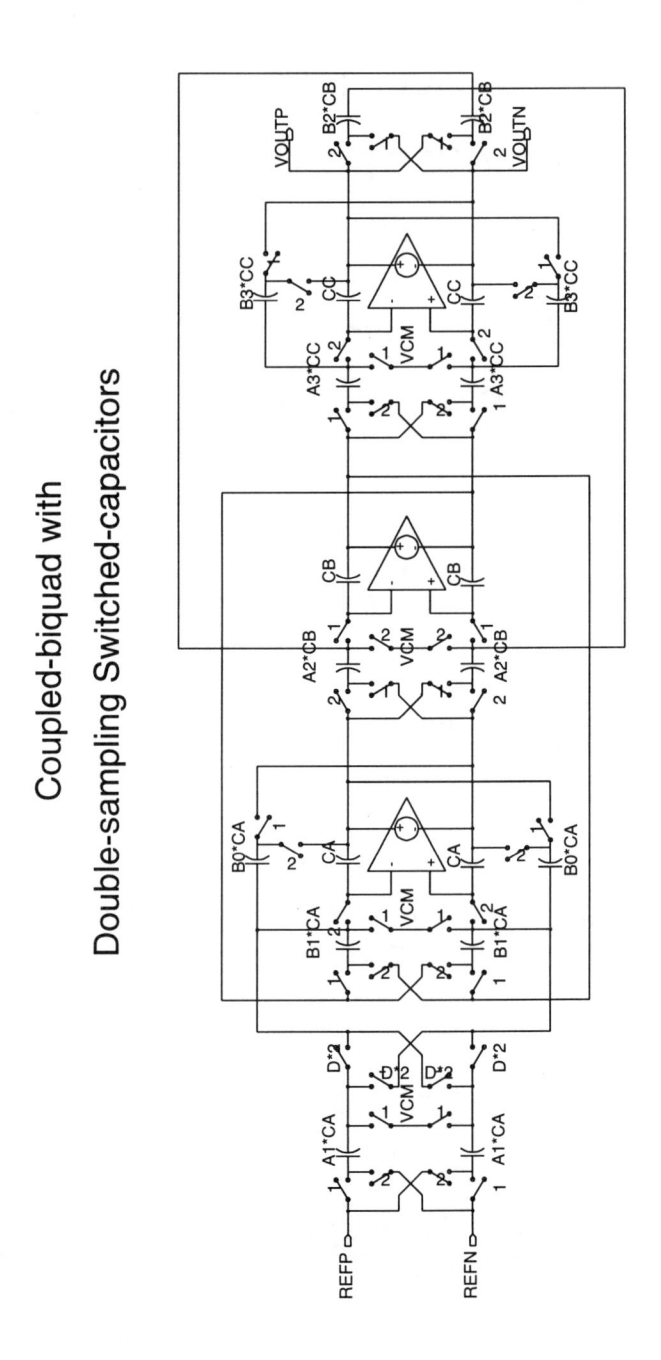

Figure 3: A schematic of a third-order coupled-biquad filter implemented using double-sampling SC's.

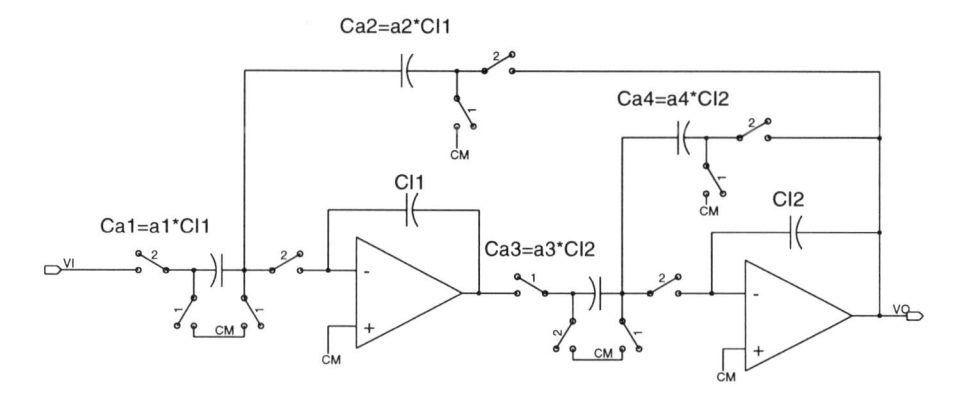

Figure 4: A schematic of a switched-capacitor biquad.

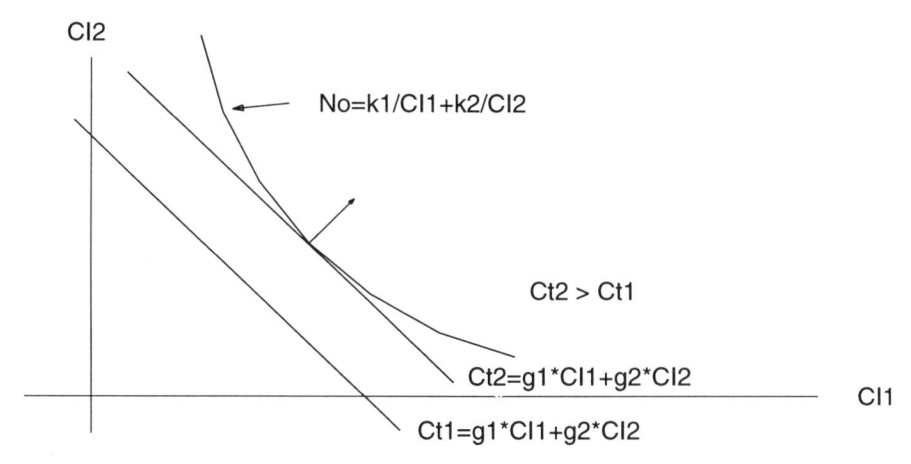

Figure 5: A plot of the KT/C noise constraint and objective function for a 2nd-order filter.

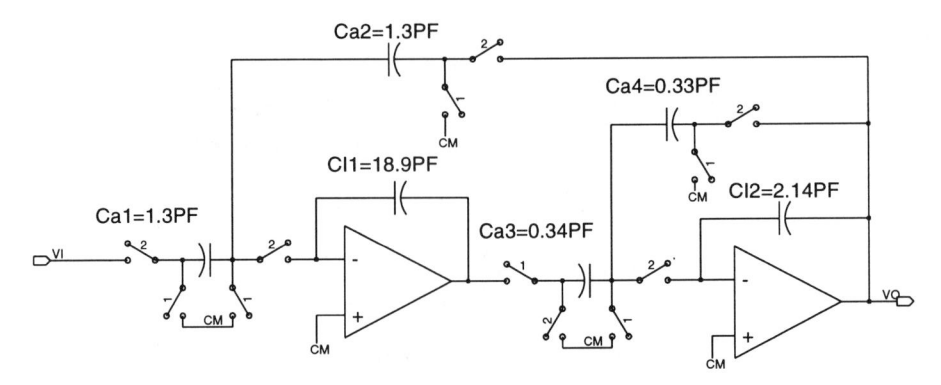

Figure 6: Biquad with damping in the second stage. Total capacitance is 24.4pF.

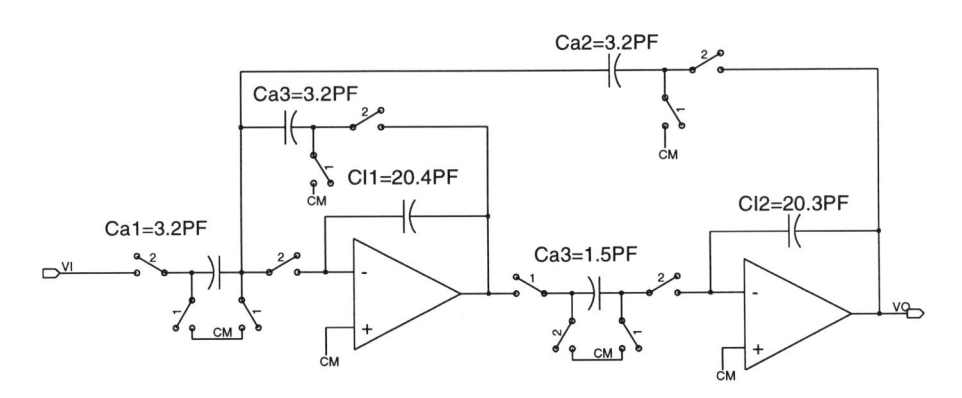

Figure 7: Biquad with damping in the first stage. Total capacitance is 51.8pF.

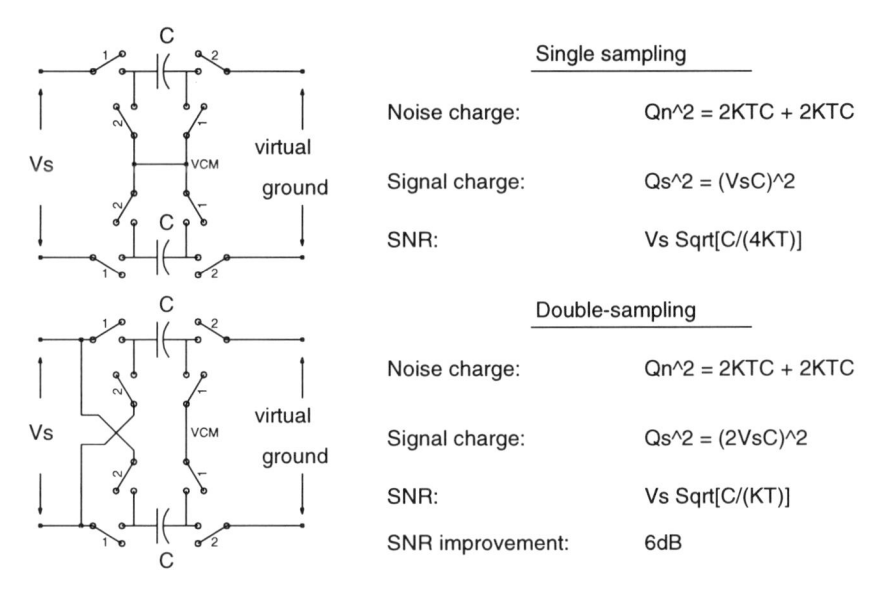

Figure 8: Comparing the SNR's of single and double-sampling switched-capacitors.

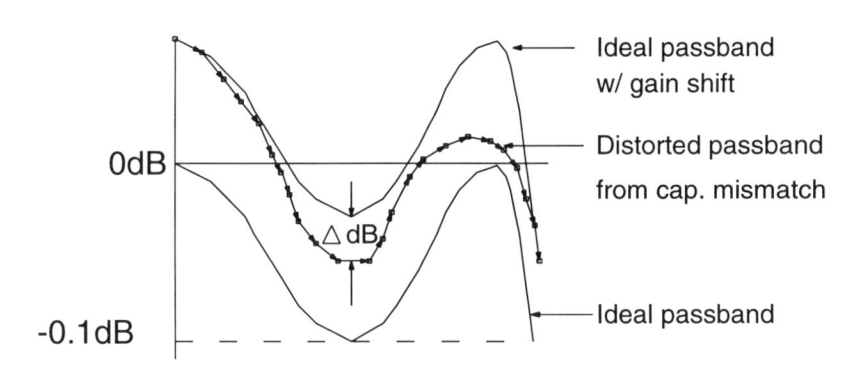

Figure 9: Definition of the transfer function sensitivity ΔdB.

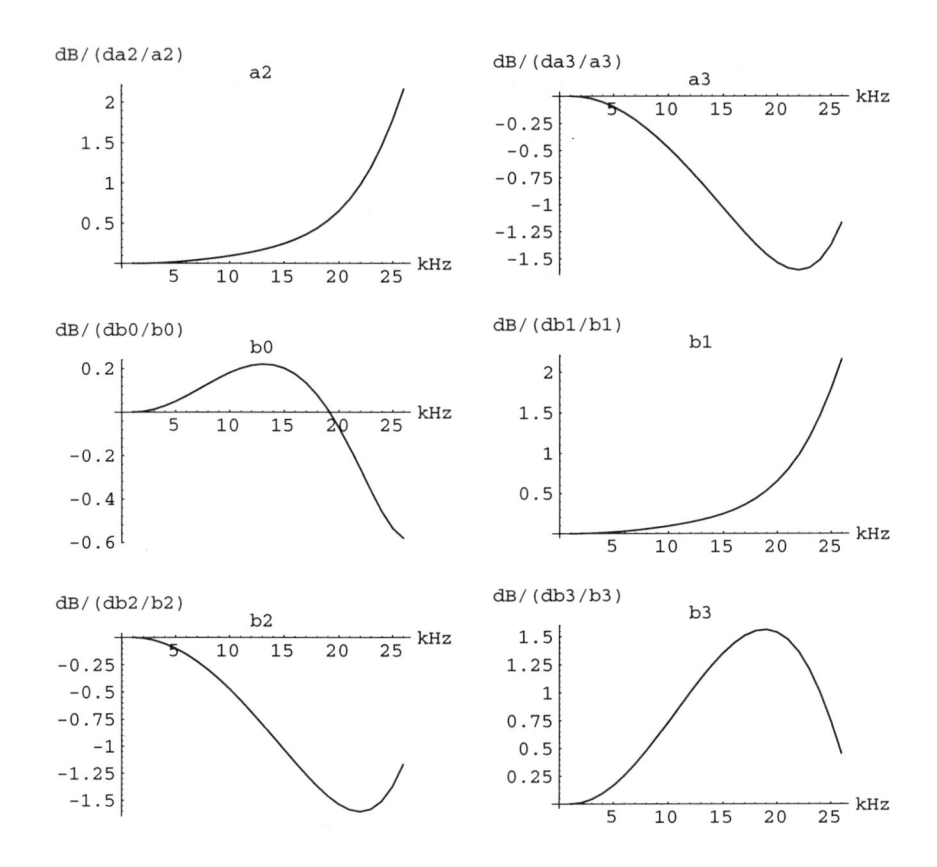

Figure 10: Transfer function sensitivities of all parameters in the "LAD2" filter structure.

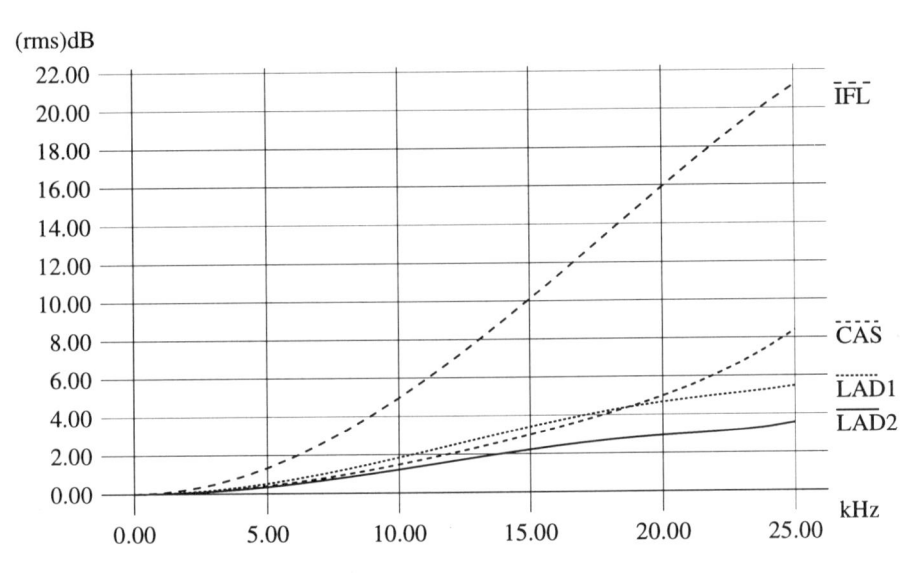

Figure 11: Plots showing the RMS value of all filter parameter sensitivities for each filter structure as a function of frequency.

References

[1] Robert Adams, "Jitter analysis of asynchronous sample-rate conversion", presented at the 95th AES convention, New York, NY, Oct, 1993.

[2] N. Sooch, J. Scott, T. Tanaka, T. Sugimoto, and C. Kubomura," "18-bit stereo D/A converter with integrated digital and analog filters", presented at the 91st AES convention, New York, NY, Oct, 1991.

[3] D. Senderowicz, S. F. Dreyer, J. H. Huggins, C. F. Rahim, and C. A. Laber, "A family of differential NMOS analog circuits for a PCM CODEC filter chip," *IEEE JSSC.*, vol. SC-17, pp. 1014-1023, Dec., 1982.

[4] D. R. Ribner and M. A. Copeland, "Biquad alternatives for high-frequency switched-capacitor filters," *IEEE JSSC.*, vol. SC-20, pp. 1085-1094, Dec., 1985.

[5] A. Baschirotto, F. Montecchi, and R. Castello "A 15 MHz 20mW BiCMOS switched-capacitor biquad operating with 150 Ms/s sampling frequency," *IEEE JSSC.*, vol. SC-30, pp. 1357-1366, Dec., 1995.

[6] J. H. Fischer, "Noise sources and calculation techniques for switched-capacitor filters," *IEEE JSSC.*, vol. SC-17, pp. 742-752, Aug., 1982.

[7] H. Walscharts, L. Kustermans, and W. Sansen, "Noise optimization of switched-capacitor biquads," *IEEE JSSC.*, vol. SC-22, pp. 445-447, June, 1987.

[8] A. Kaelin, J. Goette, W. Guggenbuhl, G. S. Moschytz, "A novel capacitance assignment procedure for the design of sensitivity and noise-optimized SC-filters," *IEEE CAS*, vol. 38, pp. 1255-1268, Nov., 1991.

[9] P.E. Gill, W. Murray and M. H. Wright *Practical Optimization*, Academic Press, N.Y., N.Y., 1981.

[10] A. Sedra and P. Brackett, *Filter Theory and Design: Active and Passive*, Matrix publishers, Champaign, Illinois, 1978.

[11] *Mathematica*, Wolfram Research Inc., Champaign, Illinois.

Tools For Automated Design of ΣΔ Modulators

F. Medeiro, J.M. de la Rosa, B. Pérez-Verdú
and A. Rodríguez-Vázquez

*Centro Nacional de Microelectrónica-Universidad de Sevilla
Edificio CICA, C/Tarfia sn, 41012-Sevilla, SPAIN
Phone #34 5 4239923, FAX #34 5 4231832
email: angel@cnm.us.es*

Abstract

We present a set of CAD tools to design ΣΔ modulators. They use statistical optimization to calculate optimum specifications for the building blocks used in the modulators, and optimum sizes for the components in these blocks. Optimization procedures at the modulator level are equation-based, while procedures at the cell level are simulation-based. The toolset incorporates also an advanced ΣΔ behavioral simulator for monitoring and design space exploration. We include measurements taken from two silicon prototypes: 1) a 17bit@40kHz output rate fourth-order low-pass modulator; and 2) a 8bit@1.26MHz central freq@10kHz bandwidth band-pass modulator. The first uses SC fully-differential circuits in a 1.2µm CMOS double-metal double-poly technology. The second uses SI fully-differential circuits in a 0.8µm CMOS double-metal single-poly technology.

Footnote

This work has been supported by the CEE ESPRIT Program in the framework of the Project #8795 (AMFIS)

1. Introduction

The performance of a $\Sigma\Delta$ converter IC is ultimately limited by its analog circuitry: the $\Sigma\Delta$ modulator front-end. Thus, efforts to enhance the performance or widen the application range of these converters concentrate mostly on the modulator and follow two parallel and largely correlated directions: exploration of high-order architectures and/or multibit quantizers and, pushing the specifications of analog cells used in the modulators at their performance edges [1][2][3]. The confrontation of these issues poses significant difficulties to IC designers. Some of the problems encountered are general of analog IC design: large number of specifications, complicated relationships between specifications and design parameters, involved analysis, critical specifications significantly sensitive to mismatch, etc. Others are specific to $\Sigma\Delta$ modulators; in particular, its accurate simulation is costly due to its highly non-linear dynamics and the necessity to use long time-series of data for evaluation purposes [4][5][6]. These difficulties render the design of $\Sigma\Delta$ modulator ICs a time- and resource-consuming process, and have prompted the development of tools which can help to increase designer productivity and, thus, reduce time-to-market and production cost of forthcoming generations of $\Sigma\Delta$-based mixed-signal ASICs.

This chapter presents a set of CAD tools for computer-aided design of CMOS *switched-capacitor* (SC) and *switched-current* (SI) $\Sigma\Delta$ modulators for *low-pass* and *band-pass* applications. These tools use *optimization* at the modulator and cell levels, advanced *behavioral simulation* at the modulator level, and include the capability of fast design space exploration of modulator architectures. The tools are vertically integrated to support top-down design of $\Sigma\Delta$ modulators, from the high-level specifications to the sizes of the cells. Their use is demonstrated in the paper through two fully-differential silicon prototypes in CMOS technologies: a fourth-order two-stage SC $\Sigma\Delta$ modulator and a fourth-order band-pass SI $\Sigma\Delta$ modulator.

2. Tool Diagram

Fig. 1 shows the design flow of $\Sigma\Delta$ modulators. It comprises top-down *synthesis* tasks:

1. **Topology selection**, i.e., to identify the best suited modulator architecture

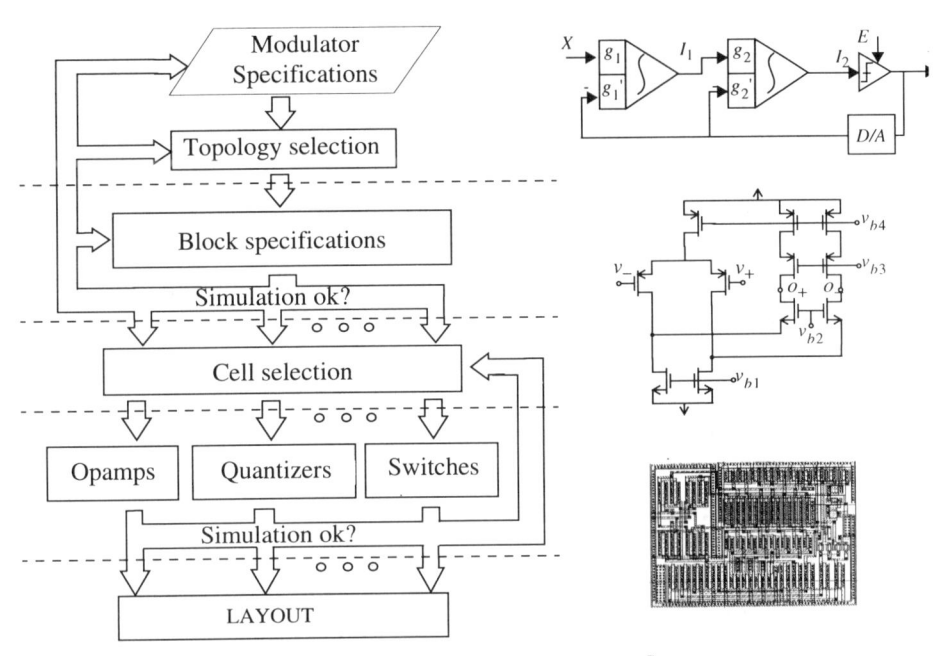

Fig 1: :Modulator design operation flow

for the required high-level converter specifications (signal baseband, resolution, etc.);

2. **Modulator sizing**, i.e., to map these high-level specifications into specifications of the basic building blocks (such as gain-bandwidth product of the opamp, slew-rate, comparator hysteresis, etc.);

3. **Analog cell selection**, i.e., to choose the cell schematics according to the specifications;

4. **Cell sizing**, i.e., to map the cell specifications into values of their components;

5. **Layout**.

These top-down tasks are complemented with bottom-up *analyses*:

6. **Modulator simulation** at the architectural level, to verify correctness of the results of high-level synthesis. Due to the large circuit complexity, and the need for long time series at the modulator output, this analysis is more conveniently handled through dedicated behavioral simulation.

7. **Cell simulation**, to verify synthesis at the *electrical* level using SPICE-like simulators [7].

8. ***Extracted layout simulation*** at electrical level -- very costly in CPU time and memory resources. Thus, it is typically used just to check connectivity and evaluate block performance degradation due to layout parasitics (not shown in Fig. 1).

The selection tasks (either modulators or cells) involve *knowledge* issues. The proposed tools include procedures to help designers in gaining insight about the operation of different modulator architectures, and hence guiding their selection. On the other hand, the sizing tasks involve principally *optimization* issues -- realized in our design framework through the use of *statistical* optimization techniques. Fig. 2 is a flow diagram of the top-down vertical integration of the set of tools described in this chapter [8].

3. Design Equation Database

These represent the behavior of typical modulator architectures realized with *switched-capacitor* (SC) and *switched-current* (SI) building blocks. The database has been conceived to be easily extendible and includes *single-loop* as well as *cascaded* architectures, *low-pass* and *band-pass*, *single-bit* and *multi-bit*. Equations describing these architectures are classified in three categories:

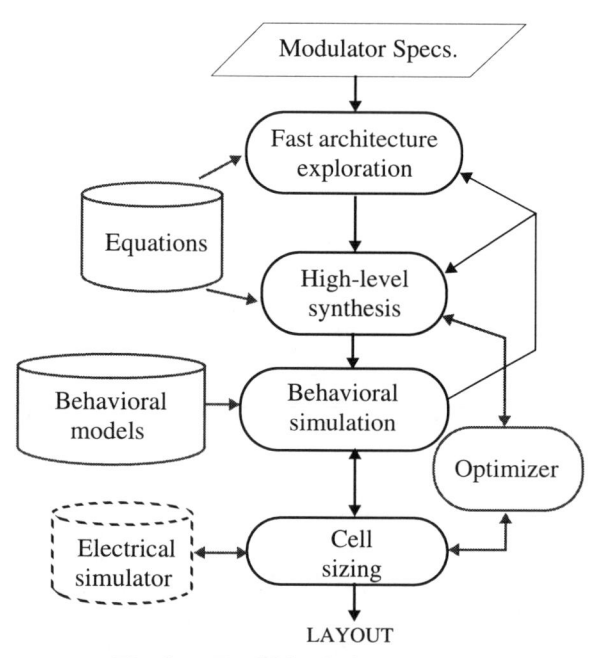

Fig 2: : Tool block diagram

- **Architecture-related.** These represent the *quantization noise* as a function of the non-idealities that affect its shaping. Their analytical expressions may be largely different for the different modulator architectures [8].
- **Circuit-related.** These represent noise sources other than quantization: *thermal noise*, incomplete *settling*, harmonic *distortion*, etc [8].
- **Fundamental limits.** They cover trade-offs between power consumption, resolution, speed, etc., for the different topologies. They are used to guide modulator selection [1][9][10][11].

Table 1 summarizes the noise and distortion contributions covered in the two first categories above, together with the responsible building block. As an example, Table 2 shows the approximate expressions for the power of noise and harmonic distortion of a cascaded SC 2-2 architecture (see Fig. 10(a)[9][12]).

Building Blocks		Non-idealities	Consequences
SC integrator	Opamps	DC-gain, finite and non-linear	Increased quantization noise, harmonic distortion.
		SR, limited	Settling noise, harmonic distortion.
		GB, limited	Incomplete settling noise.
		O.S. limited	Overloading.
		Thermal Noise	White noise.
	Switches	ON-resistance, feed-through	Settling noise, white noise, harmonic distortion.
	Capacitors	Non-linear, mismatching	Increased quantization noise, harmonic distortion.
Clock		Jitter	Jitter Noise.
Comparators		Hysteresis, resolution time	Increased quantization noise.
Quantizers		Non-linearity	Harmonic distortion.

Table 1: Non-idealities considered in the Tool

4. Sizing

Specifications contemplated for sizing include *constraints* on the performance parameters and design *objectives*. Their meaning is clarified considering for instance an opamp with the following specifications: DC-gain > 70dB; gain-bandwidth product > 5MHz; phase margin > 60 degree; input equivalent noise < 3µV; with minimum power consumption and silicon area occupation. We call constraints to the four first specifications that include > or < symbols, and design objectives to the last two, whose goal is to maximize or minimize some magnitude.

Quantization noise	$\frac{\Delta^2}{12}\left\{\frac{4\pi^2}{3M^3}\mu^2 + \frac{\delta_A^2\pi^4}{5M^5} + d_1^2\frac{\pi^8}{9M^9}\right\}$
Incomplete settling noise	$\frac{\Delta^2}{9M}\left(1 + \frac{C_p}{C_1}\right)^2 \varsigma^2 \exp\left(-\frac{g_m}{C_{eq}}T_s\right)$
Thermal noise	$\left(1 + \frac{C_{12}}{C_{11}}\right)\frac{kT}{4MC_{11}} + \left(1 + \frac{C_{12}^2}{C_{11}^2}\right)\left(\frac{kT}{6MC_i} + \frac{kTg_m R_{on}}{2MC_i}\right)$
Non-linear capacitor distortion	$\frac{\alpha^2}{8}\left(\frac{\Delta}{2}\right)^4 + \frac{\beta^2}{32}\left(\frac{\Delta}{2}\right)^6$
Non-linear opamp dc-gain distortion	$\frac{|\alpha_1|^2(1+k_1)^2 k_2^4}{4A_0^2}\left(\frac{\Delta}{2}\right)^4 + \frac{|\alpha_2|^2(1+k_1)^2 k_2^6}{8A_0^2}\left(\frac{\Delta}{2}\right)^6$
Jitter noise	$\left(\frac{\Delta}{2}\right)^2 \frac{(2\pi f_b \sigma_t)^2}{2M}$

Table 2: Approximate error power for a cascaded architecture 2-2

Sizing itself is performed through a sequence of movements in the design space until a cost function reaches a minimum. The main related issues are *cost function* formulation, and the generation of *movements* and the *management* of the optimization procedure.

A. Cost Function Formulation

The sizing of modulators and cells are formulated as *constrained* optimization problems,

$$\text{minimize } [\Phi(\mathbf{x})] \text{ subject to} \quad\quad\quad (1)$$
$$\phi_r(\mathbf{x}) \leq 0 \; ; \; 1 \leq r \leq R$$

where $\mathbf{x} = (x_1, x_2, ..., x_L)^T$ is the vector of design parameters, which defines a L-dimensional parameter space. From (1) an equivalent *unconstrained* problem is defined using different strategies for modulators and cells.

Cost function for modulator sizing

Calculated block specifications must fulfill the modulator specifications and, at the same time, be the best suited for implementation. For instance, the DC-gain and GBW of the opamps should be the lowest among the set of values which yield feasible modulators.

The modulator specifications are mapped onto a single constraint:

$$P_N(\mathbf{x}) - P_{N,max} \leq 0 \qquad (2)$$

where $P_N(\mathbf{x})$ is the total in-band output noise power at the modulator output, and $P_{N,max}$ is the maximum power that guarantees the modulator specifications: resolution, bandwidth and maximum input level. The cost function is given by:

$$\Psi(\mathbf{x}) = \begin{cases} -\sum_{j=1}^{N} K_j \log\left(\dfrac{x_j}{x_{j,norm}}\right) & \text{if } \mathbf{x} \in A \\ \log\left(\dfrac{P_N(\mathbf{x})}{P_{N,max}}\right) & \text{if } \mathbf{x} \notin A \end{cases} \qquad (3)$$

where x_j represents the value of the j-th block specification. The sign of the weight parameters K_j, indicates if the objective must be maximized or minimized. On the other hand, the normalization factors,

$$x_{j,norm} = \begin{cases} x_{j,min} & \text{if } K_j > 0 \\ x_{j,max} & \text{if } K_j < 0 \end{cases} \qquad (4)$$

are used to cope with large variations of the absolute values of different block specifications.

Logarithms in (3) renders the cost function smoother and thus, enable the trajectory to escape from local minima. The example of Fig. 3 illustrates the benefits of using logarithms. It corresponds to,

$$f(x) = K \cdot \min\left\{ -e^{-\xi \sum_{k=1}^{N}(x_k - d)^2} \prod_{k=1}^{N} \cos(x_k - d),\ -e^{-\xi \sum_{k=1}^{N}(x_k - d)^2} \prod_{k=1}^{N} \cos(x_k + d) + \gamma \right\} \qquad (5)$$

where K, ξ, d and γ are constants. Fig. 3(a) depicts (5), which has the absolute minimum in $(x, y, z) = (4, 4, 0.5)$, and many local minima in its neighborhood. The values of the function at these minima are compressed into the interval [0.4, 0.5]. On the other hand, Fig. 3(b) depicts the result of taking the logarithm of (5), where the minima are more clearly separated. Functions like this are commonly found in modulator and cell sizing [8].

a)   (b)

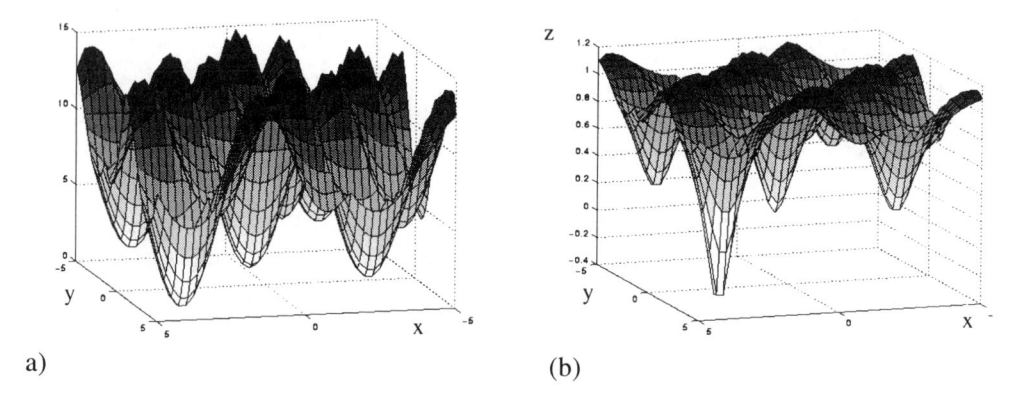

Fig 3: : Using logarithm to facilitate the search of the global minimum

Cost Function for Cell Sizing

Constraints related to cell sizing are classified in two groups:
- **Strong restrictions:** These are specifications whose fulfillment is considered essential by the designer; for instance, the phase margin of an opamp must be larger than 0 for stability. No relaxation of the specified value is allowed.
- **Weak restrictions:** These are the typical performance specifications required for analog building blocks, i.e. opamp DC-gain > 80dB. Unlike strong restrictions, weak restrictions allow some relaxation of their targets, making circuits which do not exactly meet the targets acceptable.

The costs function for cell sizing must reflect both types of constraints:

$$\text{minimize} \quad y_{oi}(\mathbf{x}) \quad , 1 \leq i \leq P$$
$$\text{subject to} \begin{cases} y_{sj}(\mathbf{x}) \geq Y_{sj} \quad or \quad y_{sj}(\mathbf{x}) \leq Y_{sj} \quad , 1 \leq j \leq Q \\ y_{wk}(\mathbf{x}) \geq Y_{wk} \quad or \quad y_{wk}(\mathbf{x}) \leq Y_{wk} \quad , 1 \leq k \leq R \end{cases} \quad (6)$$

where y_{oi} denotes the *i*-th design objective; y_{sj} y y_{wk} are constrained specifications (*w* and *s* denote *weak* and *strong* respectively) and Y_{sj} y Y_{wk} are the corresponding goals. The unconstrained cost function is defined as:

$$\Psi(\mathbf{x}) = \begin{cases} \Phi(y_{oi}) & if \; \mathbf{x} \in A \\ max[F_{sj}(y_{sj}), F_{wk}(y_{wk})] & if \; \mathbf{x} \notin A \end{cases} \quad (7)$$

where *A* denotes the acceptance regions, and where the partial cost functions are given by,

$$\Phi(y_{oi}) = -\sum_i w_i \log(|y_{oi}|) \qquad , F_{sj}(y_{sj}) = K_{sj}(y_{sj}, Y_{sj})$$

$$F_{wk}(y_{wk}) = -w_k \log\left(\frac{y_{wk}}{Y_{wk}}\right) \qquad (8)$$

w_i is the weight associated to the i-th design objective, a real positive number (alternatively negative) if y_{oi} is positive (alternatively negative); for $K_{sj}(.)$ we have

$$K_{sj}(y_{sj}, Y_{sj}) = \begin{cases} -\infty & \text{if strong restriction holds} \\ \infty & \text{otherwise} \end{cases} \qquad (9)$$

w_k is the weight associated to the k-th weak restriction -- a real positive number (alternatively negative) if the weak restriction is of $\geq$ type (alternatively $\leq$ type). These weights are used to give priority to some weak restrictions. There is no relation between the objectives and the weights of weak restrictions [13].

B. Optimization Algorithm

Fig. 4 shows a block diagram of the operation flow of the proposed iteration procedure. The updating vector, Δx_n, is *randomly* generated at each iteration. The value of the cost function is calculated at each new point of the parameter space, and compared with the previous one. The new point is accepted if the cost function has a lower value. It may also be accepted if the cost function increases, according to a *probability* function,

$$P = P_o e^{-(\Delta\Phi/T)} \qquad (10)$$

depending on a control parameter, T (temperature). This probability of acceptance changes during the optimization process, being high at the beginning (for large T) and decreasing as the system cools (decreasing T). This is the well known Metropolis algorithm [14]. Our tool enhances the basic algorithms used for *cooling schedule* (mechanism to update T) and *design parameter updating* (mechanism to generate Δx_n).

Cooling Schedule

In classical simulated annealing algorithm the cooling schedule is determined by four parameters: initial value of T, stop criterium, evolution law for T, and Markov chain length [14]. Our tool incorporates an algorithm based in the use of

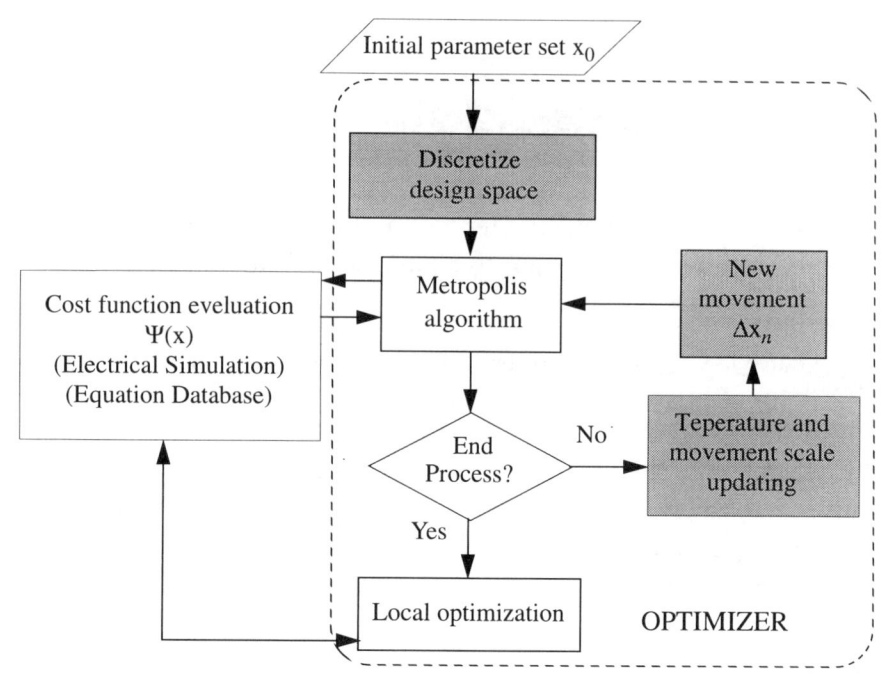

Fig 4: Flow diagram of the proposed methodology

a composed temperature [15],

$$T = \alpha(\mathbf{x}) T_o(n) \tag{11}$$

$\alpha(\mathbf{x})$ is employed to solve possible discontinuities of the cost function in the border of the acceptance region. On the other hand, $T_o(n)$ is a function of the iteration count and can vary *non-monotonically* with successive re-heatings and coolings. The tool incorporates an adaptive mechanism to automatically set the temperature and thus, keep a given *acceptance ratio*. Fig. 5 depicts the procedure. The instantaneous acceptance ratio $a[n]$ (one if the iteration has been accepted, zero if not) is low-pass filtered and the result is compared to the specified acceptance ratio

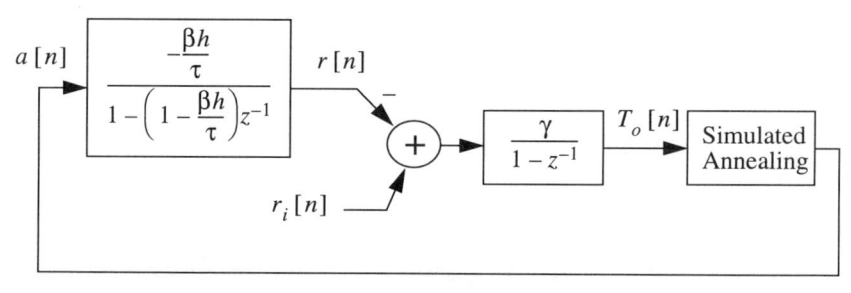

Fig 5: : Adaptive temperature flow diagram

$r_i[n]$ (commonly large at the beginning and decreasing with the iteration count). The difference between the ideal and actual acceptance ratio is integrated to obtain the new iteration temperature. The feedback loop forces the temperature to evolve such that $r[n]$ follows $r_i[n]$ -- depicted in Fig. 6. When compared with classical cooling schedules, the presence of spontaneous fast heatings and cooling has proven to be valuable to minimize complicated multi-minimum functions. In addition, the quality of the final result is only slightly dependent on the number of variables -- very convenient for analog sizing, where the number of design parameters is usually large.

<u>Design parameter updating</u>

Here the following heuristics are incorporated:
- Large amplitude movements of the design parameters are allowed at high T. On the contrary, acceptance probability decreases at low T and hence, only small movements are allowed.
- The possibility of defining logarithmic scales for independent variables with large variation range to ensure that their low range is not under-explored.
- Discretization of the design parameter space (see Fig.7). Only movements over vertices of the resulting multidimensional grid are allowed. If one

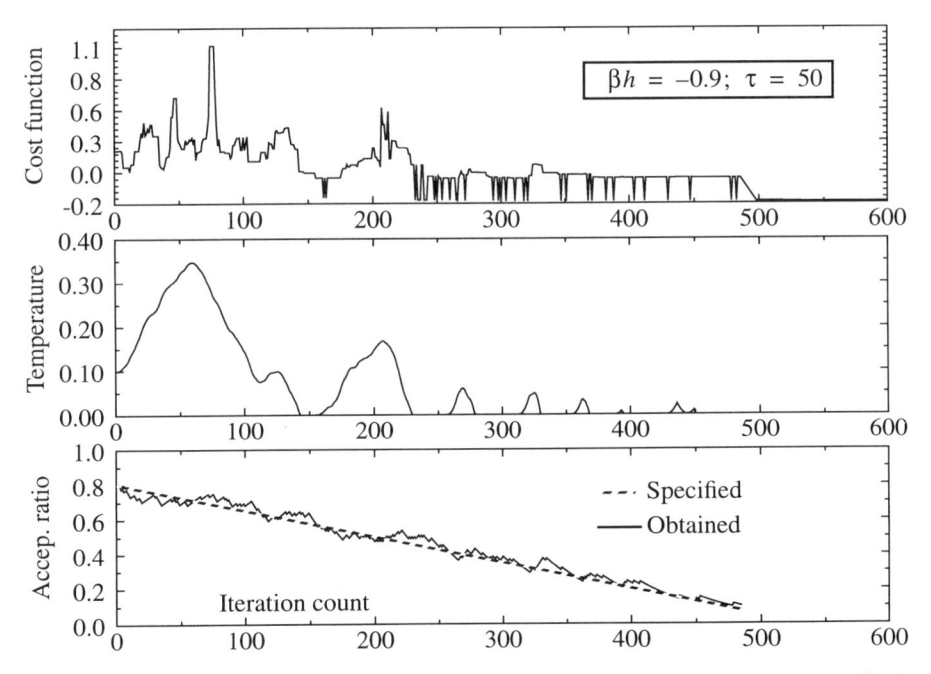

Fig 6: Example of cost function, temperature and acceptance ratio evolution.

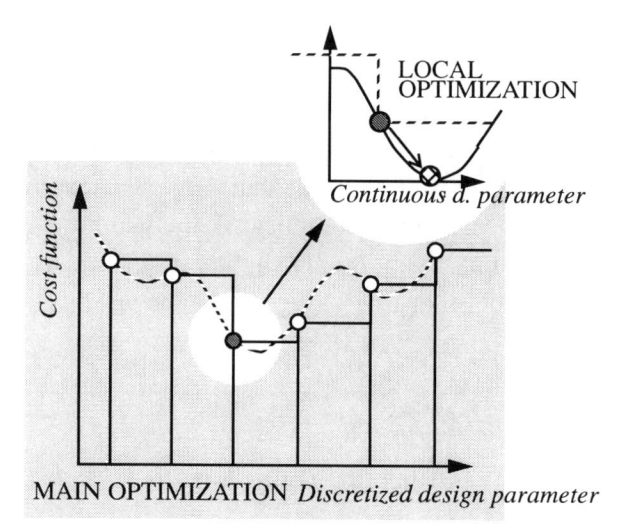

Fig 7: : Ilustrating the concept of discretized optimization

vertex is revisited during the optimization process the corresponding cost function evaluation need not be performed. When this optimization process ends, local optimization starts inside a multidimensional cube around the optimum vertex. During the local optimization it is possible to use a deterministic algorithm (Powell method) to fine tuning the design [16].

5. Behavioral Simulator

The simulator incorporated to our tool, ASIDES, starts from an input netlist containing the modulator topology and a list of non-idealities to consider during simulation, and operates in time-domain using functional descriptions of the blocks. It generates a time series which is processed using a general-purpose DSP tool, for instance *MATLAB* [17] to provide:

- Information about the ***dynamic performance*** of the modulator, including the spectrum of the converter output, graphs of the signal-to-(noise + distortion) ratio (TSNR), etc.
- Information about its ***static performance***, through evaluation of integral non-linearity, offset, gain error, etc.
- ***MonteCarlo analysis***, taking into account fluctuations of both the integrator gains and the terminal specifications of the analog cells. These fluctuations can be indicated by the user or evaluated by the tool on the basis of technological parameters and layout-related variables, for instance the capacitor size and the partition used in their layout [18]. This capability is

especially useful when cascade modulator architectures are considered because of their sensitivity to mismatching [9].
- **Parameter sweep**, for design space exploration through visualization of the impact of critical design parameters on the modulator performance.

The catalog of building blocks in ASIDES includes *voltage sources, integrators, resonators, quantizers/comparators, adders, amplifiers, preamplifiers, delays, filters,* and generic *non-linear* blocks. Each is described through a dedicated routine in C-language. To obtain time series output, these routines are invoked according to the connection specified in the input netlist. Each block output is updated using its present input (and output) and a set of non-idealities related to the electrical implementation included in an associated block model.

Fig. 8(a) illustrates this operation flow for the integrator whose corresponding routine starts a loop that involves the following calculations:
- Thermal noise originated by the on-resistance of the switches and integrator opamps. The equivalent level of white noise is first calculated based on accurate analytical expressions for standard SC multi-branch integrators, and then added in time-domain to the integrator input using a random number generator [8].
- Non-linear scaling and leakage in the integrator due to finite and non-linear DC-gain of the opamps [8].
- Capacitor non-linearity, represented as a polynomial dependence of the capacitance value on the accumulated voltage [8].
- The value of the integrator output voltage is calculated at the end of the settling period. This calculation uses either a linear or non-linear expression of the settling depending on the integrator input, the slew-rate, the gain-bandwidth product and the phase margin of the opamp and the integrator gain [10][19].
- If the *output swing* of the integrator is surpassed, the output voltage is clipped to that value.

The iterative process shown in Fig. 8(a) is necessary due, on the one hand, to the interdependency between the opamp DC-gain and the integrator output and, on the other, to the relationship among the capacitor values and their accumulated voltages. However, two or three iterations are commonly sufficient to detect convergence and complete the simulation process. The remaining primitives considered in the behavioral simulator follow descriptions similar to that for the integrator.

Fig. 8(b) shows a typical input netlist; in this case for a fourth-order 2-2 cascade ΣΔ modulator [12]. The input is a pure tone with amplitude varying from -140dBV to 0dBV in 2dBV steps. The first and second stages are second-order

(a)

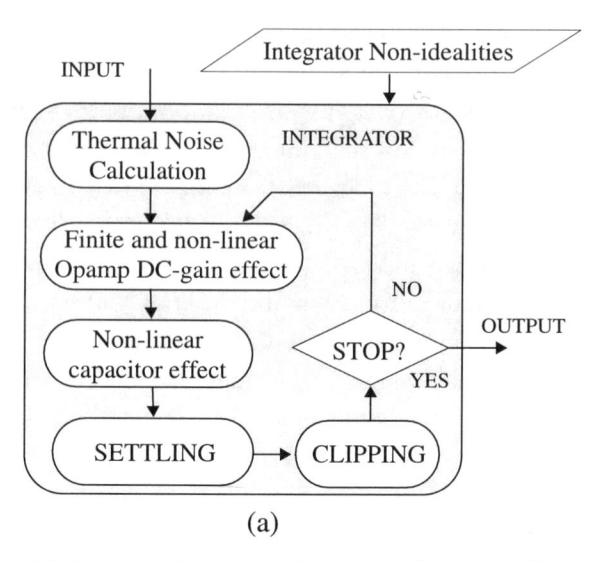

```
# 2-2 Cascade SD Modulator #
##########################
Vin inp dc=0.0 ampl=(-140 0 2) freq=1.25k;
# First Stage -> 2nd-order mod.
Comp1 out1 (oi2) real cm;
I1 oi1 (inp,0*0.5 out1,0*0.5:2) real im;
I2 oi2 (oi1,0*0.5 out1,0*0.5:2) real im;
# Second Stage -> 2nd-order mod.
Comp2 out2 (oi4) real cm;
I3 oi3 (oi2,0*0.5 out2,0*0.5:2) real im;
I4 oi4 (oi3,0*0.5 out2,0*0.5:2) real im;
# Logic for noise cancelation #
Del3 14 (out1) full;
. . .
Ad4 out (21 15) ideal;
##########################
.clock freq=35.2X jitter=1n;
.oversamp 64;
.options fullydiff mismatch;
.output snr(out) monte=30;
.output fft(out);
# Models
.model im Integrator cunit=0.25p cfb=1p cpa=1.5p cnl=50u
dcgain=70d dcgnl=20 ron=500 npwd=5n imax=800u
gm=7m osp=3 osn=-3 pm=pmp;
.model cm Comparator vhigh=1.5 vlow=-1.5 hys=50m;

.param pmp= sweep (dec 10 100 100000);
```

(b)

Fig 8: (a) Integrator operation flow. (b) Input netlist example.

modulators composed of two integrators and one comparator. All these elements are of real type and have associated models called "*im*" for the integrators and "*cm*" for the comparators. The cancellation logic, whose description is not completely printed, is formed using ideal delays and adders. The clock frequency is set to 35.2MHz with 1ns standard deviation jitter and the oversampling ratio is 64. Requested analyses include an FFT of the time series at the output node and the calculation of the signal-to-noise (SNR) curve at this node. They also include a MonteCarlo analysis of SNR with the integrator gains as random parameters. Those code lines that start with "*.model*" in Fig. 8(b) are used to specify parameters associated with non-ideal features contemplated in the block models. Note that the opamp DC-gain is not given a numerical value, but specified through the parameter "*dcg*"; this is used to sweep a range of DC-gain values -- shown in the last command line in Fig. 8(b).

Fig. 9 depicts output provided by the simulator in the case of the netlist of Fig. 8(b), presenting three graphs corresponding to simulator outputs. The contin-

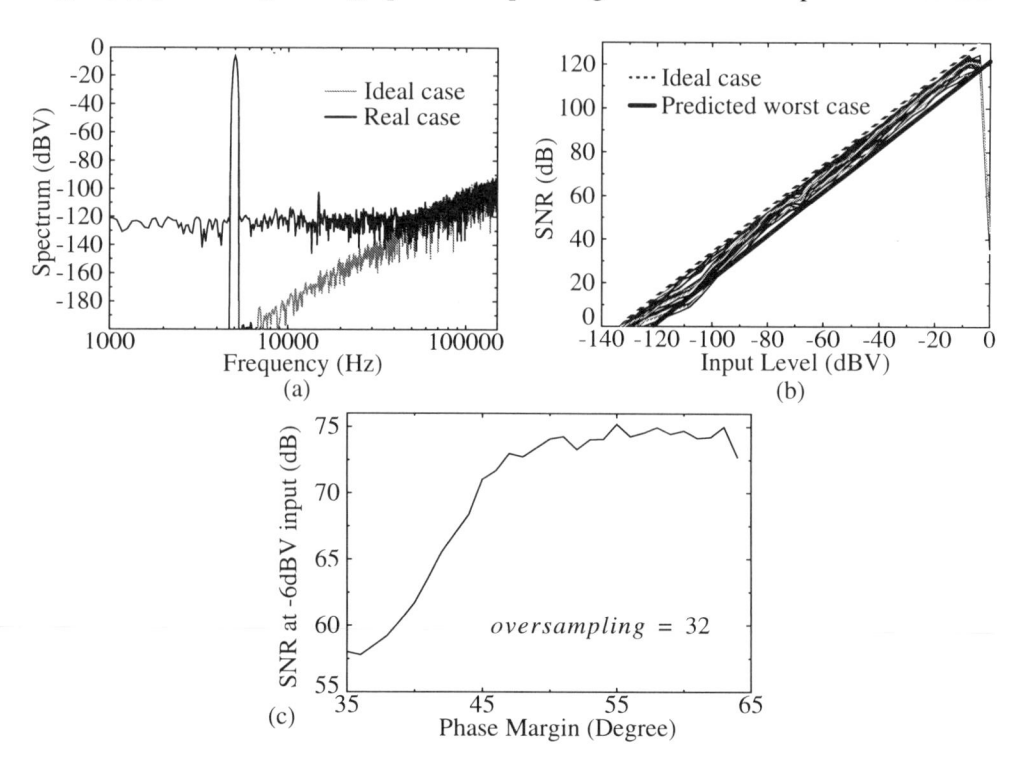

Fig 9: Three simulator outputs: (a) Real and Ideal case output spectrum for a fourth-order 2-2 cascade ΣΔ modulator. (b) SNR vs. input amplitude including integrator gain mismatching. (c) SNR for -6dBV input vs. opamp phase margin.

uous trace in Fig. 9(a) shows the simulated output spectrum, while the dashed trace shows the corresponding ideal curve. Note the presence of harmonic distortion due to non-linearity of opamp DC-gain, and an unshaped noise floor around -120dBV mainly due to thermal noise. Fig. 9(b) shows the result of MonteCarlo analysis where all error sources other than mismatch have been disconnected to highlight the influence of the latter. It is of interest to compare the results of MonteCarlo simulation with a calculation of worst-case realized in a single instance of a corresponding equation contained in the equation database used for synthesis. Fig. 9(b) includes this calculated worst-case curve which coincides with the simulation results. Finally, Fig. 9(c) shows the SNR after decimation as a function of the opamp phase margin for -6dBV@5kHz input. Based on the information contained in this graphic one concludes that, for this case, a phase margin of 50° suffices to reach maximum performance [8].

6. PRACTICAL RESULTS

The SC low-pass modulator of Fig. 10(a) has been designed to achieve 17bit@40kHz output rate, in a 1.2μm n-well double-poly double-metal CMOS technology [20]. On the other hand, the SI modulator of Fig. 10(b) has been designed for 8bit@+-5kHz bandwidth@1.26MHz central frequency, with a clock frequency of 5MHz in a 0.8μm n-well single-poly double-metal CMOS technology [21].

Table 3 displays the outcome of high-level synthesis for the SC structure. This is in the format provided by the high-level synthesis tool, which also summarizes the different noise contributions anticipated by the equations for the completed design -- displayed at the bottom in Table 3. The associated statistical optimization procedures required 30,000 iterations for the SC modulator and 20,000 iterations for the SI modulator, and lasted 11.8s CPU time and 9s CPU time, respectively, on a 100MIPs workstation. These short CPU times are a positive consequence of using equations, and render the ability to explore design spaces through iterations of the optimum high-level synthesis procedure. For both architectures the sampling capacitor was fixed to a relatively small value to evaluate the ability of the procedure to obtain feasible designs in spite of a relatively large thermal noise contribution. In particular, the thermal noise contributed by a sampling capacitor of 1pF is at the very border of feasibility for the fourth-order SC modulator (see the bottom part of Table 3). In this sense, the summary of noise contributions reported by the tool is of interest to guide design exploration if specifications are not met, and a new iteration of the high-level synthesis procedure is required.

Fig. 11 illustrates the use of the fast architecture exploration feature to evaluate

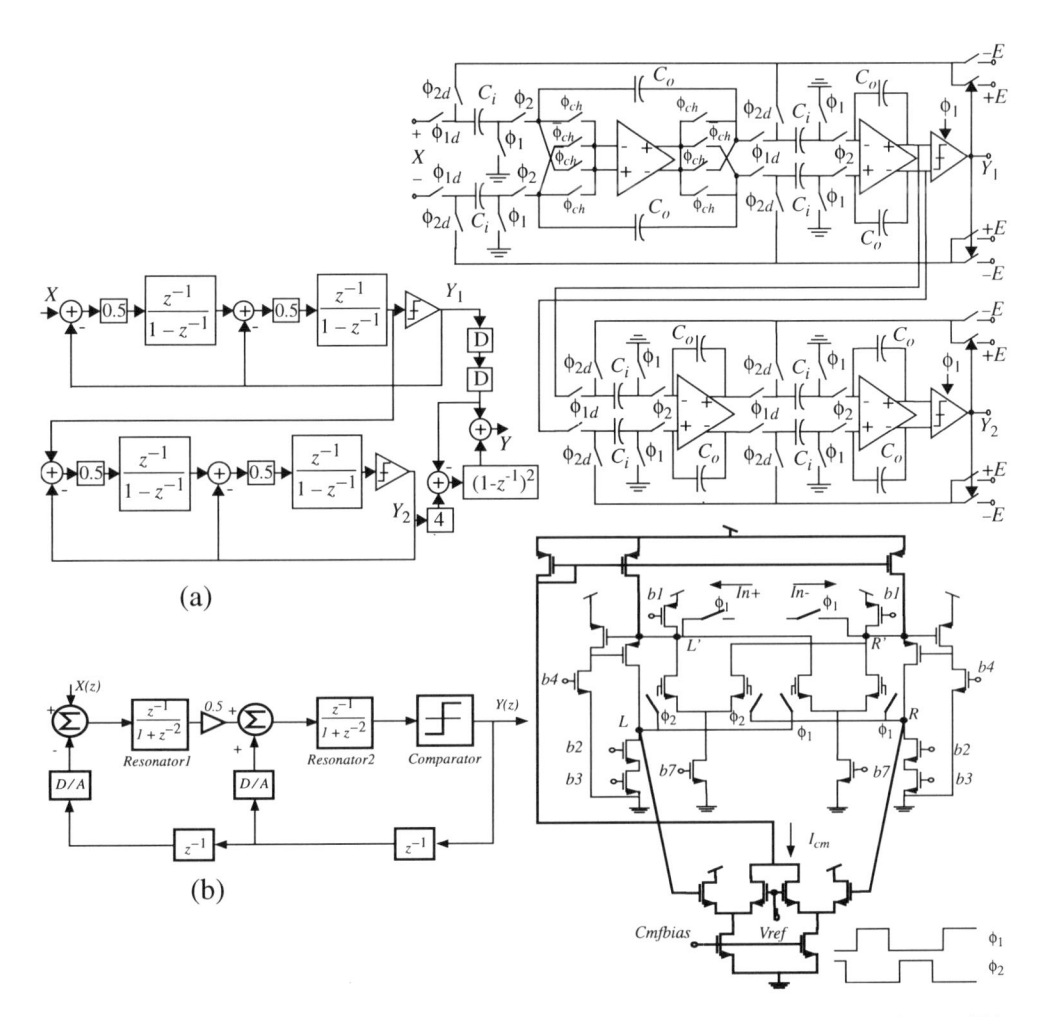

Fig 10: (a) 4th-order two-stage SC ΣΔ modulator. (b) 4th-order SI band-pass ΣΔ modulator and fully-differential regulated folded-cascode SI integrator schematic

the influence of two SI block errors in the noise transfer function (*NTF*) of the SI prototype: output-input conductance ratio error, ε_g; settling error ε_s; and the error due to changes in the feed-back loop gain in the resonator block.

Fig. 12(a) presents the schematic of the opamp used for the SC topology: a folded-cascode fully-differential OTA with degenerated mirror common-mode feedback [22]. Fig. 10(b) shows the SI integrator used to implement the resonators in the SI prototype [21]. With regards to the comparator, since speed rather than hysteresis is the more demanding specification for both modulators, we used the regenerative latches: Fig. 12(b) for SC and Fig. 12(c) for SI.

OPTIMIZED SPECS FOR:		17bit@40kHz@±1.5V
Modulator	Topology	Cascade 2-2
	Sampling frequency (MHz)	5.12
	Oversampling ratio	128
	Differential reference voltage E (V)	0.75
Integrators	Sampling capacitor C_i (pF)	1.0
	Feed-back capacitor C_o (pF)	2.0
	Unitary capacitance (pF)	≥ 0.25
	MOS switch-ON resistance (kΩ)	≤ 1.0
	Maximum clock jitter (ns)	≤ 0.9
Opamps	DC-gain (dB)	≥ 70.7
	DC-gain non-linearity (V^{-2})	$\leq 20\%$
	GB (MHz)	≥ 15.3
	Slew-rate (V/us)	≥ 57.35
	Total output swing (V)	≥ 6.0
	Input noise density (nV/sqrt(Hz))	≤ 15.0
Comparators	Hysteresis (V)	≤ 0.33
	Resolution time (ns)	≤ 24.41
Technology	Cap Non-linearity (ppm/V)	≤ 50.0
Resolution and Noise Contributions		
Dynamic range:		104dB (17bit)
Quantization noise (dB)		-118.2
Thermal noise (dB)		-104.3
Incomplete settling noise (dB)		-177.7
Jitter noise (dB)		-111.2
Harmonic distortion (dB)		-117.7

Table 3: High-Level synthesis tool output

The tool was used to automatically size the OTA, the SI integrator and the comparators to meet the specifications resulting from the high-level synthesis. The optimization process to obtain the sizes for the folded-cascode OTA required 45min CPU time, 35min for the SI integrator and 30min for the comparators. In all cases, the sizing started from scratch and no designer iteration was required. As an example, Table VI shows simulated and measured performances of the folded-cascode OTA showing good concordance with the specifications.

Fig. 13(a) shows a die photograph of the complete SC prototype with 0.94mm^2 area and power consumption of 10mW@5V. A microphotograph of the SI prototype with 0.43mm^2 core area operating with 15mW@5V is shown in Fig. 13(b).

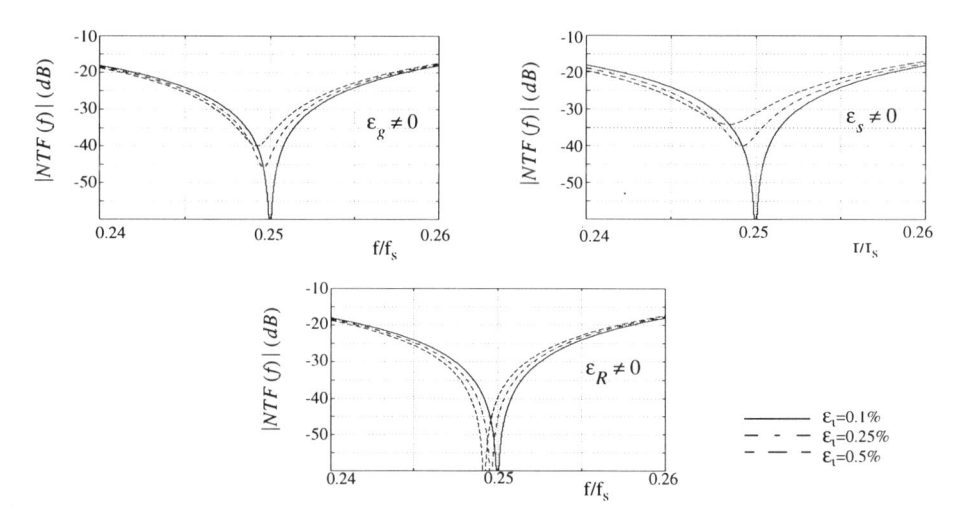

Fig 11: Influence of non-idealities on NTF(z)

To evaluate the performance of the two modulators, a test board was fabricated following the indications in [23] to reduce capacitive and inductive couplings. The modulator input was provided using a high-quality differential sinusoidal signal source (less than -100dB THD) through a simple passive low-pass filter to prevent aliasing. The output series were acquired with an HP82000 unit and transferred to a workstation for processing. The cancellation logic of the fourth-order modulator, as well as the decimation digital filters, were implemented on a workstation using

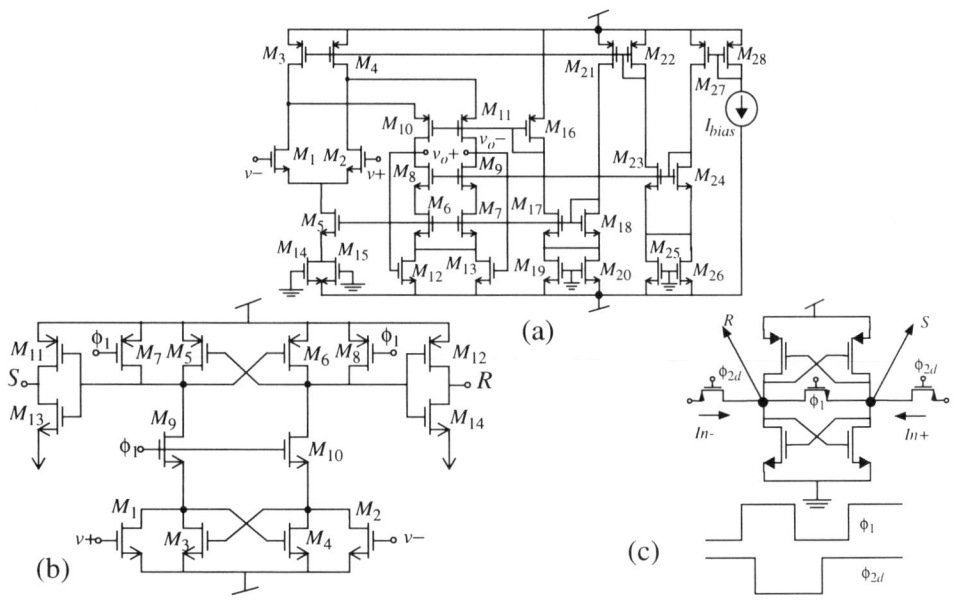

Fig 12: (a) Folded-cascode OTA. (b) and (c) Regenerative latches

(a) (b)

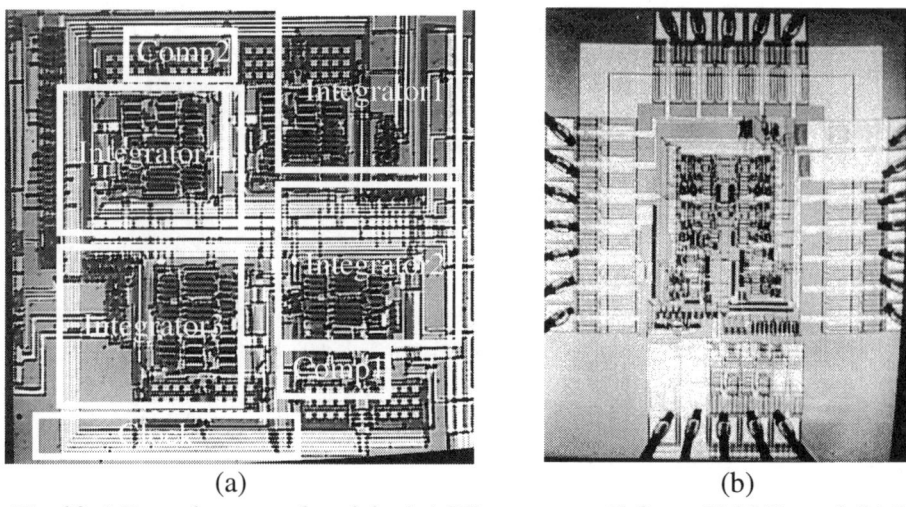

Fig 13: Microphotographs of the (a) SC prototype (1.2μm CMOS), and (b) SI prototype (0.8μm CMOS).

the same signal processor as used for simulations.

	Specs	Simulated	Measured	Units
DC-gain	≥ 71	78.52	76.01	dB
GB (1pF)	≥ 16	34.88	–	MHz
GB(12pF, 1MΩ)		4.17	4.21	MHz
PM(1pF)	≥ 60	66.28	–	Deg.
PM(12pF, 1MΩ)		87.2	86.8	Deg.
Input white noise	≤ 15	13.53	–	nV/√Hz
SR	≥ 58	74.81	70.5	V/μs
OS	≥ ±3	3.2	3.0	V
Offset	–	-	3.35	mV
Power	minimize	1.95	1.93	mW

Table 4: Simulated and measured results for the folded-cascode OTA

Fig. 14(a) presents the SNR of the fourth-order modulator as a function of the input level for three values of the oversampling ratio: 128 (nominal value), 64, and 32 which lead to 40, 80, and 160kHz digital output rate, respectively. Note that the modulator performance approaches the ideal as the oversampling ratio decreases, due to the fact that for low oversampling ratio the modulator is not thermal noise limited and thus, quantization noise dominates. The corresponding curve for the SI modulator with +-5kHz bandwidth around the central frequency is given in Fig. 14(b). Fig.15(a) presents the baseband spectrum of the SC prototype obtained through an FFT of 65,536 consecutive output samples. The input was a

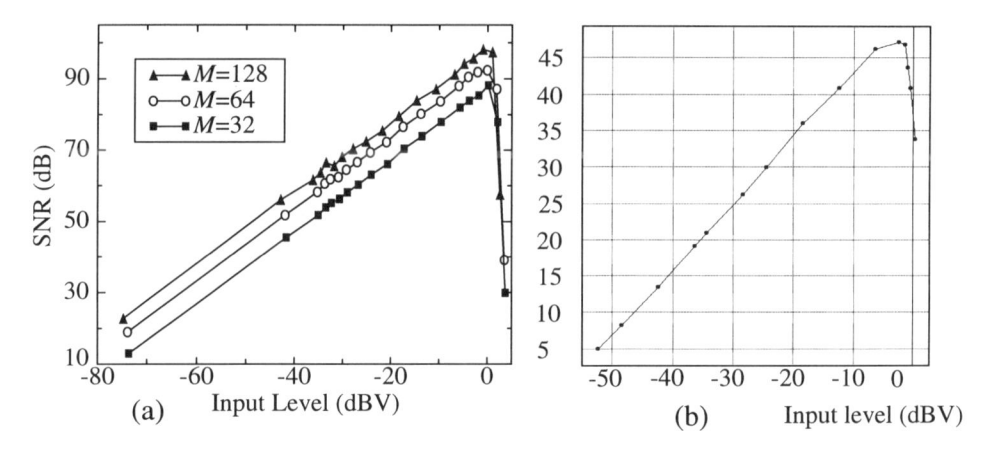

Fig 14: (a) SNR of the fourth-order prototype as a function of the input level for three values of the oversampling ratio; (b) SNR for the SI modulator.

-9dBV@4kHz sinewave sampled at 5.12MHz and compared to the output spectrum of its first stage, a second-order modulator. Differences between the two noise shaping functions are visible. However, the baseband of the fourth-order modulator is dominated by unshaped thermal noise. Our simulations show that this phenomena can be explained taking into account that the input noise power spectral density of the folded-cascode opamp was larger than expected. The output spectrum of the band-pass modulator for -6dBV@1.26MHz input tone is shown in Fig.15. Finally Table V summarizes the performance of both modulators.

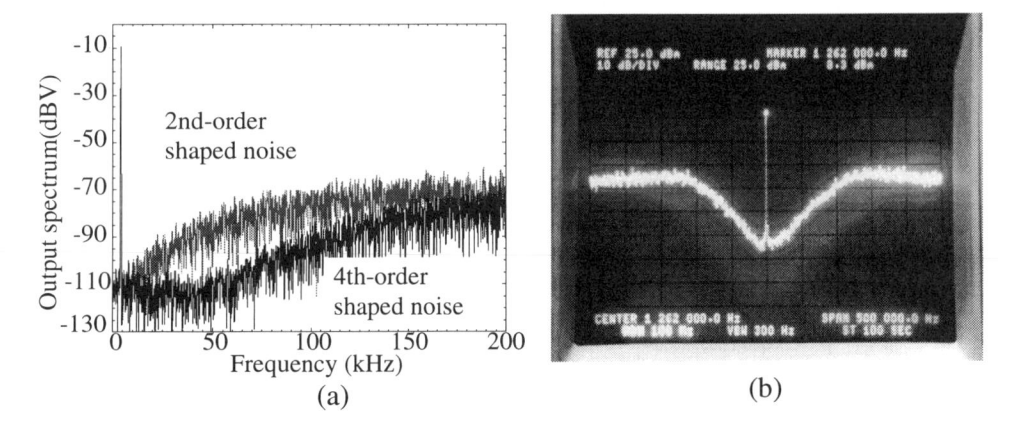

Fig 15: Output spectrums: (a) Low-pass SC modulator, (b) Band-pass SI modulator

	SC Modulator			SI Modulator
Oversampling Ratio	128	64	32	165
Resolution	16.7	15.5	14.8	8
DR	102dB	95dB	91dB	50dB
SNR-peak	98.2dB	92.5dB	88.2dB	47dB
TSNR-peak	88dB	85dB	82dB	--
Noise Floor	-110dB			--
Max. Input	1V			10µA
Max. Sampling Freq.	5.12 MHz			5 MHz
Power (Average)	10 mW			15 mW
Area (without pads)	0.94 mm^2			0.46 mm^2

Table 5: Performance of the SC and SI ΣΔ modulators

References

[1] J. C. Candy and G. C. Temes: *"Oversampling Delta-Sigma Converters"*. IEEE Press, 1992.

[2] T. Ritoniemi: "High-Speed 1-Bit ΣΔ–Modulators". *Workshop on Advances in Analog Circuit Design*, pp. 191-203. Delft, April 1992.

[3] G.C. Temes and B. Leung: "ΣΔ Data Converter Architectures with Multibit Internal Quantizers". *Proc. 11th European Conference on Circuit Theory and Design* (H. Dedieu, Ed.), Vol. 2, pp. 1613-1618, Davos, 1993.

[4] S. R. Norsworthy, I. G. Post and H. S. Fetterman: "A 14-bit 80-kHz Sigma-Delta A/D Converter: Modeling, Design and Performance Evaluation". *IEEE Journal of Solid-State Circuits*, Vol. SC-24, pp. 256-266, April 1989.

[5] C. H. Wolff and L. Carley: "Simulation of Δ-Σ Modulators Using Behavioral Models". *Proc. ISCAS'90*, pp. 376-379, 1990.

[6] V. F. Dias, V. Liberali and F. Maloberti: "TOSCA: a User-Friendly Behavioral Simulator for Oversampling A/D Converters". *Proc. ISCAS'91*, pp. 2677-2680, 1991.

[7] "HSPICE: *User's Manual*". Meta Software Inc., 1988.

[8] F. Medeiro: *"Automated Design of SC ΣΔ Modulators"*. PhD dissertation, Univ. Seville, 1996.

[9] D. B. Ribner: "A Comparison of Modulator Networks for High-Order Oversampled ΣΔ Analog-to-Digital Converters", *IEEE Transactions on Circuits and Systems*, Vol. 38, pp. 145-159, February 1991.

[10] V. F. Dias, G. Palmisano, P. O'Leary and F. Maloberti: "Fundamental Limitations of Switched-Capacitor Sigma-Delta Modulators", *IEE PROCEEDINGS-G*, Vol. 139, pp. 27-32, February 1992.

[11] B. E. Boser and B. A. Wooley: "The Design of Sigma-Delta Modulation Analog-to-Digital Converters". *IEEE Journal of Solid-State Circuits*, Vol. 23. pp. 1298-1308, December 1988.

[12] H. Baher and E. Afifi: "Novel Fourth-Order Sigma-Delta Convertor". *Electronics Letters*, Vol. 28, pp. 1437-1438, July 1992.

[13] F. Medeiro, R. Rodríguez-Macías, F. V. Fernández, R. Domínguez-Castro, J. L. Huertas and A. Rodríguez-Vázquez: "Global Design of Analog Cells Using Statistical Optimization Techniques". *Analog Integrated Circuits and Signal Processing*, Vol. 6, pp. 179-195, November 1994.

[14] P.J.M. van Laarhoven and E.H.L. Aarts: *"Simulated Annealing: Theory and Applications"*, Kluwer Academic Pub., 1987.

[15] F. Medeiro, B. Pérez-Verdú, A. Rodríguez-Vázquez and J.L. Huertas: "A Vertically Integrated Tool for Automated Design of $\Sigma\Delta$ Modulators". *IEEE J. Solid-State Circuits*, Vol. 30, pp. 762-772, July 1995.

[16] P. R. Brent: *"Algorithms for Minimization without Derivatives"*, Englewood Cliffs, N. J.: Prentice-Hall, 1970.

[17] "MATLAB: *User's Guide*". The MathWorks Inc., 1991.

[18] J-B. Shyu, G. Temes and F. Krummenacher: "Random Error Effects in Matched MOS Capacitors and Current Sources", *IEEE Journal of Solid-State Circuits*, Vol. SC-19. pp. 948-955, December 1984.

[19] F. Medeiro, B. Pérez-Verdú, A. Rodríguez-Vázquez and J. L. Huertas: "Modeling OpAmp-Induced Harmonic Distortion for Switched-Capacitor Sigma-Delta Modulator Design", *Proc. ISCAS'94*, Vol. 5, pp. 445-448. London, June 1994.

[20] F. Medeiro, B. Pérez-Verdú, A. Rodríguez-Vázquez and J. L. Huertas: "Design Consideration for a Fourth-Order Switched-Capacitor Sigma-Delta Modulator". *Proc. 11th European Conference on Circuit Theory and Design* (H. Dedieu, Ed.), Vol. 2, pp. 1607-1612, Davos, 1993.

[21] J.M. de la Rosa, B. Pérez-Verdú, F. Medeiro and A. Rodríguez-Vázquez: "CMOS Fully-Differential BandPass $\Sigma\Delta$ Modulator Using Switched-Current Circuits". *Electronics Letters*, Vol. 32, pp. 156-157, February 1996.

[22] . F. Duque-Carrillo: "Control of the Common-Mode Component in CMOS Continuous-Time Fully Differential Signal Processing", *Analog Circuits and Signal Processing*, Vol. 4, pp. 131-140, 1993.

[23] J. L. LaMay and H. J. Bogard: "How to Obtain Maximum Practical Performance from State-of-the-Art Delta-Sigma Analog-to-Digital Converters", *IEEE Trans. on Instrumentation and Measurements*, Vol. 41, pp. 861-867, December 1992.

TRANSLINEAR CIRCUITS

Johan Huijsing

Introduction

It was Barrie Gilbert who coined the name "Translinear Circuits" (TC) and proposed a classification in 1975. The function of TC's depended on the connection of the input terminals of exponential devices in loops. Later, in 1991, Evert Seevinck and Remco Wiegerink proposed a broader definition also including loops of Quadratic devices.

In this classification a distinction has been made between Gilbert's translinear loops with (TL) devices having a transconductance proportional to the signal current and voltage translinear loops (VTL) with devices having a transconductance proportional to the driving voltage. The families of translinear circuits became extremely useful in building analog linear functions with variable parameters. In the following six papers these principles will be elaborated.

In the first paper Barrie Gilbert presents us classic and new aspects of translinear amplifier design in bipolar technology.

In the second paper Max Hauser particularly elaborates on variable-gain techniques for high-frequency applications in bipolar and CMOS technology.

In the third paper Evert Seevinck shows the family CMOS translinear circuits.

In the fourth paper Remco Wiegerink shows a systematic approach to the design CMOS translinear circuits.

In the fifth paper Klaas-Jan de Langen applies translinear principles to the input and output stages of low-voltage operational amplifiers.

Finally in the sixth paper Rinaldo Castello applies translinear principles to the design of low-voltage continuous-time filters.

ASPECTS OF TRANSLINEAR AMPLIFIER DESIGN

Barrie Gilbert
Analog Devices Inc.
1100 NW Compton Drive
Beaverton, OR, 97006, USA

INTRODUCTION

The translinear principle has become quite familiar to IC designers during the past twenty years. Originally conceived within the narrow framework of bipolar, wideband, fixed- and variable-gain current-mode amplifiers employing *closed loops* of junctions [1,2]—now called TL cells, in which input and output signals and biases are all in pure current form—the scope of the concept has gradually broadened to include any circuit in which the *essential function depends directly on a precise exponential relationship* existing between the current at one terminal of a suitably-biased three-terminal active device and the voltage applied across the remaining two terminals. A translinear cell not including any directly closed loops has more recently [3] been called a *translinear network* (TN).

For the BJT, this key relationship exists between the collector current I_C and the base-emitter forward-bias voltage V_{BE}. It will be apparent that SiGe heterojunction bipolar transistors (HBTs) exhibit the same essential exponential behaviour and therefore can be used in all translinear applications with little if any modification to the theory. The absolute value of the band-gap voltage—hence, the $I_S(T)$—does not appear in the final equations of TL circuits, nor in many TN circuits; thus, GaAs HBTs (having a V_{BE} of over a volt) may be used, or even pure-germanium transistors (usefully having a V_{BE} of about half that of silicon), if such might ever be fabricated in monolithic form.

This idea has more recently been applied to MOS devices operating in the subthreshold (weak inversion) domain [4]. But the original formulation of the principle, since expounded at greater length by Seevinck [5], cannot be applied to translinear-loop cells using enhancement-mode MOS transistors operating in strong inversion, because, at least according to simple theories, the channel current I_{DS} of an MOS transistor bears a *quadratic* relationship to the gate-source voltage. Consequently, Seevinck and Wiegerink have proposed an alternative formulation [6] of the 'translinear idea', based on the observation that, in contrast to the BJT, which (for a V_{CE} greater than about 200mV) exhibits a linear relationship between the transconductance $g_m = \partial I_C / \partial V_{BE}$ and the collector *current*, MOS devices in strong inversion (and with $V_{DS} > V_{GS}$) exhibit this linear relationship between the transconductance $g_m = \partial I_{DS} / \partial V_{GS}$ and the *excess voltage* V_{GS} above the threshold voltage, V_{TH}.

This extension of the original principle has been called 'MOS translinear', or MTL. However, the mathematical relationships are very different, and are much less tractable than the simple 'repeated product' form of the translinear-loop principle based on exponential junction behaviour. Furthermore, the 'quadratic-I_{DS}' assumption is only an approximation, even for long-channel devices, with serious non-idealities in practice, due to channel-length modulation below 1μm, as well as back-gate effects; these are rarely addressed with adequate realism in the literature. In fact, for modern sub-micron MOS transistors, it is the I_{DS}—not the g_m—which is an almost *linear* function of V_{DS} above V_{TH}, and the cell behaviour errs very significantly from that predicted by 'MTL' theory.

To avoid going beyond the spirit of the original definition of 'translinear', and risking ambiguity about the intended meaning and application of the term in MOS applications[1], its use without an adjectival qualifier should be reserved for those cells invoking *exponential* device behaviour, in recognition of long-standing and familiar usage. The strong-inversion idealization should be termed *voltage-translinear*, or VTL, since the proposed acronym MTL, when read as 'MOS-translinear', could refer to either the VTL mode or to classical translinear operation (either TL or TN) using MOS transistors in subthreshold[2].

It is likely that the increasing utilization of CMOS in analog applications will gradually soften the dependence on translinear techniques, which have yielded an impressive portfolio of bipolar integrated circuits, and continue to be of value, either in new applications (or the rediscovery) of classical cell topologies, or in more subtle ways. On the other hand, little use can be made of the idea in the CMOS domain. A noteworthy exception is the implementation neural networks [7,8], using MOS devices at very low currents, where their bipolar-like behaviour can be exploited; a new development in this field is the use of floating-gate cells [9] to perform summing of exponential arguments, hence multiplication, in an otherwise classical translinear modality. Such applications generally place very modest demands on accuracy, so considerable deviation from the presumed device 'law' is of little consequence. Similarly, modest accuracy requirements allow VTL cells to be employed in specialized non-demanding applications [10].

1 EARLY TRANSLINEAR CELLS

The basic idea of a translinear circuit was conceived by the author in 1967, in the context of high-bandwidth (500MHz) electronically-variable gain cells, for use in oscilloscope vertical amplifiers at Tektronix Inc. Out of this initial work came many developments, one of which was a monolithic doubly-balanced modulator (or mixer, a nonlinear multiplier), which was then *linearized* by the addition of another pair of BJTs to realize a wideband four-quadrant analog multiplier, reported in February 1968 at one of the earliest International Solid-State Circuits Conferences in Philadelphia [1] and later described in full detail in two seminal papers [11,12] which anticipated many of the translinear circuits that would later be turned into commercial products, including high-accuracy and wideband multipliers [13,14], RMS-DC converters [15], an analog array processor [16], as well as a variety of other interesting and useful nonlinear circuit concepts, such as vector sum and vector difference cells [17]. These pioneering products, many of which remain in full-scale production, some nearly three decades later, all had at their core small circuit cells sharing certain common features:

1) They required the use of *monolithically integrated* bipolar transistors, since isothermal operation and matching of device geometry and doping levels were essential. While such circuits were conceivable in the mid-'sixties, the full realization of their potential had to await the availability of 'analog-quality' process technologies, today refined to a high art.

2) They were strikingly elegant, being comprised of little more than *DC-coupled bipolar transistors and current sources*, with essentially no dependence on ancillary passive components (resistors and capacitors). Their transistor-intensive schematics were more suggestive of current-mode logic, quite unlike the contemporary analog circuits in which each transistor was typically supported by several passive elements.

3) The transistors were arranged in *closed loops*, each containing at least two base-emitter junctions (as in a simple current mirror) but often four, six, eight, or even more junctions. Overlapping loops were frequently used.

4) Rather than the voltages used almost universally in analog circuits, the 'signals'—the cell inputs, outputs and control biases—were *currents*. Whatever voltages arose across the junctions were of only *incidental* importance; furthermore, the *full-scale* voltage swings were very small—typically only a few tens of millivolts (that is, comparable with kT/q).

5) Because the junction and signal voltages were small, operation at *very low supply voltages* was often possible, down to 1V in certain cases, where the junctions are not stacked, but rather alternate in polarity[3].

6) The inherently-minimal branch impedances associated with this mode of operation resulted in *high bandwidth*, often limited by the required V-I and I-V interfaces. The minimal *internal* voltage swings under all signal conditions largely eliminated the distinction between small-signal and large-signal operation. Slew rate limitations were essentially absent.

7) Unlike prevalent high-frequency analog signal-processing cells, these novel circuits exhibited highly-predictable, *fundamentally exact* and *temperature insensitive*, linear and nonlinear relationships between the signal variables.

8) They were particularly useful in implementing a wide variety of *continuous algebraic functions*, using very few transistors, including: squaring; square-rooting; multiplication and division; vector addition and subtraction; the direct computation of amplitude ratios in an array; polynomial, trigonometric and implicit-form function generation. The large-signal response delays were often only a nanosecond or two, unprecedented at the time.

In the pre-microprocessor world, the high functional capacity, versatility and speed of TL circuits represented assets of outstanding practical value.

260

Below are shown four of the seven pages of schematics that appeared in an early patent (U.S. 3,689,752), from which it is apparent that numerous possibilities for novel current-mode topologies were generated by the translinear point of view.

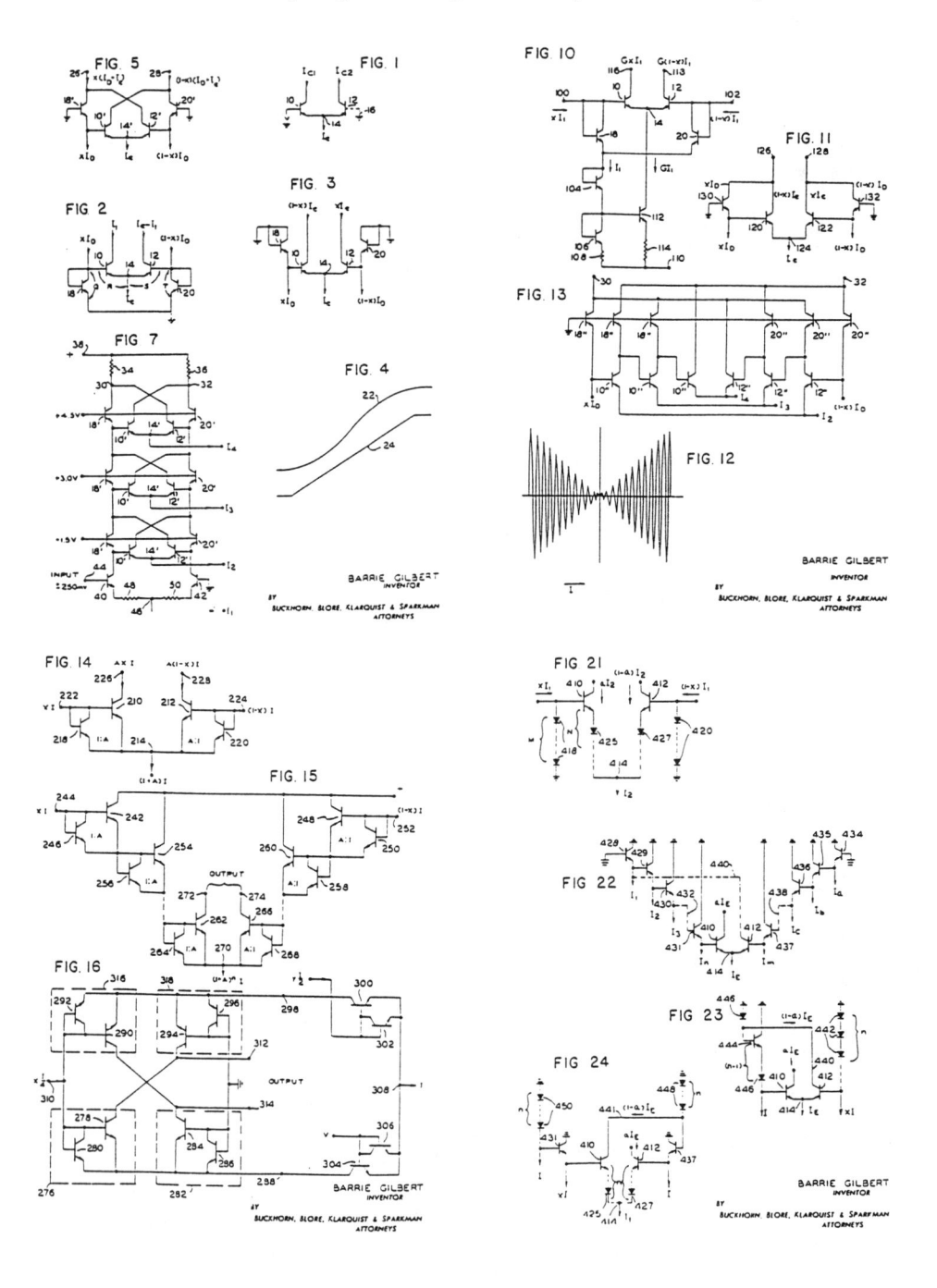

A simple comparison will serve to point out the differences between classical analog circuit design and translinear design. Consider the prosaic amplifier cell shown in **Figure 1.1** (which was popular in the 'sixties, and might still be found in contemporary textbooks), in the light of the above eight points:

1A) Discrete transistors can be used; matching between them is *not* an issue.

2A) There are eight passive components for two transistors.

3A) Though loops in the topology can be traced, these are not the central feature of the cell, nor do they involve only junctions.

4A) Signal inputs and outputs are voltages.

5A) Operation from a 1.2V supply is possible (R2=∞), but only with significant performance compromises, particularly with regard to dynamic range.

6A) The parasitic capacitance at the collector node of Q1 usually determines the bandwidth; in those cases where the amplifier can support an output signal swing of several volts, slew-rate limitations will be evident.

7A) Significant nonlinearities arise due to the variation in incremental emitter resistance (the r_e of both transistors) with the instantaneous values of V_{IN} and V_{OUT}, causing distortion; there may be various types of temperature sensitivities, for example, in the gain and input/output impedance.

8A) The circuit is limited to essentially linear-signal applications.

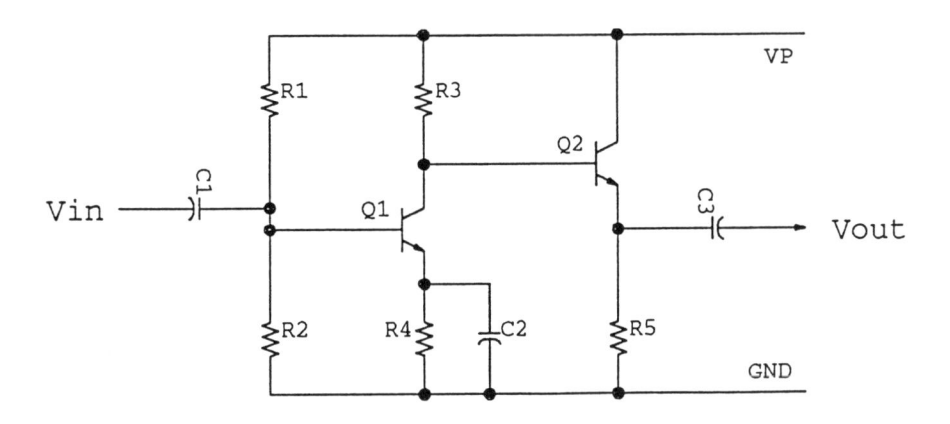

Figure 1.1 A Traditional BJT Amplifier Cell

Now consider the circuit of **Figure 1.2**, a *pure current-mode amplifier*, in the important sense that no 'use' is made of voltages at any point in the circuit. It was described in the classic 1968 paper [11], which focused on the design of wideband translinear amplifiers. The first thing to note is the large number of transistors used, an appropriate concession to the monolithic medium. Secondly, note the use of a differential signal path; this, too, was already becoming a strong feature of monolithic circuits, beginning with the advent of the two-transistor differential amplifier stage.

While not a complete design (the currents have to be eventually generated within the supply constraints), the possibility of operation at 1.2V is evident; Figure 1.2 shows the incidental voltages for a nominal V_{BE} of 0.8V and the assumption is made that a minimum V_{CE} of 0.2V is acceptable. If we again go through the eight points listed above, we now find:

1B) Close matching of the transistors (particularly between pairs, and particularly in their saturation current I_S, determining the absolute V_{BE}) and isothermal operation are essential, demanding monolithic realization.

2B) There are no passive components in the amplifier, and none are required *in principle*, though in practice (solely because current sources are not as plentiful as voltage sources), one or two resistors may be needed to define the bias currents, convert a voltage-mode input signal to an input current, and convert the current-mode output currents to voltages.

3B) There are N loops of junctions for an N-stage amplifier (three in this example, identified on the figure), each embracing four transistors.

4B) As noted already, all signals and biases are currents, and the peak voltage swings for *full-scale* conditions are typically only ±50mV.

5B) A low supply voltage has little bearing on the dynamic range, which is now largely[4] defined by the maximum signal currents—which may range from a few nanoamps to many milliamps—and by the shot-noise currents.

6B) The large- and small-signal bandwidth of each cell is defined predominantly by the f_T of the transistors, operating under the particular bias conditions (rarely if ever the *peak* f_T, of course); and there is no slew-rate limitation even at full-scale signal levels.

7B) The signal path is linear, and the LF current gain—which can be as high as β_0^N, while still being accurate[5]—is independent of temperature.

8B) The variable gain of this amplifier is a nonlinear function of I_G.

Such amplifiers are manifestly not universal in their domain of application, but they nevertheless point to a quite different paradigm for amplifier design.

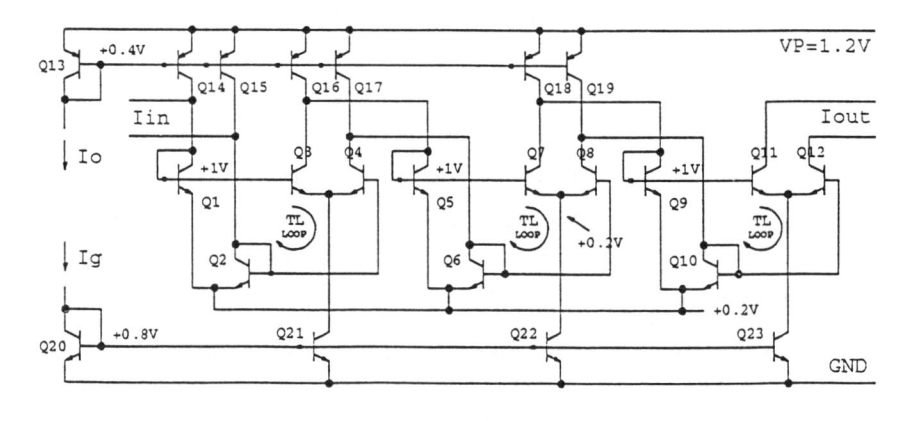

Figure 1.2 An Early Translinear Amplifier

1.1 BJT 'Translinearity'

Following the early work on current-mode amplifiers, a proliferation of other current-mode cells were conceived, built in monolithic form, and proven. Before the term 'translinear' was coined, the common theme utilized in all these cells was informally referred to as *'The Pervasive Principle'*, a reflection on the fact that a general technique had been discovered with which one could effortlessly devise and analyze a large class of linear and nonlinear analog cells [18].

A new word (proposed much later [2], in 1975) captured the essence of this principle, distilled in that most quintessential property of the BJT, namely, that its transconductance, g_m, is linearly proportional to its collector current, I_C. This relationship arises in turn from the *exponential* relationship between I_C and V_{BE}, which, for a modern bipolar transistor, is dependable over six to eight decades. Bipolar designers are quick to identify this as *the chief distinction between the BJT and all kinds of field-effect transistors*. Thus, the BJT can be accurately modeled as a simple voltage-controlled current-source (VCCS):

$$I_C = A_E J_S(T) \, exp \, (V_{BE}/nV_T) \qquad (1.1)$$

where A_E is the emitter area, $J_S(T)$ is the saturation current density, n is the 'emission coefficient' (typically very slightly greater than unity over the broad central current range[6]) and $V_T = kT/q$. This relationship is reliable over a current range of at least a million to one (for example, 1nA to 1mA) over which range n is reasonably constant. I_C is also sensibly independent of the collector bias, V_{CB}. A plot of $log(I_C)$ vs. V_{BE} reveals extraordinary linearity over many decades.

The saturation current $I_S(T) = A_E J_S(T)$ can be viewed as a *current scaling* parameter, determined, amongst other things, by the doping levels and doping profiles, base width and the band-gap energy E_{GO}. For a transistor having a room-temperature V_{BE} of 800mV at 100μA, it would be only 3.6×10^{-18}A. It is rarely possible to measure $I_S(T)$ directly; it is usually deduced from measurement of the V_{BE} of a transistor operating at a known collector current $I_C = I_R$ and temperature $T = T_R$. If measured with zero collector bias ($V_{CB} = 0$), at the default system temperature (usually 27°C) it then can be used as the SPICE parameter IS.

The magnitude of $I_S(T)$ exhibits notorious temperature sensitivity, varying by a factor of about 10^{13} from −55°C to +125°C, from typically 10^{-24}A to 10^{-11}A. The collector current I_C, being proportional to both $I_S(T)$ and $exp(T_R/T)$, would vary enormously if V_{BE} were held at a fixed value. For example, applying a B-E bias of 650mV to a BJT having $I_S = 5 \times 10^{-17}$A at 300K would result in an I_C of from approximately 1nA to 1mA over this temperature range. It is therefore hardly surprising that the early users of discrete BJTs strenuously avoided 'hard voltage biasing' of the E-B junction, because it seemed so inappropriate. Instead, the notion of a 'current-controlled current-source'—the 'beta view'—was emphasized, since beta varies only mildly over temperature. As we shall see, TL and TN circuits are fundamentally immune to this immense variation in I_S. Note that, 'turned around', the $V_{BE}(T)$ for constant I_C varies by only about −0.25%/°C.

From (1.1) we find

$$\frac{\partial I_C}{\partial V_{BE}} = g_m = \frac{I_C}{nV_T} \qquad (1.2)$$

While this property of the BJT is very widely known to be 'useful' in general circuit design, it is absolutely pivotal to the *translinear* view of the BJT. What is so remarkable about this result is that all of the details that go into the theory of transistor behaviour—the material type (Ge, Si, SiGe, GaAs, etc.), the bandgap voltage, the polarity type (PNP or NPN), the doping concentrations and their profiles, the effective mass of electrons and holes, mobilities, carrier diffusion coefficients, etc., all of which play a role in determining $J_S(T)$—have vanished[7].

Even the emitter junction area A_E is unimportant to g_m, and, though perhaps not immediately apparent, neither does the g_m depend on many such 'incidentals' as the Early voltage, the collector bias $V_{CB} \geq 0$ or the beta[8]. While all these will determine the bias conditions required to establish a certain precise I_C, only the latter is important in determining g_m, and this remains true over a current-density range so wide that only moderate attention needs to be paid to absolute device sizing, with quite different considerations in mind for optimization.

It is this sublime attribute, this *translinearity*, rather than its 'speed' or 'high current-handling', or any other claims to 'high-performance', that is the central appeal of the BJT, leading to great certainty of outcome in design and providing a sure foundation for robustness and manufacturability. From (1.2) it is obvious that to render a g_m (for example, that of either Q1 or Q2 in Figure 1.1) stable over temperature, I_C needs to be proportional to absolute temperature (PTAT). This ploy is very well-known, and can be achieved with the utmost simplicity and predictability using further BJT transistors[9], in a ΔV_{BE} cell, which also exploits translinear properties; the g_m is accurately traceable to a resistor in this cell.

1.2 MOS 'Translinearity'

One only has to look at the basic MOS equations to see how very different the situation is. For an enhancement-mode device, operating in 'moderate to strong' inversion, and in 'saturation' it is said that:

$$I_{DS} = \kappa \frac{W}{2L}(1+\lambda V_{DS})(V_{GS} - V_{TH})^2 \qquad (1.3)$$

from which we get

$$g_m = \frac{\partial I_{DS}}{\partial V_{GS}} = \kappa \frac{W}{L}(1+\lambda V_{DS})(V_{GS} - V_{TH}) \qquad (1.4)$$

where κ and λ are functions of carrier mobility, substrate doping and oxide thickness amongst other things [19,20]. In (1.4) we appear to have something like the voltage-translinear (VTL) form of the 'pervasive principle', namely, that the transconductance of an enhancement-mode MOS device is linearly proportional to the gate-source voltage above threshold. Unfortunately, even if the (quadratic) modeling were correct, the equations that thereafter arise in the analysis of cells containing *loops of transistors analogous to their BJT prototypes* are complicated and mathematically awkward, which is quite different to the original TL case.

There are many more qualifiers needed to adequately describe MOS operation for analog design purposes. To begin with, there is no broad region of I_{DS} over which a single set of modeling equations apply, except perhaps the bipolar-like weak-inversion region which, for moderate channel widths and typical gate lengths, may only extend up to a few microamps. Then the device enters a transition region, in which neither bipolar-like exponential behaviour nor the quadratic behaviour described by (1.3) prevails. At higher gate voltages, another transition occurs and eventually the device enters the region of very strong inversion, where another set of equations takes over.

Second, the factor κ is not in the nature of a fundamental constant (like kT/q), but as noted depends on carrier mobility and oxide thickness; the former has a strong (but imprecise) temperature variation, and the latter varies significantly from one production lot to another. Thus, κ generally has a range of values, and if g_m is to be determined accurately, replica-biasing techniques must be invoked, in which a real resistor sets the g_m, as for the bipolar case. Note, however, that CMOS circuits are generally characterized by a very sparse use of resistors, and the available materials are often not optimal for high-precision analog design.

Third, the choice of the width and length of the channel—that is, the device sizing—affects the g_m, and, unless undesirably wide and long channels are used, production uncertainties in these quantities significantly affect the absolute g_m. There are also initial uncertainties and temperature-dependencies in the threshold voltage V_{TH}, causing further variability in g_m, which is more strongly affected by the drain-source voltage V_{DS}, than is true of the bipolar transistor and V_{CE}. Additional influences on the channel current are due to the substrate or well bias.

Finally, and most importantly, the power of two, shown in (1.3) and (1.4), is not correct for sub-micron MOS devices, invariably being closer to one. This is, of course, a very desirable artifact in amplifiers, but becomes a serious defect in nonlinear applications. Unfortunately, the bulk of the literature on such things as MOS analog multipliers places implicit reliance on that quadratic assumption.

1.3 BJT vs. MOS as Amplifier Technologies

The foregoing may sound like a condemnation of MOS for the realization of analog functions, but this is not what is intended. In the first place, it would be foolhardy to ignore all the excellent work that has been done, and continues to be done, in this regard, or to imagine that MOS technologies (by which, of course, we invariably mean CMOS) are ill-suited to analog design. The observation that BJT design is straightforward[10] because "the transconductances are very predictable" scarcely provides an adequate rationale for its use. Such may make design a bit faster and reduce time-to-market, but that is largely a matter of familiarity, experience and having a strong repertoire of tried-and-trusted cells.

On the other hand, the case usually forwarded for CMOS is its *lower cost*, and this may not be an entirely valid argument. Modern CMOS processes are very complex; they use a large number of masking steps, and wafer costs are not too dissimilar to pure-bipolar processes. Their yields are generally higher, but while this is crucial in the manufacture of large dice, such as imagers, microprocessors, memories, DSP products and other VLSI, the impact of yield on product cost is not severe for moderate scales of integration using bipolar technologies. For example, Analog Devices has mixed-signal products built on an in-house 25GHz process, incorporating over 30,000 active transistors[11]. In RF applications, the transistor count may drop to a few thousand for a 900MHz cellular phone transceiver with on-board frequency synthesizer, or to a mere handful of devices in simple functions such as a 2GHz LNA/Mixer or I/Q Modulator. At this point, yields are very high, and die cost is of diminishing importance, being a small fraction of the overall *cost of delivery*, which must include product development, manufacturing overheads, testing, documentation, advertising and field support.

We need to find more compelling reasons for migrating to CMOS for analog functions. Many of the new graduates, more skilled in MOS design than bipolar, are demonstrating solid (and sometimes stunning) achievements in their medium, often by appealing to radically different architectural approaches. For example, the upsurge of interest in bandpass sigma-delta A/D converters is a response to the difficulties of achieving broadband operation in a conventional converter, the need to integrate more of the system into a common technology, to lower overall power consumption, and even to achieve some of the signal filtering. But it is becoming apparent that one reason for considering CMOS may be its *analog performance*. This is probably a novel idea for the seasoned bipolar designer.

That is, there are a growing number of cases where the *fundamental device limitations* of the translinear BJT are painfully apparent, and where majority-carrier transistors promise valuable benefits. A not-so-trivial advantage, of course, is their essentially-zero gate current. This certainly is one place where CMOS affords significant design simplification, sometimes in subtle ways. For example, one often needs to provide several current-sources sharing a common bias line. In the BJT case, if any one of these sources should saturate ($V_{CE} \to 0$), the bias line will often be pulled, and all currents change their value, while the MOS version will be quite benign in a similar scenario.

But let's return to this matter of g_m. The high values which can be attained using BJTs (about 40mA/V at I_C=1mA), and the often-valuable exponential junction law may not always be desirable. Indeed, there are many situations in which this traditional 'strength' stands like a fundamental road block to performance improvements. Not surprisingly, one such situation is that of high-linearity fixed-gain RF amplifiers.

Consider the simple low-noise amplifier shown in **Figure 1.3**, a type often used in receivers operating in the 1-10GHz range. To simplify the analysis, a biasing scheme is used that might be chosen in a monolithic realization: the larger Q1 is mirror-biased by Q2 to I_C=MI_O, which establishes a nominal $r_e = V_T/I_C$. R_{B2} may be made slightly larger than MR_{B1} to provide beta compensation. The LNA will for the present illustrative purposes be presumed to operate between equal impedances[12] of Z_O. The shunt feedback resistor R_F (more generally, an impedance Z_F) determines the matching impedance at the two ports.

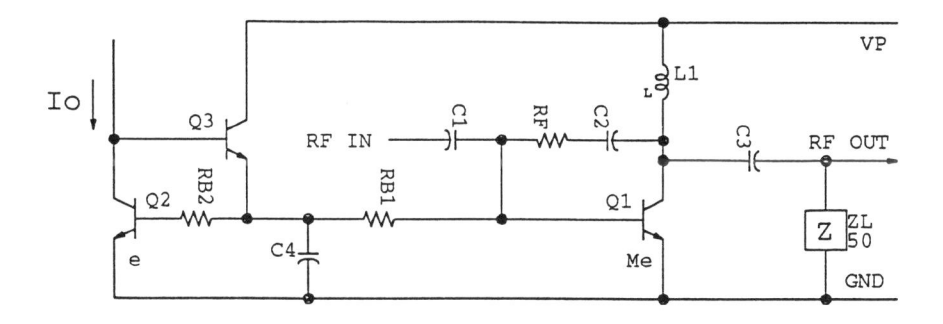

Figure 1.3 A Bipolar Low-Noise Amplifier with Translinear Biasing

For an ideal BJT, having a sufficiently high f_T, there is a unique value of R_F to achieve this double match:

$$R_F = Z_O^2/r_e \qquad (1.5)$$

As a philosophical whimsy, we may note that this result can be obtained without analysis, using the following *reductio ad absurdum*: (1) There are only two given quantities, Z_O and r_e, each of which have the dimension of resistance. (2) The required result must be a simple ratio. (3) It can therefore only have the above form *or* the form r_e^2/Z_O. (4) The latter form leads to the conclusion that for a very high g_m ($r_e \rightarrow 0$) and/or a high Z_O, R_F would need to approach zero, so that the 'amplifier' would collapse to a diode, with input (B) and output (C) shorted together, leading to the conclusion that the gain would be zero for high bias currents. (5) Therefore, the form shown in (1.5) is the correct form.

The nonlinearity of this LNA may be captured in the 1dB compression power referred to the input (P1dB), by the input-referred two-tone third-harmonic intercept power (IIP3), or by various other metrics. One may be forgiven for supposing that the ability to handle large inputs ought to improve as the bias current is increased. It is here that the exponential behaviour of the BJT becomes a liability, for *the P1dB and IIP3 are in principle independent of the bias current*. This is because once R_F is correctly chosen (for example, it would be 387Ω for r_e=6.46Ω, that is, I_C=4mAP, where the latter terminology refers to the use of a PTAT current to stabilize r_e over temperature) and a power match is achieved, then the input voltage at the node B is *precisely half* the generator voltage V_{GEN}. Thus, the nonlinearity for a given source amplitude is *independent of the bias*.

It is easy to show that the P1dB of this LNA, in a 50Ω system, is −18.7dBm and the two-tone IIP3 is −8.8dBm. This is quite inadequate for many receiver applications. The solution is to include *emitter degeneration*, that is, the use of series feedback by the insertion of an impedance in the emitter branch. This could be simply a resistor, but the ohmic value required to effect useful linearization will usually be too high to maintain a sufficiently low noise figure. Thus, it is necessary to use an inductor, which may be either on-chip or (sometimes incidentally) provided by the bond wires and the package.

In this way, one is essentially thwarting the translinearity of the BJT, which clearly is *not at all desirable in this application*. Other things being equal, one would here prefer a device having a more linear g_m, which all majority-carrier devices *do* have. As already noted, for short-channel MOS transistors the g_m is almost independent of the channel current, which is obviously very desirable, leading to much higher values for P1dB and IIP3, that is, lower intermodulation.

Now let us examine the noise performance. For an ideal BJT, there is a shot noise component $\sqrt{(2qI_C)}$, which, working into r_e, generates an equivalent voltage noise spectral density across the base-emitter junction, that is, across the matched input port, which evaluates to

$$S_N = \frac{0.463 \text{nV}/\sqrt{\text{Hz}}}{\sqrt{I_C}} \quad (1.6)$$

at T=300K, where I_C is expressed in mA. The corresponding noise figure for a 50Ω system is shown as the lower curve in **Figure 1.4**. The first observation to be made here is that the power gain for this particular amplifier crosses below 0dB for I_C=1.04mAP, where r_e=25Ω. However, this is easily addressed by a more suitable output matching scheme. Second, in order to minimize the noise figure, there is no way[13] to avoid the use of high bias currents. Even assuming this is not of major concern (as in base stations) there is still a *lower limit* on noise figure, imposed by the shot noise in the base, which operates on the source impedance. But this is an essentially *practical*, not a fundamental, limitation of the BJT, since it is possible to achieve a high current gain at moderate frequencies, and various novel device structures promise further improvements[14].

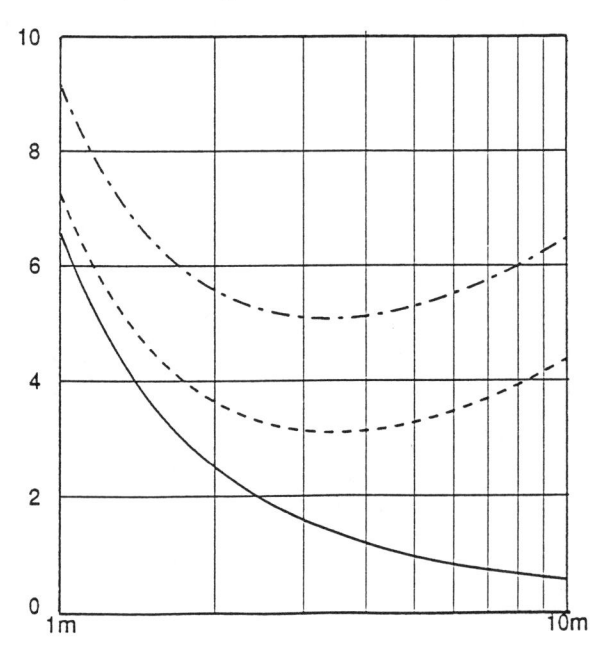

Figure 1.4 Noise Figure of a Bipolar LNA

However, the base shot noise in the frequency range of interest is unrelated to the DC beta. Instead, we need to use the value of beta at the signal frequency, f_S. This is essentially just the ratio f_T/f_S; the base current shot noise is thus $\sqrt{(2qI_Cf_S/f_T)}$. For a device operating at an effective f_T of 20GHz, the AC beta $\beta(f)$ at f_S=2GHz is only 10. The middle curve in Figure 1.4 shows the effect on noise figure when the base noise for this effect is included, using $\beta(f)$=10. There is a further practical problem, and that is the Johnson noise due to the base resistance, $r_{bb'}$. This evaluates to

$$0.129\text{nV}/\sqrt{\text{Hz}} \times \sqrt{r_{bb'}} \qquad (1.7)$$

and adds directly to the input noise. The final noise figure using a transistor with a base resistance of 15Ω is shown as the top curve in Figure 1.4.

These results clearly point to some rather basic—if not *entirely* fundamental—limitations of the BJT in low-noise high-frequency applications. The base shot noise can be lowered only by increasing the f_T. Recent developments in SiGe and GaAs heterojunction transistors[15] show that it is now possible to achieve peak f_Ts of the order of 100GHz. HBT technologies allow base resistance $r_{bb'}$ to be lowered through the use of heavier base doping; the emitter injection efficiency is restored by the addition of up to 10% germanium. The $r_{bb'}$ may then be further reduced by the use of larger emitter-base regions, since one may be able to allow some reduction in operating-f_T to effect an overall improvement in noise figure.

It will be clear from the foregoing that simultaneously achieving state-of-the-art linearity and noise performance from bipolar transistors poses some taxing challenges. Thus, ironically, it is possible that designers of high-performance RF front-ends may choose *not* to use the BJTs that are available in a BiCMOS process, ostensibly provided at their insistence for precisely such purposes! Of course, each case must be considered on its merits. It is unlikely that the CMOS devices available in *most* BiCMOS processes are poised to eclipse the BJT. But whether in advanced BiCMOS or pure-CMOS, future RF ICs are certain to lean less heavily on BJTs, and benefit from the use of majority-carrier devices.

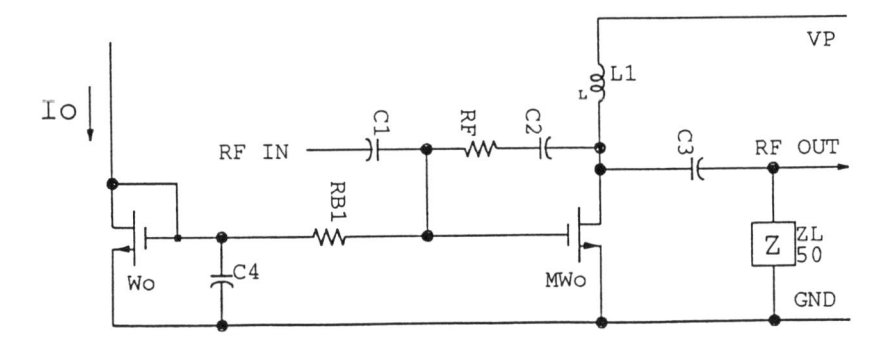

Figure 1.5 An NMOS Matched-Impedance LNA

We can re-cast the amplifier in NMOS form as shown in **Figure 1.5**, using the same basic topology and biasing technique (though probably not optimal). The g_m is no longer solely dependent on the bias current, and by choosing a technology providing very short channel lengths and thinner gate oxides, resulting in increased values of κ in (1.4), and by increasing the width of the channel, one can use *device geometry* to regain some of the g_m, which is invariably less than that of a BJT at the same current. The matching criterion is essentially the same; it may now be written

$$R_F = Z_O^2 g_m \qquad (1.5a)$$

But this is a g_m which is inherently more linear, particularly for deep sub-micron devices. Consequently, the all-important P1dB and IIP3 can be much higher than for the BJT case. On the other hand, these numbers no longer have a fundamental quality, but depend on the properties of a particular technology.

There is essentially no shot noise associated with an FET[16]. Instead, we may attribute the noise entirely to the effective noise-resistance of the channel. Using a high W/L ratio, this resistance can be lowered, in principle without limit, though of course, in practice the effective f_T of the device will eventually be impaired. It follows that, in principle, low noise figures can be achieved *without the use of large bias currents*. This is a major advantage of MESFET and MOS amplifiers. There is of course no equivalent to base current noise for the MOS device; neither is there a direct equivalent to $r_{bb'}$, though the resistance of the polysilicon gate can generate a similar Johnson noise component. This can be addressed using silicided gates, and by designing the device to have multiple short gate fingers. The performance of CMOS amplifiers in RF applications at frequencies in the low GHz range has not generally reached that of bipolar circuits, but it is surely only a matter of time before that interesting milestone is reached.

2 VARIABLE-GAIN CELLS

One of the strongest areas of application of translinear circuits has been in providing accurate electronic control of gain. Both linear and exponential (linear-in-dB) control functions are readily provided, and other forms (such as hyperbolic or square-law) are just as readily accommodated. Perhaps the greatest appeal of translinear variable-gain amplifier (VGA) cells is the ease with which many such variants can be devised. It is of course beyond the scope of this paper to provide a comprehensive survey. We will touch on just a few ideas that show the general value of translinear design techniques in RF and IF applications.

An excellent survey of VGAs, including translinear types, was prepared by Max Hauser, fifteen years ago [21], and the latest chapter—which perhaps might be entitled "Transcending Translinearity"—in this on-going compendium was added at the 1996 Advances in Analog Circuit Design workshop [22]. No attempt is made here to address the topic of CMOS alternatives to bipolar VGAs. Many promising alternatives exist, though often not direct replacements; however, in the pursuit of contemporary product design, one must always ask whether it is necessary to seek *exact* functional equivalents in the development of new systems.

Almost the simplest BJT cell, the differential pair, immediately provides a simple and practical approach to gain control. **Figure 2.1** shows the essential elements of a voltage-mode amplifier cell, whose small-signal gain is a *linear* function of the tail current I_T:

$$G_O = \frac{R_L I_T}{2V_T} \qquad (2.1)$$

(We assume a high beta, so that $I_C \approx I_T/2$, since it is actually the *collector* current that determines g_m). In principle, the gain will be stable with temperature when I_T is PTAT. Note that this VGA is a *direct application* of the BJT's translinearity. As for the LNA discussed above, a major source of noise is that due to shot mechanisms. The input-referred voltage noise-spectral density at T=300K is quite closely given by

$$S_N = \frac{0.93 \text{nV}/\sqrt{\text{Hz}}}{\sqrt{I_T}} \qquad (2.2)$$

where I_T is in mA, assuming negligible noise due to the source/base resistance and/or base currents. The 1dB gain-compression voltage, V1dB, can be shown to be 28.5mV RMS (times T/300K), independent of I_T. Thus, the dynamic range for a 100MHz bandwidth, defined as the ratio V1dB/$S_N\sqrt{\Delta F}$, is about 3,000, or nearly 70dB, for I_T=1mAP, and 60dB for I_T=100µAP. The distortion is predominantly third-harmonic, and is −40dBc (1%) for an input amplitude of 18mVP. Methods to improve the linearity and dynamic range will be presented in a moment.

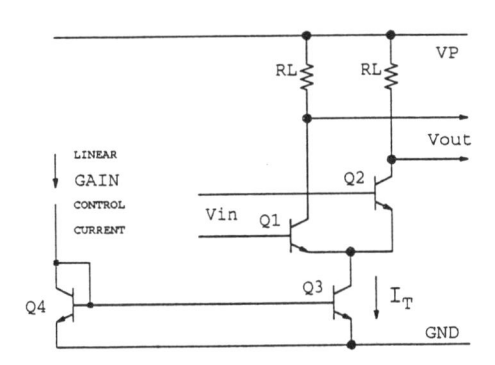

Figure 2.1 The BJT Differential-Pair as a VGA Cell

Any power-law for the gain-control can be achieved by simply cascading linear-law stages of this sort. Such cascading is often useful for other reasons, which include increasing overall bandwidth, minimizing changes in frequency response with gain, apportioning gain so as to minimize noise and power, improving isolation, etc. In many applications, it is desirable to provide exactly exponential, or linear-in-dB, gain control. This is, of course, the kind of thing that BJTs take in their stride. **Figure 2.2** shows one way in which this can be achieved. Q3 and Q4 form a current mirror, aided by the emitter-follower Q5, which is biased by R_E. When the gain control current I_G is zero, the main bias I_T is simply replicated in Q3.

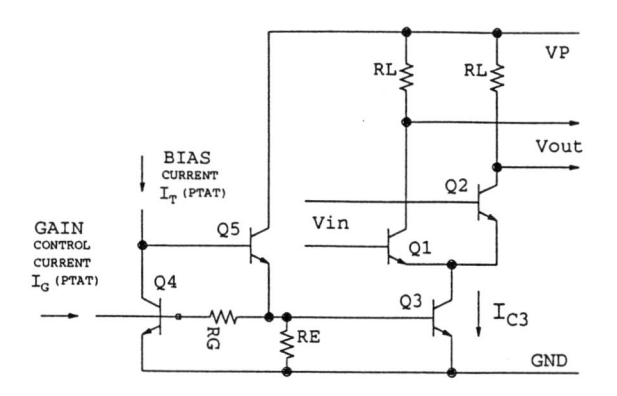

Figure 2.2 A Linear-in-dB Modification of the VGA

For $I_G=0$, R_G has a negligible effect on I_{C3}, but it may usefully be chosen to provide beta-compensation. For finite I_G, the voltage $V_G=I_G R_G$ introduced across R_G lowers the V_{BE} of Q3, resulting in a collector current of

$$I_{C3} = I_T \exp(-V_G/V_T) \qquad (2.3)$$

Referring the decibel gain to the maximum value (when $I_G=0$), we have

$$G_{dBc} = 20 \, lgt \{\exp(-V_G/V_T)\}$$
$$= -0.336 \text{ dB/mV of } V_G \text{ at } T=300K \qquad (2.4)$$

Gain variations of 60dB are practicable at moderate bandwidths (that is, up to about 10% of f_T). Of course, one of the advantages of direct-decibel control is that the law is unaffected by the number of gain cells connected in cascade. Thus, the gain variation used for each stage will often be much less than 60dB. Incidentally, this is a good example of a mixed translinear cell, combining a TN section (the differential amplifier), with a modified TL section (the current-mirror with the addition of I_G and R_G).

In precise gain-control applications, I_G needs to be rendered PTAT. Since the controlling variable will invariably be stable with temperature, this requires the inclusion of a simple one-quadrant multiplier, to generate $I_G = I_{dB} I_P / I_R$, where I_{dB} is the primary controlling current (proportional to an external voltage), I_P is a first reference current which is proportional to absolute temperature and I_R is a second reference current which is stable with temperature. This can all be done very elegantly using a TL multiplier cell. A simplified schematic of this cell is shown in **Figure 2.3**. With care in design, the production error in the law-conformance can be held to ±0.1dB over a 80dB range, and the absolute gain error held to well under 1dB (both over the full −55°C to +125°C temperature range). **Figure 2.4** shows typical results for the AD607, a single-chip 500MHz superhet receiver with I/Q demodulation, which uses this type of amplifier to provide very accurate RSSI measurements[17].

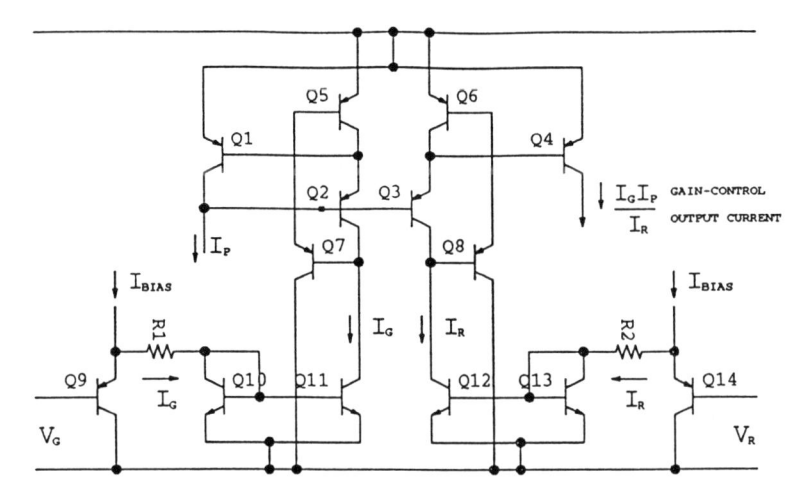

Figure 2.3 A One-Quadrant Multiplier for use with the Linear-in-dB VGA

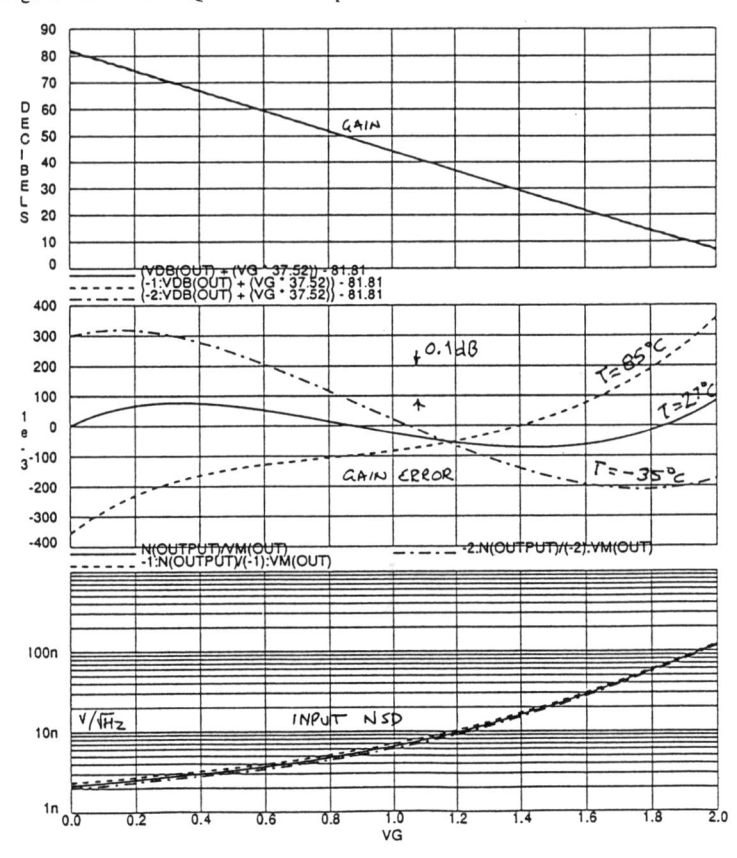

Figure 2.4 Typical Gain, Gain Linearity and Noise for Cascaded Linear-in-dB VGA Cells

While excellent linearity is possible in the *gain* function, the same cannot be said of the *signal channel*, whose large-signal transfer function has the well-known hyperbolic tangent form:

$$V_{OUT} = R_L I_T \tanh(V_{IN\_DC}/2V_T) \qquad (2.5)$$

The small-signal gain is thus a strong function of the instantaneous value of the input voltage:

$$\frac{\partial V_{OUT}}{\partial V_{IN\_DC}} = G_O \, sech^2(V_{IN\_DC}/2V_T) \qquad (2.6)$$

The gain is reduced to 50% of G_O for V_{IN_DC}=45.57mV at T=300K. Clearly, we cannot use the 'emitter degeneration' ruse here, introduced to thwart the nonlinearity of the LNA (Figure 1.3), since the translinear behaviour is essential to the variable-gain function of this cell. One obvious and straightforward solution is to use a linearized V/I interface and interpose a translinear (TL, current-mode) two-quadrant multiplier cell, as has been done in numerous cases in the past. The output section of such a TL cell can then be biased by either of the previous current sources to achieve simple-linear or linear-in-dB gain control. With a little ingenuity, such schemes can be designed to operate from a 1.2V supply, although some solutions that have been proposed preclude operation down to low temperatures, are not very linear, and have serious beta sensitivities; however, each of these errors will yield with attention to the details.

2.1 Multi-tanh Cells

An alternative approach, providing an extended linear range with little extra complexity, is the use of a method named by the author the multi-*tanh* technique, a concept which has been known and taught since the mid-seventies but which has only recently been widely utilized in commercial analog ICs. The basic idea is to use *two or more pairs* of BJT differential pairs, operating in simple parallel at both input and output, and arranged to be offset along the voltage axis. In general terms, these circuits synthesize the function

$$I_{OUT} = \sum_{n=1}^{N} I_n \tanh \left\{ \frac{V_{IN} + V_n}{2V_T} \right\} \qquad (2.7)$$

where I_n is the tail current to the n-th stage, and V_n is the base offset voltage associated with that stage. The net g_m of these N stages is

$$g_m = \sum_{n=1}^{N} I_n \, sech^2 \left\{ \frac{V_{IN} + V_n}{2V_T} \right\} \qquad (2.8)$$

There are several ways in which the necessary offsets can be achieved. The simplest method is to use emitter-area ratios; alternatively, they can be introduced using PTAT-biased emitter-follower input stages with small emitter resistors to generate the discrete voltage offsets. A two-section cell is called a *multi-tanh doublet*; it provides a rapid improvement in linearity. *Triplets* and higher-order N-*tuplets* have been used, but the benefits diminish rather rapidly with N.

Figure 2.5 shows a linear-in-dB VGA based on a multi-*tanh* doublet (N=2), implemented as two differential pairs in parallel. The junction area of one emitter in each pair is arranged to be A times larger than the opposite emitter; this will invariably be implemented by the replication of unit emitters. In a monolithic embodiment, the common bases and collectors of Q1-Q3 and Q2-Q4 respectively allow the cell to be made very compactly using just two pairs of emitters instead of one inside common collector-base regions (though this possibility is less often used today, since the modeling of such special structures is usually not readily provided in a modern simulation system).The effect of the emitter area ratio A is easily shown to shift the peak of each g_m by an equivalent offset voltage

$$\pm V_n = \pm V_T \ln A \quad \text{or} \quad A = exp(abs(V_n/V_T)) \qquad (2.9)$$

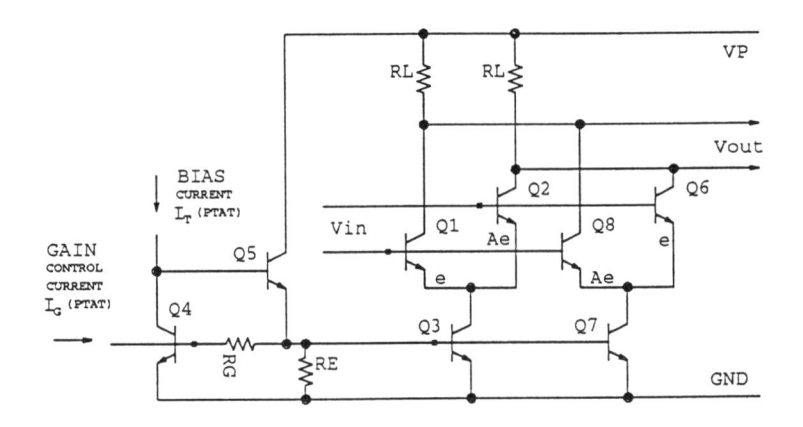

Figure 2.5 Linear-in-dB VGA using Multi-*tanh* Doublet

Thus, for A=4, the Q1-Q2 pair shift about 36mVP in one direction while the Q3-Q4 pair shift by the same amount in the other, with the result that the two transconductance segments, each having a *sech²* form, add in such a way as to result in an overall g_m that is much flatter than the simple differential pair, with a resulting improvement in linearity, as shown in **Figure 2.6**.The zero-signal g_m is reduced according to

$$g_m = g_{mo} \frac{4A}{(1+A)^2} \qquad (2.10)$$

where g_{mo} is the g_m that results for A=1, that is, when the cell is collapsed to a simple differential pair, biased by the full tail current I_T. The input noise due to shot mechanisms increases as

$$E_N = \frac{0.93 nV/\sqrt{Hz}}{\sqrt{I_T}} \frac{1+A}{2\sqrt{A}} \quad (I_T \text{ in mA}) \qquad (2.11)$$

However, overall, an improvement in dynamic range can be achieved; for A=4, the 1dB compression point increases from −18.7dBm to −11.8dBm (6.9dB), while the noise increases by only ×1.25 or about 1.9dB, an SNR extension of some 5dB.

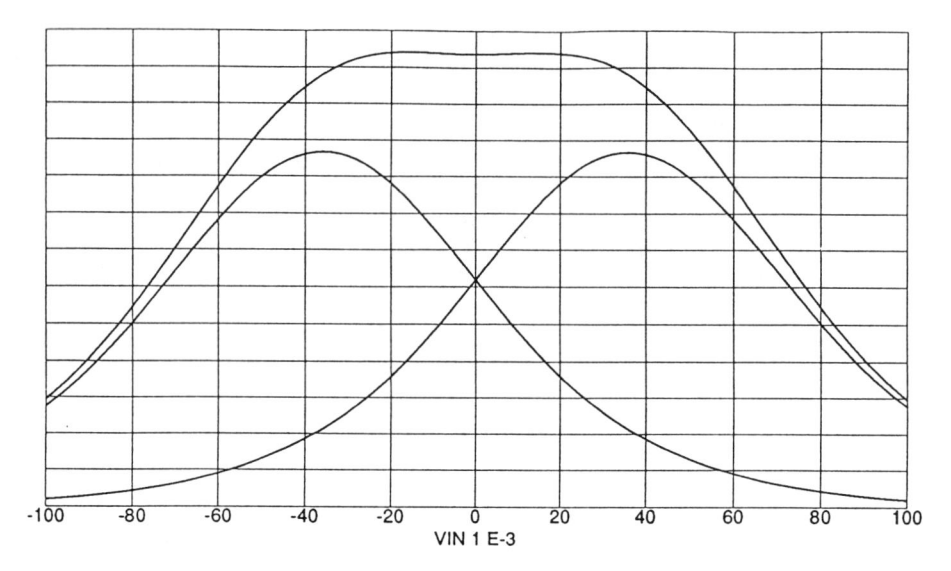

Figure 2.6 Separate and Composite G_m's (Linear Scale) of the Multi-*tanh* Doublet, A=4

2.1.1 Minimum-Distortion Criteria

It is of interest to determine the conditions for achieving a maximally-flat g_m-vs.-V_{IN_DC}, resulting in minimum distortion *for very small-amplitude signals* (say, V_{AC}). This means that the first derivative and as many as possible of the higher-order derivatives of the g_m function (2.8) must be set to zero. For the generalized N-section multi-*tanh* cell, the solution is daunting. Using the simplifying notation $x=V_{IN\_DC}/2V_T$ and $\lambda_n=V_n/2V_T$, we have:

$$g_m = \sum_{n=1}^{N} I_n \, sech^2(x+\lambda_n) \tag{2.12}$$

$$g_m' = -2 \sum_{n=1}^{N} I_n \, sech^2(x+\lambda_n) tanh(x+\lambda_n) \tag{2.13}$$

$$g_m'' = -2 \sum_{n=1}^{N} I_n \{sech^4(x+\lambda_n) - 2sech^2(x+\lambda_n) tanh^2(x+\lambda_n)\} \tag{2.14}$$

and so on. It will be apparent that finding the optimal solution for general parameter values is tedious when N is large. But for low values of N, and using symmetric values of I_n and λ_n, the problem is tractable. Thus, for the doublet, with only a single offset parameter, λ, and at $V_{IN\_DC}=x=0$, (3.1.6) becomes

$$g_m' = 0 = sech^2(\lambda)tanh(\lambda) + sech^2(-\lambda)tanh(-\lambda) \tag{2.15}$$

But this is true for all values of λ, due to the symmetry of the *tanh* function, so we need to consider the second derivative.

Setting it to zero

$$g_m'' = 0 = sech^4(\lambda) + sech^4(-\lambda)$$
$$- 2\{ sech^2(\lambda)tanh^2(\lambda) + sech^2(-\lambda)tanh^2(-\lambda) \} \quad (2.16)$$

which simplifies to
$$0 = sech^4(\lambda) - 2\,sech^2(\lambda)tanh^2(\lambda) \quad (2.17)$$
the solution to which is
$$\lambda = sinh^{-1}(1/\sqrt{2}) = 0.65848 \quad (2.18)$$

The corresponding offset voltage is $2\lambda V_T$, or 34.043mVP. This can be generated using A=3.732... (see 2.9), which can be closely approximated using an emitter-area ratio of 15/4; the consequences of using the nearest integer value of 4 are very slight. **Figure 2.7** shows how the gain and HD3 of the doublet vary with A.

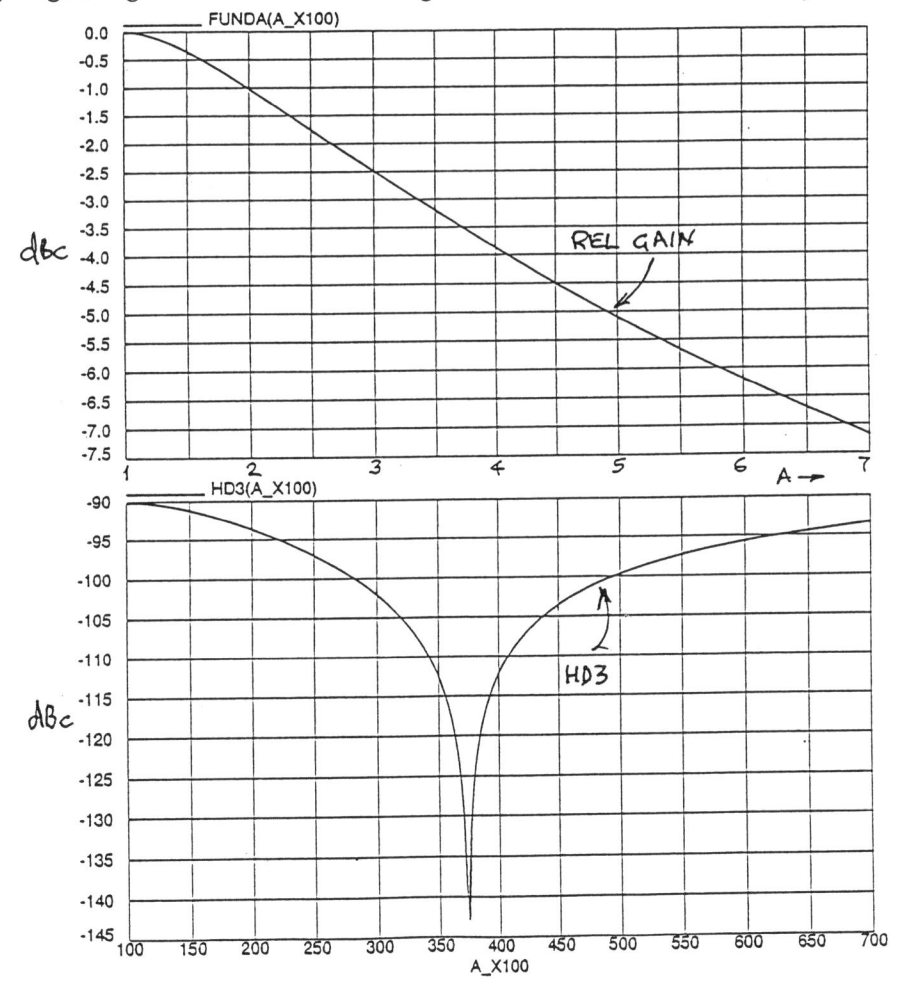

Figure 2.7 Relative Gain and Third-Harmonic Distortion of the Multi-*tanh* Doublet (V_{AC}=1mV)

Thus, the doublet—yet another application of the translinear perspective—provides an immediate benefit in dynamic range, compared to the simple differential pair (A=1), as shown in the following table. Note that the dBm power used in defining P1dB here is relative to 50Ω; thus, 0dBm corresponds to a single-tone sine amplitude of V_{AC}=316mV, or 223.6mV RMS.

TABLE I

A	HD3 (dB) for V_{AC} of 10mV	100mV	P_{1dB} (dBm)	ΔP_{1dB} (dB)	(1+A) $2\sqrt{A}$	Col. 5 (in dB)	ΔDR (dB)
1	−50.33	−15.84	−18.67	0.00	1.000	0.000	0.00
2	−53.79	−16.90	−17.13	1.54	1.061	0.512	1.03
3	−62.11	−18.73	−14.50	4.17	1.155	1.252	2.92
4	−100.25	−20.88	−11.87	6.80	1.250	1.938	4.86
5	−60.09	−23.40	−9.72	8.95	1.342	2.555	6.40
6	−55.96	−26.48	−8.04	10.63	1.429	3.100	7.53

To extend the linear range further, one may also use derivatives of the multi-*tanh* concept in which the offset sub-cells are connected in series, rather than parallel. **Figure 2.8** shows one such series-connected doublet. It will be apparent that this circuit has exactly twice the signal capacity as the parallel-doublet, but it is not so obvious that, provided the source impedances at both input nodes are equal, there is no net noise contribution from either the Johnson noise of the base resistors, R_B, or from the shot-noise currents from Q2 and Q3 which sum into the center base node. Consequently, the dynamic range (SNR) performance of this cell is the same as for the parallel doublet. The reason is simply that the cell is left-right symmetric, so any *common-mode* noise at this node causes equal but opposite-phase noise currents in the inner transistors. Even a moderate mismatch in the source impedances leaves a substantial amount of noise cancellation.

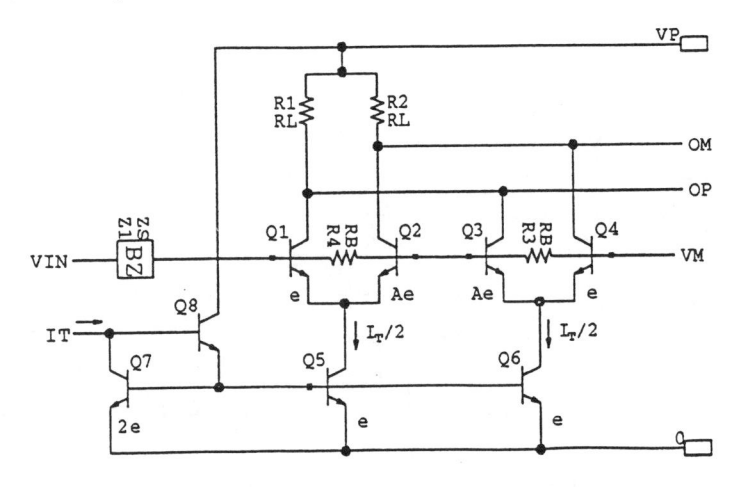

Figure 2.8 A Series-Connected Multi-*tanh* Doublet

2.1.2 Higher-Order Multi-tanh Cells

The multi-*tanh* triplet, using three differential pairs, with the central pair balanced and the two outer pairs offset, is even more attractive. **Figure 2.9** shows one of many possible implementations. In this case, PTAT voltages V_{OS}, developed across resistors R_{OS}, are used to introduce the offsets, to avoid the use of very large transistors; the effective area ratio is $A=exp(V_{OS}/V_T)$. Thus, for A=13, a V_{OS} of 66.3mVP would be used. Provided that the resistors are fairly small, the additional noise they introduce may be negligible when compared to that of the g_m cells. The emitter followers usefully raise the input impedance, though not without a noise penalty. Obviously, one can also use *emitter-area ratios* in a pair of EFs to introduce some, or even all, of the triplet offsets with less noise degradation than using resistors. Many such schemes have been used over the years. A different gm-control scheme is used here: I_G now raises the gain, calling for a much lower I_T (and/or larger Q1) than used in the previous examples. Again, combinations of these two gain-control arrangements have been used.

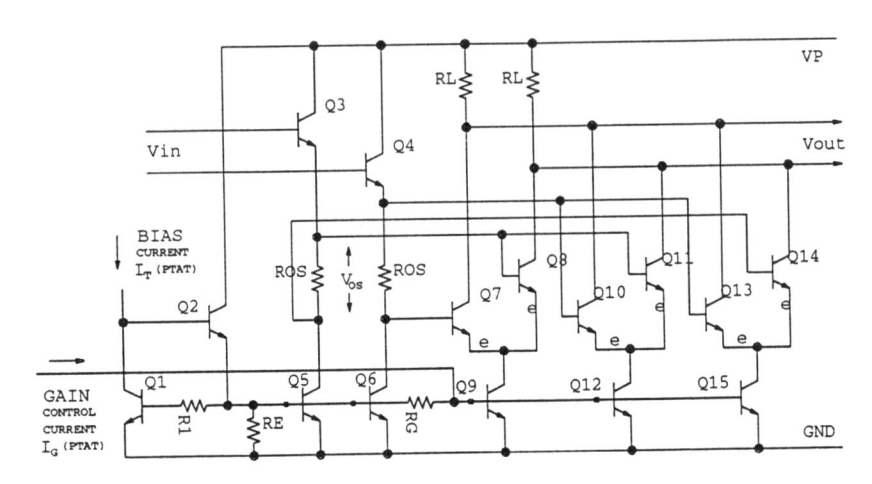

Figure 2.9 Multi-*tanh* Triplet with Alternative Linear-in-dB Scheme

The optimization of the triplet is a little more complicated, since now there are two variables: A, and the ratio of the inner to outer tail currents, K. (Note that Q12 would have an emitter-area ratio of K times that of Q9 and Q15). The performance for two typical optima, A=13, K=3/4 (Case 1) and A=21, K=10/12 (Case 2) is shown in Table II below. For Case 1, the ripple in the small-signal gain is within ±0.04dB for instantaneous inputs up to ±56.2mV (−15dBm); for Case 2, it is within ±0.25dB for inputs up to ±84mV (−11.5dBm). The zero-signal g_m is reduced by 6.64dB and 7.6dB respectively. The noise spectral density figures are for the ideal translinear case, and assume a total tail current of 1mA/300K. Again using the P1dB to define the upper extremity of the dynamic range, it is apparent that the triplet offers a modest improvement of about 7.5dB. Higher-order *n-tuplets* provide further dynamic range extension, but with diminishing returns and with increasing complexity of optimization.

TABLE II

CIRCUIT FORM	P_{1dB} (dBm)	ΔP_{1dB} (dB)	NSD nV/√Hz	NSD in ΔdB	ΔDR (dB)
Diff pair, A=1	−18.67	0.00	0.93	0.00	0.00
Doublet, A=4	−11.87	6.80	1.16	1.94	4.86
Triplet-1	−8.15	10.52	1.35	3.28	7.24
Triplet-2	−7.09	11.58	1.43	3.78	7.80
Quadlet	−4.00	14.67	1.65	5.03	9.64
Quinlet	−3.30	15.37	1.73	5.44	9.93

One may wonder whether similar tricks can be played with CMOS cells in strong inversion, using differing W/L ratios. Unfortunately, the simple answer is "no". Lacking the essential translinearity, the g_m of the individual pairs no longer sum in the required manner. There are better ways to use CMOS devices in VGA applications, particularly by operating them in their triode region.

2.1.3 Other TL VGA Cells

Before moving on to a more recent type of VGA, using advanced translinear principles, mention will be made of two more variants of the g_m-style cell. As noted earlier, a square-law gain function can be implemented by simply cascading two linear-control sections, which may, of course, be provided by an off-the-shelf dual analog multiplier, preferably a two-quadrant type optimized for gain-control applications, such as the Analog Devices AD539. Nevertheless, there may be occasions when a simpler solution is needed, or limited performance is permissible. **Figure 2.10** shows how a square-law TL cell can be used to effect this function in a single stage. Here, I_O is a fixed (zero-TC) bias current, while I_G is the gain-control current; for equal transistor sizes in the TL section, $I_T = I_G^2/I_O$.

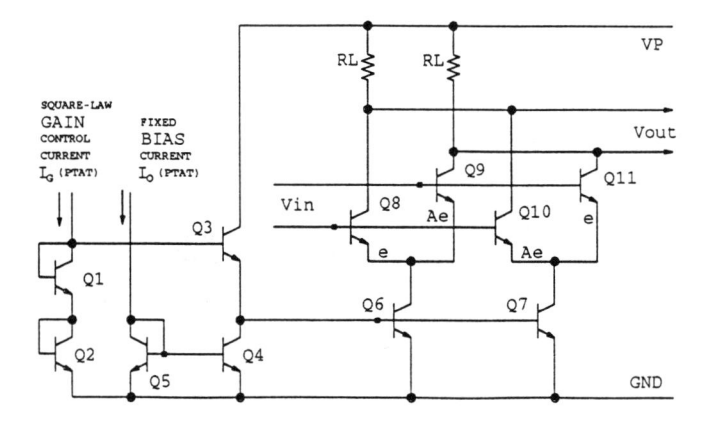

Figure 2.10 Single-Stage VGA with Translinear Square-Law Scheme

3 THE XAMP™

From the foregoing it will be evident that VGAs based on differential pairs have good noise performance but, even with the use of multi-*tanh* n-tuplets, have rather poor large-signal linearity and signal handling capacity. On other hand, current-mode (translinear-loop) multipliers, augmented by linear V/I converters can have excellent linearity, particularly when laser-trimming is used to eliminate the last vestige of emitter-area mismatch (producing even-order distortion) and the effects of junction resistance (producing odd-order distortion). For example, through meticulous attention to details, the AD734—a state-of-the-art four-quadrant multiplier—attains a THD of –80dBc on both its 'X' and 'Y' inputs. But this approach to VGA design leads to relatively high noise levels, often because of the contributions from the resistors used in the V/I stages, and their low g_m. However, the large supply voltages (±15V) and signal swings (±10V) used for that product allow a respectable dynamic range of 95dB to be achieved.

What is often needed is an amplifier offering uncompromised, state-of-the-art *noise* performance at *maximum* gain, combined with impeccable *linearity* when dealing with large signals at *minimum* gain. Further, its noise should degrade gracefully as the gain is reduced. Since VGAs are commonly used to level an input signal having a large range of amplitudes to something more nearly constant in amplitude (that is, the classical AGC function), it follows that another desirable aspect of the 'ideal VGA' would be that its *output noise is independent of gain*. These objectives can be met by placing an attenuator ahead of a very quiet fixed-gain amplifier, as shown in **Figure 3.1**. Of course, it is understood that this sketch shows a highly generic concept. For such a scheme to provide the VGA (electronically-variable gain) function, the 'slider' must be somehow controlled by another voltage (or possibly current) input, and the attenuator must have very low noise. The XAMP™ solves these problems in an interesting way. Several products, for example the AD600, AD602 and AD603, have proved the value of this technique in critical instrumentation such as medical ultrasound, and even in advanced IF/AGC applications [30].

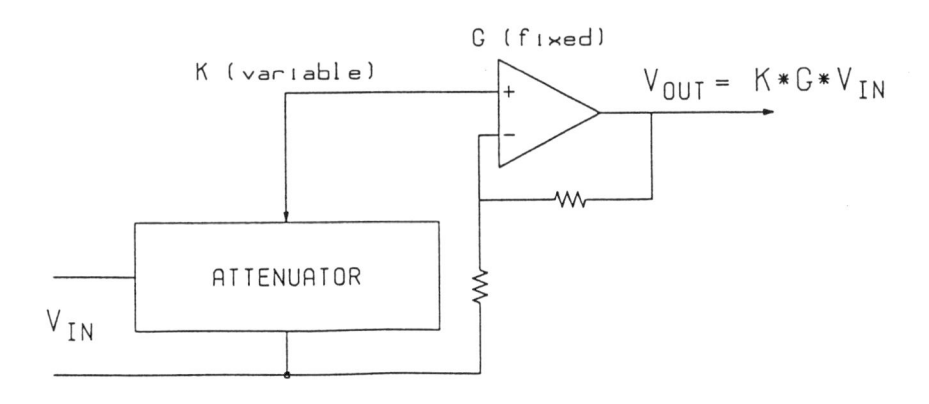

Figure 3.1 The XAMP™ Concept

The key question is: how is an electronically variable attenuator realized? (This is the central challenge in most VGA design). **Figure 3.2** shows how this was achieved in the first-generation solution. The ladder network—made of real resistors (in practice, thin-film SiCr is used)—provides a tapped attenuator, which *automatically* guarantees an exponential decay of signal amplitude from left to right. That is, the useful linear-in-dB gain function now springs, *not from translinearity*, but from an even more precise principle, based on a network composed of nothing more than very-closely matched resistors.

Using the 'binary' R-2R ladder of the sort found in classical DACs and ADCs, the attenuation is 6.0206dB/tap. There is no special reason to use that ratio; one can use an R-R network, providing 8.36dB/tap. In practice, the network will always be made using integer replications of a unit resistor, but many dB/tap variants can readily be obtained.

Nonetheless, translinearity plays an important role in the *overall* design of the XAMP™ (which is a contraction of *exponential amplifier*). Note in Figure 3.2 that there are N g_m stages. These act as linear multiplier cells, looking very much like the VGA of Figure 2.1, being controlled by currents I_1 through I_N. One by one, as they are activated, they become the input g_m cell for a composite op-amp. The control currents overlap, so that in general two or more stages are active at any gain setting. In the process, they generate a 'phantom' tap point that moves from left to right across the top layer of resistors R1. Translinear multiplication is essential to ensuring smooth gain interpolation. The full design details go well beyond the scope of this paper, which is mainly intended to show the many ways in which BJT translinearity is useful. It will be apparent that one cannot simply substitute MOS devices for the g_m cells: the VTL principle is fundamentally unsuited to this application[18].

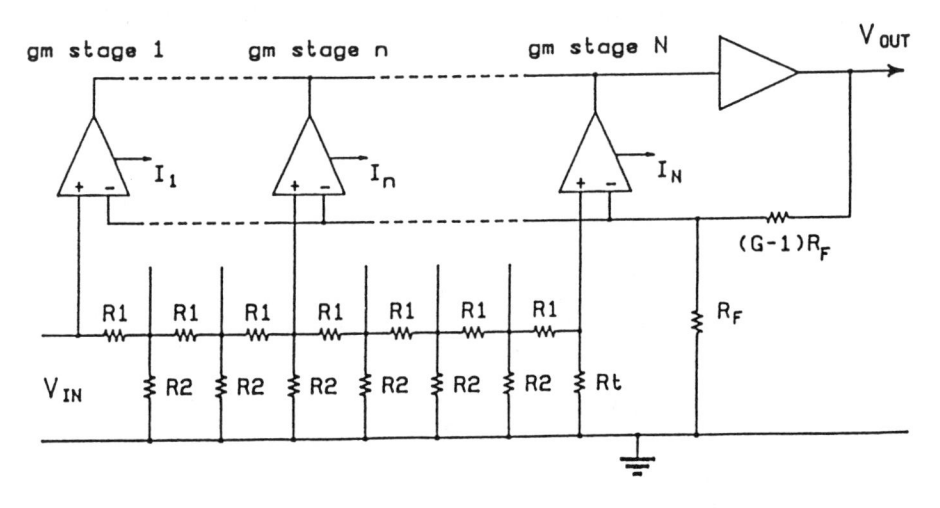

Figure 3.2 Linear-in-dB Scheme Based on a Ladder Network

The gain-control currents I_1 through I_N are provided by another translinear cell, shown in **Figure 3.3**. This circuit generates a series of N current pulses (spaced along the 'voltage' axis) each of which is essentially Gaussian in shape; these provide the bias currents for the N g_m stages in the previous figure. The operation of this circuit can be explained by noting that the base voltages take on a parabolic form due to the currents I_B delivered to the internal base nodes, these currents flowing outward through the resistors R_B to the external nodes at which the differential bias voltages $-V_B/2$ and $+V_B/2$ (provided by a gain-control interface) are applied. The product $I_B R_B$ is temperature-stable, resulting in a stable gain function, while I_E is PTAT, resulting in a stable g_m in the input cells.

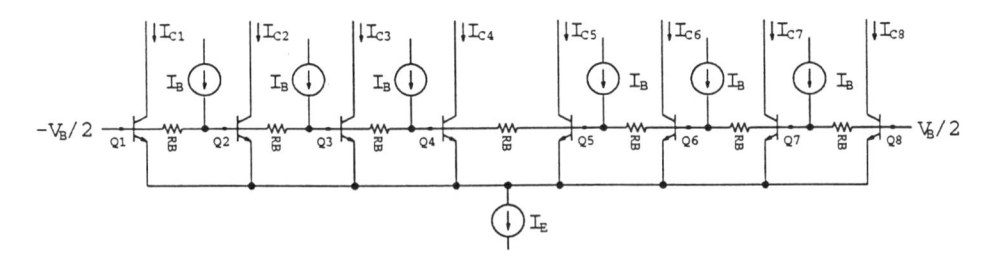

Figure 3.3 The Gaussian Interpolator

When $V_B=0$, the network is fully balanced in the left-right direction, and the tail current I_E flows equally in the two central transistors, where the local base voltages are highest. Because of the parabolic voltage distribution along the bases, and the high g_m of the BJTs, very little current flows in the other transistors. As V_B increases, the left-hand bases become more positive, and I_E is 'steered' into these transistors. The shape of the variation of the collector currents versus V_B is essentially Gaussian, as can be seen from the e^{-x^2} form of the numerator in the equation for I_{C_n}, where x is defined as shown below; the denominator is fairly close to a constant value except for large values of x, for moderate values of Q.

$$I_{C\_n} = I_E \frac{exp\{-Q(x-x_n)^2\}}{\sum_{n=-(N-1)/2}^{+(N-1)/2} exp\{-Q(x-x_n)^2\}} \quad (3.1)$$

where
$$Q = I_B R_B / 2V_T \quad (3.2)$$

$$x = V_B / \{(N-1)I_B R_B\} \quad (3.3)$$

$$x_n = n - (N+1)/2 \quad (3.4)$$

Once again, we are seeing how very important those exponential terms are in forming complex, though eminently practical functions. **Figure 3.4** shows typical current pulses for $I_B=200\mu A$, $R_B=715\Omega$ (Q=2.76), $I_E=1mA$; note that the peaks are a little under the full tail current, due to the slight overlap in the 'tails'.

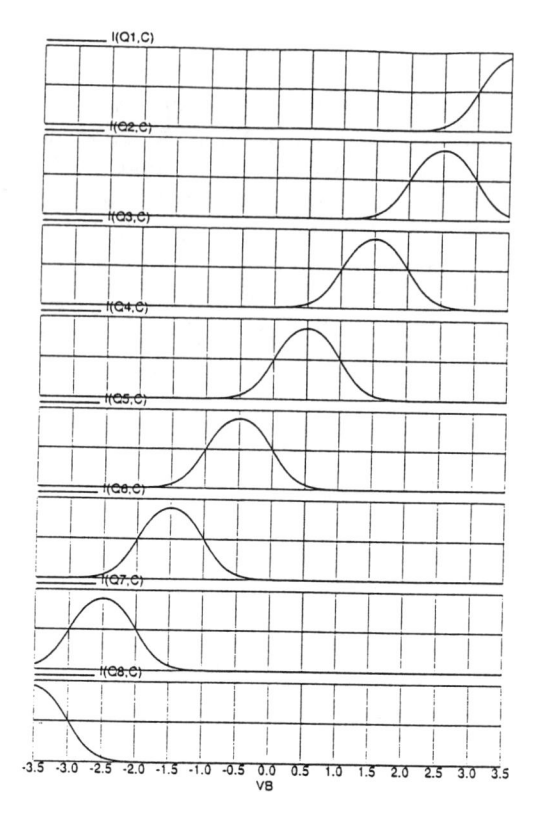

Figure 3.4 The Currents in Q1 through Q8 vs. V_B

Incidentally, these same principles were used in the AD639, a Trigonometric Microsystem, in which they provide a highly-accurate basis for the generation of the sine function (to a production accuracy of 0.02%), as well as all other normal and inverse trigonometric functions [25], providing a further example of the wide range of applications of the translinear perspective. It is probably not incorrect to assert that this level of functional versatility and the fully-practical high accuracy would be quite impossible in CMOS.

In the XAMP™, the Gaussian form for these currents is convenient because it means that the over gain function exhibits a smooth form, having an essentially sinusoidal deviation from the ideal linear-in-dB law. A linear interpolator would introduce 'cusps' in the gain function as it is handed over from one g_m cell to the next. Nevertheless, it is an easy matter to build a linear interpolator, that is, one generating a precisely triangular series of current pulses to bias the g_m stages. **Figure 3.5** shows one such possibility. Unlike the last circuit, which was a translinear network (TN) using the BJT's exponential properties at a basic level, this is a strict translinear-loop (TL) circuit, in which the voltages are of merely incidental interest. The detailed operation is left as an exercise for the reader, except to note that once again the current I_E is 'steered' from left to right, this time under the control of the modulation factor X, and that the current I_X is arranged to be $(N-1)I_O/2$. For example, for the results shown in **Figure 3.6**, N=5, $I_O=100\mu A$ and $I_X=200\mu A$. The horizontal diodes are preferable Schottky's, to minimize the voltage swings at the input nodes (which have a staircase form), in order to fit the circuit into a low supply voltage.

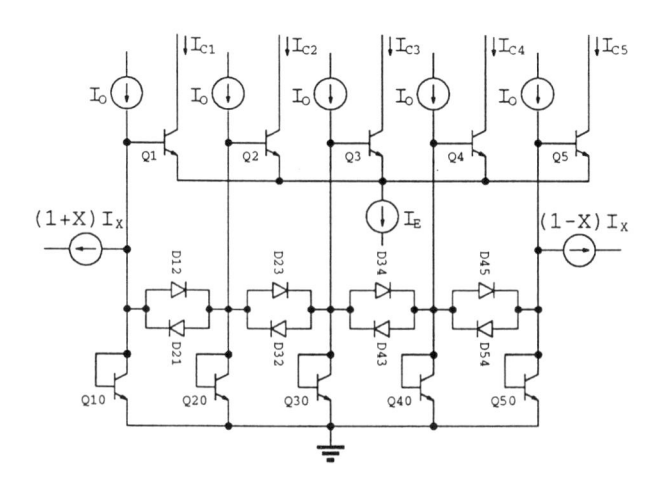

Figure 3.5 A Linear Interpolator

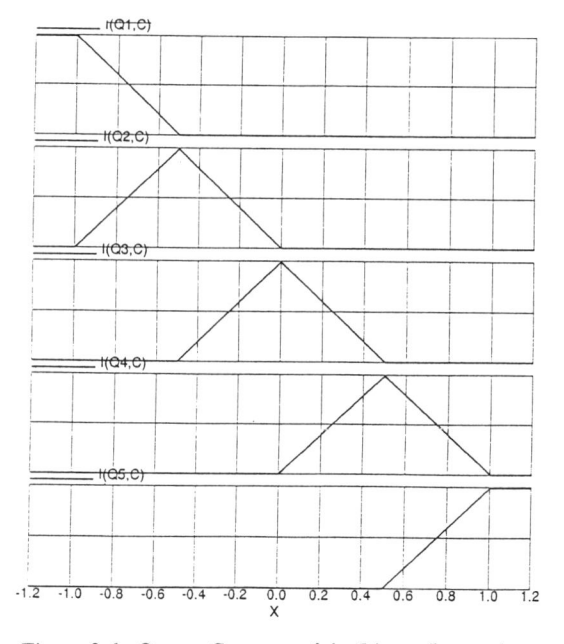

Figure 3.6 Output Currents of the Linear Interpolator

A more recent realization of the XAMP idea provides fully-differential input handling, and allows operation from a single supply. Products based on this technique (the AD604, which is a dual VGA, and the single-channel AD605), though not yet fully utilizing all of the possibilities of this new technique, combine a state-of-the-art input-referred noise-spectral density of about 0.75nV/√Hz with an input impedance of several megohms, provided by a pre-amplifier [26].

Several important ideas are embodied in the 'DSX' (Differential, Single-Supply XAMP), shown in **Figure 3.7**, the key elements of which are:

1) A fully-balanced attenuator is used, still having integer ratios of resistors, but providing slightly more than 6dB per tap point.

2) An 'Active Feedback' scheme (see [13]) closes the loop around the main gain stage, resulting in fully-separated input and output grounds, allowing summing to the output and providing greater flexibility with regard to positioning the gain range using external adjustments.

3) The Gaussian interpolator uses a PTAT bias (I_B, in Figure 3.3), resulting in a 'Q' in Eqs (3.1) and (3.2) that is temperature stable; this prevents changes in the 'gain ripple' over temperature.

4) A two-quadrant translinear multiplier cell generates $V_B = V_G V_P / V_R$ to provide a PTAT drive to the interpolator from the (temperature-stable) gain-controlling voltage, V_G, which may also be differential.

5) The chip is enabled when V_G exceeds 50mV; thus, assuming a DAC is used to provide this voltage, the digital null-word controls the power; when 'OFF', the attenuation is very high.

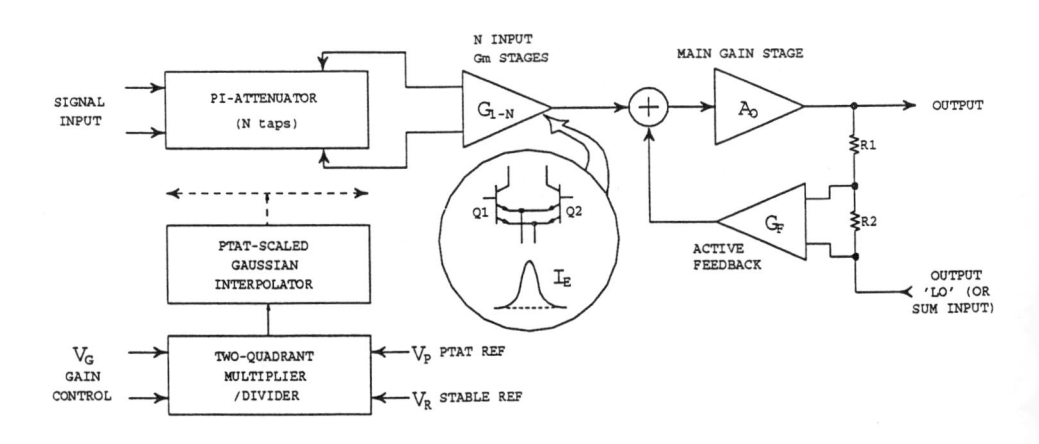

Figure 3.7 The 'DSX' — A Differential, Single-Supply XAMP

4 SUMMARY

Translinear circuits began as a new way to make wideband amplifiers. They were different from other amplifiers of the time in that the signals were all in current form. In fact, these circuits were originally called 'current-mode' amplifiers. They marked an important turning point in the way signal-processing was implemented in a monolithic medium, inasmuch as the conversion of the signals into currents resulted in merely *incidental* voltages, which were quite nonlinear, even though the overall (I_{OUT}/I_{IN}) transfer was linear. The basic idea at the outset was to pre-distort the signal using a logarithmic transform, a principle that has recently been invoked in the form of log-domain filters [27-29]. These amplifiers were not only very linear: the linear range extended right up to the extremities of the available range, followed by abrupt clipping. This was a further novelty. Variable-gain and multiplication cells were immediate spin-offs of these ideas.

During the intervening thirty years since the 1966 conception of these 'strict-translinear' circuits, it gradually became apparent that it was really the fundamental 'translinearity' of the bipolar junction transistor that was of such pervasive value, and its key strength. The broader class of circuits in which this property is exploited has been called 'translinear networks'; it includes numerous specialized nonlinear circuits performing almost every type of function, of one, two or many input variables (even hundreds, in such operations as analog array processing [16]). This translinearity is independent of the size of the transistor or the particular style of technology; thus, all the circuit cells developed over the years are immediately transferable to SiGe and GaAs HBT's.

The question now arises: is this almost-magical exponential behaviour of the BJT *always* a strength? By studying very basic amplifier cells, such as LNAs for high-performance radio applications, where the minimization of intermodulation poses taxing design challenges, we now realize that it is not. The better g_m-linearity of silicon MOS and GaAs MESFET devices, the absence of shot-noise in the channel and of both DC and shot-noise current in the gate electrode all point to their use in such applications. The suitability of CMOS for mixed-signal applications, and the diminishing gate-lengths, leading to even more linear transconductance, suggest that our approach to amplifier design will change rather significantly in the next few years. Somewhat ironically, the fact that the g_m of a short-channel MOS transistor tends to be almost independent of channel current will mean that many 'voltage-translinear' (VTL) cells developed using older MOS technologies, which depended on an assumed quadratic I_{DS}-V_{GS} relationship, will not be transferable; this includes the many CMOS variable-gain cells and multipliers which have been proposed.

It is likely that the fundamental benefits of the BJT in amplifier design, a few of which have been mentioned here, will continue to be of value for the foreseeable future. Of course, BiCMOS technologies, including CBCMOS, may offer the best of both worlds. Their use will gain momentum only when they are promoted and manufactured as a company's mainstream technology, with a view to trading off the higher wafer costs against overall improvements in efficiency of utilization.

ACKNOWLEDGMENT

The author is indebted to Monica Cordrey for her vigilant proof-reading of this material, and for her patient assistance in bringing the MS to completion.

NOTES

1. As has happened in the past between BJT and MOS terminology; for example, the term 'saturation' means precisely the opposite for the two types!
2. At the AACD '96 workshop, Seevinck proposed the alternative term 'quadratic translinear' (QTL), on the grounds that 'voltage translinear' may be misunderstood as referring to voltage-mode signal-processing.
3. This seems to have only recently been rediscovered, but the potential has always been there. The full realization of contemporary low-voltage translinear circuits is greatly facilitated by the availability of a complementary process.
4. Not entirely, since the base resistance $r_{bb'}$ will introduce Johnson noise, which, when multiplied by the relatively high transconductances of the closed-loops of junctions, can be troublesome. For a BJT operating at $I_C=100\mu A$, the noise is increased by 3dB for $r_{bb'} \approx 130\Omega$.
5. In [3] the gain cell form used in Figure 1.2 is called 'beta-immune', since the current-gain is substantially independent of the finite beta. This is believed to be a unique property of this cell.
6. Its actual value is invariably unimportant in TL cells, but affects the *absolute* g_m in more general TN (translinear network) cells.
7. This is only one of the ways in which the BJT manifests profoundly fundamental behaviour. Another is the way in which it is possible to use a single junction to determine temperature with absolute calibration accuracy.
8. However, ohmic junction resistances such as $r_{ee'}$ and $r_{bb'}$ will lower g_m at all frequencies, and the junction capacitances and base transit time will impact the effective g_m at high frequencies.
9. Even more simply, by applying a voltage equal to the bandgap value to a series-connected junction and resistor; thus, in Figure 1.1, the bias current of Q1 can be rendered PTAT by setting the voltage across R2 to this value, then choosing R3 appropriately.
10. Particularly when using complementary processes, even better, when the transistors are also dielectrically-isolated and augmented by low-TCR thin-film resistors and metal-oxide-metal capacitors, such as in Analog Devices' XFCB Process.
11. By contrast, their most advanced DSP products, using the SHARC architecture, have a great deal of memory on board, and use nearly 30 million transistors, *one thousand times* as many as the largest bipolar circuits.
12. This is not essential, and it is a simple matter to re-design the LNA to match to, say, a higher output impedance. In fact, lower noise figures can be achieved by so doing.
13. This needs a little clarification. The assumption here is that the 50Ω generator is matched directly by the amplifier. If we interpose a suitable network between the source and the amplifier, which results in a voltage increase, it is quite easy to achieve an improvement in noise figure, but only at the cost of a proportionately lowered P1dB and IIP3.
14. Experimental high-speed bipolar transistors have been fabricated using tunneling emitters, having DC current gains of tens of thousands, and simultaneously having extremely high Early voltages.
15. However, the current density required to realize the peak f_T in GaAs HBTs is about ten times higher than in SiGe devices.
16. This is not completely true, since there is a shot-noise component which becomes dominant in sub-threshold operation of enhancement-mode MOS devices, and is never entirely absent even in strong inversion.
17. In many cases, where only an AGC function is to be provided, temperature stability may be unimportant, allowing the temperature-compensation means to be eliminated.
18. However, there are variants of the XAMP™ using CMOS techniques.

References

[1] B. Gilbert, "A DC-500MHz Amplifier/Multiplier Principle", *ISSCC Digest of Technical Papers*, February 1968, pp. 114-115].

[2] B. Gilbert, "Translinear Circuits: A Proposed Classification", *Electronics Letters*, Vol. 11, No. 1, January 1975, pp. 14-16; and errata, Vol. 11, No. 6, p. 136.

[3] B. Gilbert, *Current-Mode Circuits from a Translinear Viewpoint: A Tutorial*, Chapter 2 of *Analogue IC Design: The Current-Mode Approach*, edited by C. Toumazou, F. J. Lidgey and D. G. Haigh, IEE Circuits and Systems Series, Vol. 2, Peter Perigrinus Press, 1990.

[4] A. G. Andreou and K. A. Boahen, "Translinear Circuits in Subthreshold MOS", *Analog Integrated Circuits and Signal Processing*, Vol. 9, No. 2, March 1996.

[5] E. Seevinck, *"Analysis and Synthesis of Translinear Integrated Circuits"*, in the series Studies in Electrical and Electronic Engineering, Elsevier, 1988.

[6] E. Seevinck, & R. J. Wiegerink, "Generalized Translinear Principle", *Journal of Solid-State Circuits*, SC-26, No. 8, August 1991, pp. 1198-1102.

[7] B. Gilbert, "Nanopower Nonlinear Circuits Based on the Translinear Principle", *Workshop on Hardware Implementation of Neural Networks*, San Diego, CA, January 1988.

[8] C. A. Mead, *Analog VLSI and Neural Systems*, Addison-Wesley, 1989.

[9] B. A. Minch et al, "Translinear Circuits Using Subthreshold Floating-Gate MOS Transistors", *Analog Integrated Circuits and Signal Processing*, Vol. 9, No. 2, March 1996.

[10] R. J. Wiegerink, "Computer Aided Analysis and Design of MOS Translinear Circuits Operating in Strong Inversion", *Analog Integrated Circuits and Signal Processing*, Vol. 9, No. 2, March 1996.

[11] B. Gilbert, "A New Wideband Amplifier Technique", *Journal of Solid-State Circuits*, SC-3, No. 4, December 1968, pp. 353-365.

[12] B. Gilbert, "A Precise Four-Quadrant Multiplier with Subnanosecond Response", *Journal of Solid-State Circuits*, SC-3, No. 4, 1968, pp. 365-373.

[13] B. Gilbert, "A High-Performance Monolithic Multiplier Using Active Feedback", *Journal of Solid State Circuits*, Vol. SC-9, No. 6, pp. 364-373, December 1974

[14] B. Gilbert and P. Holloway, "A Wideband Two-Quadrant Multiplier", *ISSCC Digest of Technical Papers*, February 1980, pp. 200-201.

[15] B. Gilbert, "A Monolithic RMS-DC Converter with Crest-Factor Compensation", *ISSCC Digest of Technical Papers*, February 1976, pp. 110-111.

[16] B. Gilbert, "A Monolithic 16-Channel Analog Array Processor", *IEEE Journal of Solid-State Circuits*, SC-19, No. 6, December 1984, pp. 956-963.

[17] B. Gilbert, "High-Accuracy Vector-Difference and Vector-Sum Circuits", *Electronics Letters*, Vol. 12, No. 11, May 1976, pp. 293 -294.

[18] B. Gilbert, "Translinear Circuits: An Historical Review", *Analog Integrated Circuits and Signal Processing*, Vol. 9, No. 2, March 1996.

[19] Y. P. Tsividis, *"Operation and Modeling of the MOS Transistor"*, McGraw-Hill, 1987.

[20] D. Divekar, *"FET Modeling for Circuit Simulation"*, Kluwer Academic, 1988.

[21] M. W. Hauser, *"Large-Signal Electronically Variable Gain Techniques"*, MIT Master's Thesis, December 1981.

[22] M. W. Hauser, "Variable-Gain Techniques: Translinear or Not?", *Advances in Analog Circuits and Devices*, Lausanne, 1996.

[23] B. Gilbert, *"Precision Nonlinear Techniques"*, *Integrated Circuit Techniques Conference*, Katholieke Universiteit Leuven, Heverlee, Belgium, May 1977.

[24] B. Gilbert, "A Low-Noise Wideband Variable-Gain Amplifier Using an Interpolated Ladder Attenuator", *ISSCC Technical Digest*, February 1991, pp. 280-281.

[25] B. Gilbert, "A Monolithic Microsystem for Analog Synthesis of Trigonometric Functions and Their Inverses", *Journal of Solid State Circuits*, Vol. SC-17, No. 6, December 1982, pp. 1179-1191.

[26] E. Brunner, "An Ultra-Low-Noise Linear-in-dB Variable-Gain Amplifier for Medical Ultrasound Applications", *Wescon Conference Record*, November 1995, pp. 650-655.

[27] R. W. Adams, "Filtering in the Log Domain", Preprint #1470, presented at the 63rd Conference of the Audio Engineering Society, New York, May 1979.

[28] E. Seevinck, "Companding Current-Mode Integrator: A New Principle for Continuous-Time Monolithic Filters", *Electronics Letters*, Vol. 26, No. 24, November 1990, pp 2046-2047.

[29] D. R. Frey, "Log Domain Filtering: An Approach to Current-Mode Filtering", *IEE Proc. Part G*, Vol. 140, No. 6, December 1993, pp. 406-416.

[30] B. Carver, "A High-Performance AGC/IF Subsystem", *QST* (the journal of the American Amateur Radio League), May 1996, pp. 39-44

Variable-Gain, Variable-Transconductance, and Multiplication Techniques: A Survey

Max W. Hauser
Linear Technology Corporation, Milpitas, CA, USA

Eric A. M. Klumperink
MESA Research Institute
University of Twente, Enschede, The Netherlands

Robert G. Meyer
Department of Electrical Engineering and Computer Science
University of California, Berkeley, CA, USA

William D. Mack
Philips Semiconductors, Sunnyvale, CA, USA

1. Introduction

Continuous electronic variability of a memoryless transfer characteristic (current gain, voltage gain, resistance, etc.) is the shared theme of variable-gain amplifiers or attenuators, electronically variable resistances and transconductances, and analog multipliers. Such functions have a long history, predating solid-state devices. The need for variable gain *per se* has arisen for many decades in such applications as automatic gain control (AGC) [1-5]. Analog multipliers were originally important in analog computers, and more recently in analog implementation of certain signal-processing arithmetic. Variable transconductance (G_m) and resistance

(R) circuits, the heart of many variable-gain amplifiers and multipliers, have acquired further importance as elements of tunable monolithic continuous-time filters. Much overlap exists between the circuit techniques for these various functions, embodying relatively few basic circuit principles. This chapter reviews those principles together (switched gain control, a related class, is outside the scope of this survey). Some upshots are summarized from a broader study of this subject matter as of 1981, which is also available [60]. Certain uses of nonsaturated FETs in particular have been reinvented over the decades in different applications, most recently in the application of filters.

2. Variable-gain basics

Certain generic classifications apply. *Small-signal* variable-gain circuits (a unifying principle despite diverse practical forms) exploit the change in some small-signal gain parameter such as an incremental transconductance (g_m) in a device or circuit that is inherently nonlinear under large signals. ("Large signals" means excursions of voltage or current approaching the quiescent DC values in the device.) Any circuit element with a smooth large-signal nonlinearity can provide some range of small-signal gain control. In contrast, high-linearity or *large-signal* variable-gain circuits, the main focus of this chapter, must exploit one of the limited number of electrical situations that realize an accurate signal multiplication.

Multipliers are formally a subclass of variable-gain circuits having a linear gain-control law. The application requirements tend to be implicitly different, however, the multipliers demanding the additional linearity on the control port, but less stringent signal-port performance over a wide gain range than in typical "variable-gain" applications. Moreover, a linear control of gain is inconvenient in a gain-control application with a wide gain range, and for other reasons. Feedback-type AGC systems demand specific nonlinear gain control laws to achieve desirable performance in the control loop, an old topic surveyed elsewhere [1]. The best-known example of this is with integrating AGC loops: requiring the incremental dynamics of the loop to be independent of absolute signal amplitude leads to a differential equation, whose solution is an *exponential* gain-control law for the variable-gain element. Other classes of AGC loop filters lead to different forms of desirable nonlinearity in the gain control, as demonstrated for example by Oliver in 1948 [2], Victor and Brockman in 1960 [3], McFee and Dick in 1965 [4], and Vol'pyan in 1968 [5].

Narrowband applications of gain control are common in RF communications, as compared with *baseband* applications, whose signal bandwidth extends to DC, or at least overlaps with the control bandwidth. These two application classes differ in their sensitivity to *control feedthrough*, the residue of the gain control appearing at the signal output of a variable-gain element. In narrowband applications, control residue tends to be outside of the frequency range of interest and therefore to be suppressed by tuned circuits in the signal path. Also, baseband variable-gain circuits specify noise and distortion in such terms as signal-to-noise ratio (SNR) and harmonic distortion (HD). Narrowband applications are more concerned respectively with noise figure (NF) and with nonlinearity expressed as gain compression and intermodulation (IM) distortion (specifically the signal level for third-order IM intercept), since these numbers measure errors within the frequency band of interest.

Special-purpose solid-state components have sometimes competed successfully with the bipolar junction transistor (BJT) and field-effect transistor (FET) variable-gain circuit techniques described in this chapter. These special-purpose components include solid-state photoelectric [6] and thermoelectric [7,8] systems, and P-I-N diodes [9,10,11]. Some of these alternatives are innovative and instructive. Székely's thermoelectric circuit [7] was completely monolithic. On the other hand, the simplest AGC system ever devised (intended for voiceband applications, where the inherent time constant fits the AGC requirements) is a carefully applied two-terminal thermistor where self-heating regulates the gain of the signal. Kornev reported effective AGC over a 40 dB range with this method [8]. P-I-N diodes, which deliver high-frequency large-signal impedance control and can handle large voltage and power levels [9,10,11], are another class of two-terminal gain-control device, and share with all such devices the property that the same port receives both signal and control, which must therefore be distinguished by their frequency ranges.

3. Bipolar-transistor large-signal variable-gain mechanisms

Figure 1 shows a basic current-steering core [12-22]. This may be the single most frequently used solid-state large-signal gain control element, as well as one of the oldest; usually it appears in fully differential form (Figure 2). The utility of these circuits stems from the fundamental ability of an emitter-coupled pair of BJTs, with a base voltage difference V_C, to steer a fraction of the total collector current to one side, this fraction being controlled by V_C but independent of the magnitude of the current. Pairs of non-exponential devices such as saturated MOSFETs, configured as in Figure 1, lack this property: a large-signal input I_{IN} modulates the fractional current steering and yields a distorted output[1].

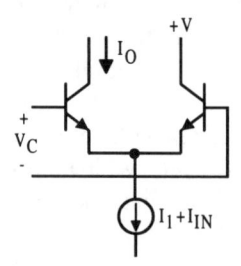

Figure 1: Basic BJT current-steering pair.

Quantitatively, the basic behavior of the current-steering pair in Figure 1 follows from the basic BJT large-signal forward-active model for collector current,

$$I_C = I_S \exp(\frac{qV_{BE}}{kT}) \tag{1}$$

(where I_S is a saturation-current parameter). Neglecting base currents, the sum of collector currents equals the total emitter current ($I_1 + I_{IN}$). It is easy to show then that for identical transistors, and neglecting other second-order effects such as resistive parasitics,

$$I_O = \frac{(I_1 + I_{IN})}{1 + \exp\left(\frac{-qV_C}{kT}\right)}. \tag{2}$$

---

[1] The proportional-current-steering property is equivalent to stipulating that the small-signal transconductance g_m of the devices be proportional to their large-signal operating (e.g., collector) current. This can be seen by considering the destination of an incremental current I_{IN} at the emitters in Figure 1. That g_m property in turn implies a differential equation for the large-signal current-voltage characteristic of each device, whose solution is that the current be an exponential function of the controlling voltage (V_{BE}).

A scaled component of I_{IN} appears in the output current, with a gain factor from zero (as V_C is large and negative) to one (as V_C is large and positive). Note that this is large-signal gain control: (2) does not presume that $|I_{IN}|$ is any smaller than the value that would cause a BJT to leave forward-active operation and invalidate (1), a magnitude of $|I_{IN}|$ equal to the DC bias value I_1. A signal-independent current proportional to I_1 and modulated by the gain control also appears in I_O; this is an example of control feedthrough, as mentioned in Section 2.

Figure 2 is the fully-differential form with complementary tail currents. From (2), the differential output current in Figure 2 is

$$I_{O1} - I_{O2} = \frac{2I_{IN}}{1+\exp\left(\frac{-qV_C}{kT}\right)}. \tag{3}$$

The differential form cancels not only the V_C-dependent control feedthrough (the I_1 term in (2)) but also even-order distortion that results from resistive parasitics in a single BJT pair [15,19,21].

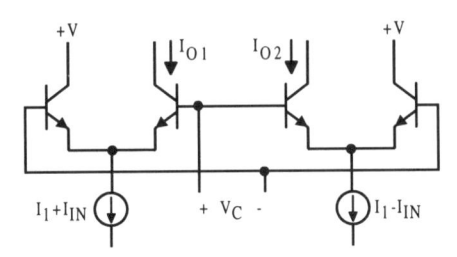

Figure 2: Fully differential BJT current-steering gain control.

Davis and Solomon in 1968 [14] and Sansen and Meyer in 1974 [17] were among those to build practical wideband variable-gain circuits around Figure 2, implementing the tail current sources with an additional emitter-coupled pair below the transistors in Figure 2. Sansen and Meyer introduced the refinement of operating the current-generating transistors (the lower pair) at higher currents than the current-steering transistors (the upper quad). This decoupling of the operating currents of upper and lower transistors permits the input-referred voltage noise to be reduced without sacrificing linearity, and is still of practical importance [118].

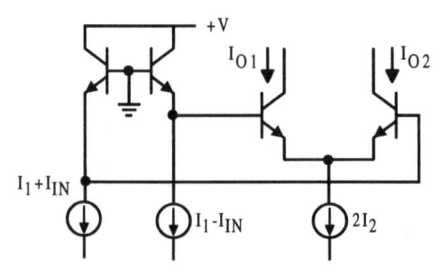

Figure 3: "Gain-cell" form of BJT gain control.

Figure 3 illustrates the other major BJT large-signal variable-gain mechanism, the linearized base-driven pair or "gain cell" [12]. This circuit, which like Figures 1 and 2 inherently requires its signal input in the form of a current, can achieve variable current gains greater than unity, unlike Figures 1 and 2, at some cost in noise and high-frequency signal feedthrough [22,60]. Figure 3 admits many published variations, with the common principle that two pairs of junction devices operate with the same voltage difference but different total currents [60]. Because of the proportional-current-steering property noted earlier, each of the two pairs (in Figure 3, the left and right pair) experiences the same *ratio* of collector currents, so that the differential current gain through the cell in Figure 3 is I_2/I_1, for magnitudes of I_{IN} less than I_1. It is easy to show this from (1).

Combining Figure 2 and Figure 3 (replacing the right-hand pair in Figure 3 with the two pairs of Figure 2, and cross-connecting the four collector currents from Figure 2) yields Gilbert's famous "six-pack" multiplier [13,18]. This is a current-input current-output core whose inputs and output are differential; it can realize an accurate device-parameter-independent four-quadrant multiplication.

4. Translinear and related circuit classes

The gain cell of Figure 3, and the "six-pack" multiplier derived from it, epitomize what are now called translinear circuits [23]. Briefly, such circuits share an underlying structure: a group of series V_{BE} drops that equals the V_{BE} sum in an opposing series, since altogether the V_{BE}s form a Kirchhoff voltage loop. Since each V_{BE} is proportional to the logarithm of a collector current (more generally a current density), the Kirchhoff summing constraint implies one product of currents equaling another. Moreover, unpredictable process parameters and temperature cancel out in the relationship. This principle has been exploited for analog multiplication and division, polynomials and roots, variable gain, and in its trivial form (one V_{BE} on each side), current mirrors.[2]

The numerous variations of Figure 3 that have appeared include the 1966 current gain cell of Grasselli and Stefanelli [24], which is almost a translinear circuit, using BJTs in saturated operation as the input pair. The single-ended translinear current-gain cell of Hamilton and Finch [25] links the input and output pairs of a Figure-3 variant in a positive feedback loop and exhibits reciprocal reverse transmission accompanying its forward current gain.

Straightforward generalizations of the translinear idea to other types of nonlinear devices configured into Kirchhoff-type voltage loops or current sums permit process-parameter-independent large-signal linear or nonlinear analog functions from virtually any device type. The multiplicative or additive process parameters (for example I_S and kT/q for BJTs, and V_T and the transconductance parameter "K" for FETs) cancel or become ratios, and the resulting "pure" large-signal input-output relation contains only designable quantities (input and output variables and dimensionless, temperature-independent ratios). Such generalizations of the translinear principle have been explored in their own right, since at least 1981 [26], and more recently Seevinck and Wiegerink further developed this idea in the application of saturation-mode MOSFETs [27,28]. Specific nonsaturation-mode FET circuits embodying the broad idea appeared as early as 1965 [30-32]. (FET cases may therefore actually predate bipolar translinear circuits, whose chief origin of record was Gilbert's three papers in 1968.) Hutcheon and Puddefoot in 1965 [30], Abu-Zeid et al. in 1968 [31], and Abu-Zeid and Groendijk in 1972 [32], although they did not use any such general terminology, employed Kirchhoff-linked groups of nonsaturated FETs to cancel process parameters in analog multiplier-dividers. These circuits appear in further detail in the next section. They yield large-signal input-output relations such as

$$\frac{I_{D1}}{V_{DS1}} = \frac{K_1}{K_2}\frac{I_{D2}}{V_{DS2}}$$

---

[2] Although the name "translinear" came later [23], the basic principle was expounded in Gilbert's 1968 "amplifier" paper [12], along with the remark "The number of circuits that can be devised to perform functions of this kind is legion."

where K_1/K_2 is a designable ratio of device sizes. Notably, the 1972 circuit of Abu-Zeid and Groendijk [32] illustrates a topology deriving from Kirchhoff's *current* law, rather than Kirchhoff's voltage law as in the more familiar BJT translinear circuits.

Generalizations of translinearity have practical limits, and are in need of clearer nomenclature. The large-signal current-voltage relationships available from a FET in its various modes lack the inherent accuracy, over many decades of signal amplitude, possible in the V_{BE} - I_C relationship of a forward-active BJT. Gilbert, moreover, has objected to the language "translinear" or "generalized translinear" for circuits not based, as was the original principle, on devices with a *trans*conductance *linear* in current. Nevertheless an important umbrella principle does exist, yielding designable large-signal input-output relations and device-parameter cancellation in return for the proper Kirchhoff connection of devices or subcircuits. The BJT translinear class is one especially accurate and useful manifestation of it.

5. Large-signal variable gain using FETs in nonsaturation

The last 35 years have seen steady use of FETs in large-signal variable-gain or variable-G_m circuits. We first review those that operate the FETs in the nonsaturation mode, building on the natural idea of the FET channel as a "voltage-controlled resistance." Methods for linearizing this resistance are very old. This history reveals a remarkable degree of reinvention. Well-established subcircuits used for analog multipliers in the 1960s reappeared in switched-capacitor and charge-coupled-device (CCD) circuits in the 1970s, and again in continuous-time filters since the 1980s (sometimes without evident awareness of their extensive published history, even in the same journal). In continuous-time filters, a variable R or G_m combines with a monolithic capacitor to define the physical time constant that is the kernel of any continuous-time filter. Particular importance for filters attaches to these FET circuit methods. This is not only because of their compatibility with CMOS manufacturing but more fundamentally from the recognition that FET large-signal synthetic variable R and G_m circuits can, with good noise figure, support signal-voltage excursions comparable to the power-supply voltage, and consequently can achieve filter signal-to-noise ratios unattainable today by other monolithic means [46,47].

Triode, prepinchoff, "ohmic," and "linear" are some of the names used in the literature to describe a FET operated with a voltage along its channel (strongly inverted, in the case of a MOSFET) low enough that the channel does not pinch off. The terms "ohmic" and "linear" have the wrong connotations for our context, since a central issue is that the FET channel resistance is not in fact linear (*i.e.*, "ohmic"). We follow Tsividis [29] in denoting this mode of operation as "nonsaturation."

In the nonsaturation mode, FET large-signal DC drain current obeys to first order the common charge-control approximation

$$I_D \cong K\left[(V_{GS}-V_T)V_{DS} - \frac{V_{DS}^2}{2}\right] \quad (4)$$

Countless *small-signal* variable-gain circuits have exploited the approximate "voltage-controlled resistance" linking I_D with V_{DS} in a nonsaturated FET, and accepted the quadratic nonlinearity explicit in (4). Since the early 1960s, linear, large-signal variable gain has also been achieved through two methods of canceling the quadratic term in (4). Note that early FET-based variable-gain circuits used junction FETs (JFETs) rather than MOS FETs, with principles that apply to both. MOS devices appear in the illustrations below.

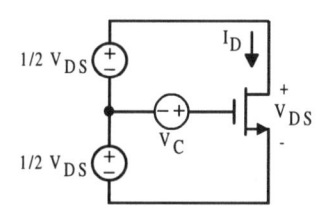

Figure 4: Linearizing nonsaturated FET channel resistance by symmetrical drive with respect to gate bias.

The first FET-linearization method adds a $V_{DS}/2$ component into a V_{GS} control voltage, which cancels the quadratic nonlinearity in (4). Another way to explain this method is that the source and drain experience symmetrical voltage excursions with respect to the gate bias. Control occurs through this gate bias. The composite circuit then presents a linearized variable resistance at the FET source and drain terminals, which can be exploited in some larger circuit. Figure 4 shows the basic configuration. For the FET in Figure 4,

$$V_{GS} = V_C + \frac{V_{DS}}{2}$$

Therefore from (4),

$$I_D = K(V_C - V_T)V_{DS} \qquad (5)$$

A large-signal conductance now links I_D to V_{DS}, linear to the accuracy of the model (4). In general this result holds for both polarities of V_{DS} and I_D, as long as the FET channel does not saturate. For gain control, this can be employed as a conductance controlled by V_C (the usual approach) or, in principle, as a transconductance where V_C is the signal input while V_{DS} is the gain control. (Since the back gate or bulk of a MOSFET introduces parasitic junction-FET behavior also, careful disposition of the bulk also requires biasing it with respect to the mid-point of V_{DS}, which is not shown in Figure 4 [60].)

This principle has a diverse and illustrious history. Attributed to Martin in 1962 [33], it was used initially with discrete junction and MOS FETs, as reported in various forms by Elliott in 1964 [34], Hutcheon and Puddefoot in 1965 [30], Todd in 1965 [35], Bilotti in 1966 [36], Abu-Zeid et al. in 1968 [31], Leighton in 1968 [38], and von Ow in 1968 and 1969 [37,39]. One practical version of it, using a voltage divider to combine half of V_{DS} into V_{GS}, appears in Figure 5.

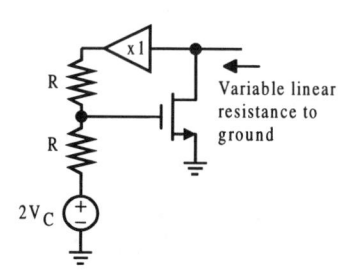

Figure 5: One practical form of Figure 4 with optional unity-gain voltage follower.

In particular, Hutcheon and Puddefoot [30] and Abu-Zeid *et al.* [31] employed linearized nonsaturated FET circuits of this kind as the basic building blocks for multiplier-divider circuits obeying the superset of the translinear principle mentioned in Section 4. (Hutcheon and Puddefoot used a time-shared form that might not be immediately recognized.) The basic idea of these circuits was to arrange two subcircuits obeying (5) with equal values of the controlling voltage V_C. This leads to a relation in which the magnitudes of the parameters K and V_T cancel:

$$\frac{I_{D1}}{V_{DS1}} = \frac{K_1}{K_2}\frac{I_{D2}}{V_{DS2}}$$

This method can be generalized, although the original authors did not suggest it, by arranging V_C's in a voltage loop, resembling the stacking of V_{BE} voltages in a bipolar translinear circuit [26]. This leads to circuits obeying relations such as

$$\frac{I_{D1}}{V_{DS1}} + \frac{I_{D2}}{V_{DS2}} = \frac{I_{D3}}{V_{DS3}} + \frac{I_{D4}}{V_{DS4}}$$

in which the variables can represent inputs or outputs by the details of the circuit arrangement.

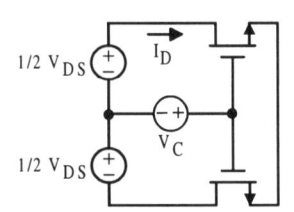

Figure 6: Circuit of Figure 4 with FET split into two series devices.

As a further variant of the linearization technique in Figure 4, note first that splitting the FET into two half-length devices yields the configuration of Figure 6. If the common source potential is maintained not by a short circuit but by a virtual short circuit from an op amp, we obtain the configuration of Figure 7, proposed in 1983 by Banu and Tsividis to provide electrically-variable input resistances in an op-amp integrator [40,41]. Figure 7 allows another degree of freedom in that the two FET currents need not be equal; the op amp can absorb a common-mode current. In Figure 7 the basic model of (4) predicts a differential FET drain current, forced then by the op amp through the feedback impedances Z_f, of

$$I_{D1} - I_{D2} = K(V_C - V_T)V_{IN} \tag{6}$$

where K is the transconductance parameter for each device, as in (4). This is the same basic result as for the single linearized FET, (5). More thorough modeling of the FETs to include further imperfections shows that the configuration of Figure 7 can cancel all of the even-order nonlinearities in the large-signal channel conductance [41]. The integrator of Banu and Tsividis and its variations have been seminal to the development of monolithic tunable continuous-time filters with good noise performance [46,47].

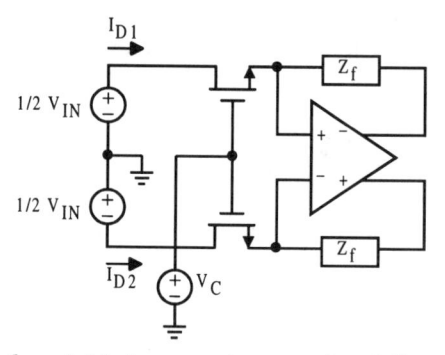

Figure 7: Linearized variable input resistances for differential op amp. After Banu and Tsividis [40].

Doubly-balanced variants of Figure 7, with further cancellation of nonlinearities, were described in multipliers by Crooke and Wegener in 1976 [42] and by Sage and Cappon in 1979 [43], and in filters by Czarnul [44] and Song [45] in 1986.

The second chief method of linearizing nonsaturated FETs for large-signal gain control is a differential transconductor with V_{GS} as the signal input and V_{DS} as the gain control. The same V_{DS} voltage appears across both FET channels in Figure 8, a fundamental difference compared with Figures 6 and 7. In practice, circuitry not shown in Figure 8 derives an output signal from the current difference $I_{D1} - I_{D2}$. The differential configuration cancels both the quadratic nonlinearity and the V_T

parameter in (4). This shares with the previous linearization technique a utility evident in frequent rediscovery.

When two matched FETs obeying (4) receive differential gate drive in Figure 8, the difference of their drain currents is

$$I_O = I_{D1} - I_{D2} = K(V_{GS1} - V_{GS2})V_{DS} \tag{7}$$

Both V_{DS} and V_{GS1}-V_{GS2} can have either polarity so (7) affords a four-quadrant multiplication capability. Also this linearization, unlike (5) and (6), entails standing currents in the two FETs, but it places no first-order constraint on the magnitudes of these currents. Thus in (7) the transconductance from differential V_{GS} to differential I_D depends on V_{DS} but not on common-mode V_{GS} or I_D. This degree of freedom is occasionally exploited.

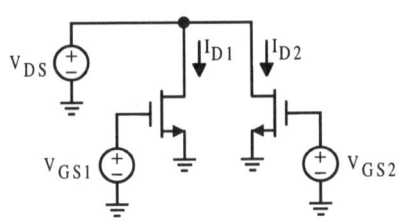

Figure 8: Linearizing channel conductance in nonsaturated FET pair by differential gate drive. Output is drain-current difference.

Highleyman and Jacob in 1962 were apparently the first to use this gain-control or variable-transconductance technique [50], and it has reappeared regularly in various applications. Taking examples only from the major Anglophonic literature, Østerfjells in 1965 [51], Abu-Zeid and Groendijk in 1972 [32], Bosshart in 1976 [52,53], Mavor *et al.* in 1977 [54], McCaughan *et al.* in 1978 [55], Enomoto *et al.* in 1982 [56], Enomoto and Yasumoto in 1985 [57], Pennock in 1985 [58], and Alini, Baschirotto, and Castello in 1992 [59] used versions of this linearization technique. Abu-Zeid and Groendijk in 1972 moreover used two of the pairs in Figure 8 in a multiplier-divider circuit embodying the generalization of the translinear principle described in section 4 [32], canceling the absolute magnitude of the remaining process parameter "K" in (7) as well as V_T, so as to yield a relationship of the form

$$\Delta V_{GS2} = \frac{K_1}{K_2} \frac{V_{DS1} \Delta V_{GS1}}{V_{DS2}}.$$

With these nonsaturation-mode circuits, as well as with the other large-signal FET techniques that follow, it is important to keep in mind the limitations of the elegant algebra that predicts a clean linear signal path from a clever configuration. The mathematical device models such as (4) that underlie all of these FET circuit techniques are idealizations, and they are is not as inherently accurate as the counterpart model for bipolar transistors, equation (1). Limitations to the linearity of the basic linearized-channel-resistance circuits due to second-order effects not modeled in (4), especially modulation of carrier mobility in the channel, have been studied extensively in the literature, and reviewed both in the context of gain control [60] and of filtering [48,49].

6. Large-signal variable gain using FETs in saturation

A FET in saturation (with a MOSFET, the usual case, we imply strong inversion as part of the "saturation" mode) obeys the approximate model

$$I_D \cong \frac{K}{2} \cdot (V_{GS} - V_T)^2 \qquad (8)$$

corresponding to (4) in nonsaturated operation. Equation (8) can be exploited in several systematic ways to achieve large-signal analog multiplication or variable G_m, as we shall summarize.

The published history of saturation-mode large-signal variable-gain circuits is more recent than that of circuits using nonsaturation. Ikeda's 1970 saturated-MOS AC-coupled "variable resistor" [61] was presented as an alternative to Bilotti's nonsaturated circuit [36] but unlike Bilotti's, it was actually a small-signal circuit, lacking correction for the device's quadratic nonlinearity. Házman in 1972 described an approximate multiplier core with one common-source pair providing the source currents to two differential pairs [62]. The 1979 two-MOSFET squaring core of Seriki and Newcomb [63] was proposed for quarter-square multiplication, a venerable analog-computer technique exploiting the algebraic identity

$$\frac{1}{4}\left((a+b)^2 - (a-b)^2\right) = a \cdot b. \tag{9}$$

These early circuits preceded considerable development of saturation-mode methods for large-signal multiplication and variable G_m, R, or gain circuits [64-108].

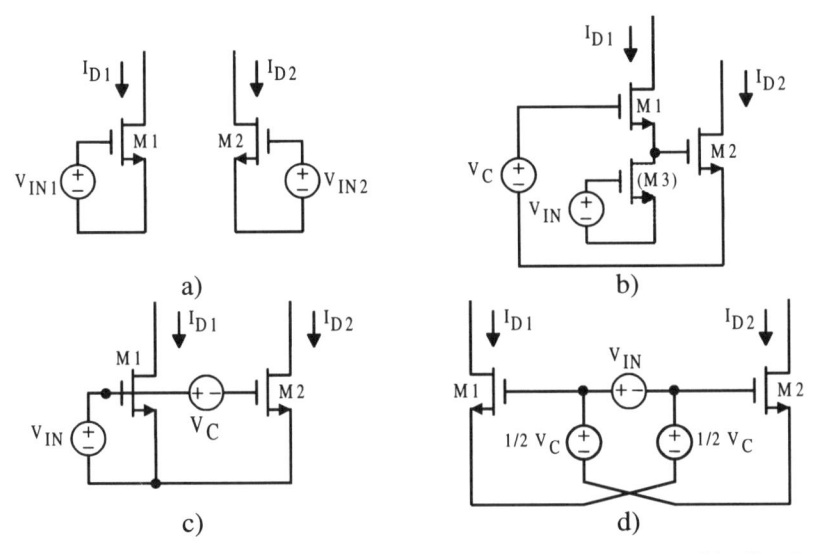

Figure 9: Examples of four different ways of forcing two Kirchhoff voltage relations among V_{GS1} and V_{GS2}, implementing a linear variable G_m. Output is the differential current ID_1-ID_2.

Many saturation-mode variable-G_m circuits have a functional kernel with two equal MOS transistors[3] obeying (8), which leads to:

$$I_{OUT} = I_{D1} - I_{D2} = \frac{K}{2}\left(V_{GS1} + V_{GS2} - 2V_T\right) \cdot \left(V_{GS1} - V_{GS2}\right) \tag{10}$$

---

[3] In some cases NMOS and PMOS FETs with $K_N \approx K_P$ are used [79,93,107], or two "CMOS pairs" with series combination of NMOS and PMOS [71,75,88,94,96,101,106].

The fundamental electronic variability here arises from the multiplication of two voltages to form a current. There are essentially four different ways to exploit this relation to obtain an I_{OUT} linear in a signal input V_{IN} and modulated by a control voltage V_C.

a) Force two V_{GS} values (as in Figure 9a) but impose a special constraint on them that prevents quadratic distortion of the signal input V_C. This occurs naturally when the V_{GS} values arise from certain preprocessing circuits, as in gain cells due to Klumperink [82,93,107] and Wang [102]. Suitable weighted sums of V_C and V_{IN} can also be created using differential pairs as proposed by Torrance et al. [67], Klumperink et al. [81,100], and Wilson and Chan [92]. Such preprocessing circuits however introduce additional errors and noise. The remaining cases apply V_C and V_{IN} to the two FETs in such a way that preprocessing circuits are not necessary.

b) Force $V_{GS1} = V_{IN}$, and the sum $V_{GS1} + V_{GS2} = V_C$ (as in the example of Figure 9b).

$$I_{OUT} = I_{D1} - I_{D2} = K(V_C - 2V_T)V_{IN} - \frac{K}{2}(V_C - 2V_T)V_C \quad (11)$$

I_{OUT} is now linear in V_{IN} but not in the control voltage V_C. The variable-G_m circuit shown in Figure 9b (and others) was proposed by Bult and Wallinga in 1987 [74,78] (transistor M3 copies V_{IN} to V_{GS1}). Cheng and Toumazou used this as the heart of a so-called composite MOSFET (COMFET) [98]. Szczepanski et al. employed two such circuits in a cross-coupled configuration [101].

c) Force $V_{GS1} = V_{IN}$ and the difference $V_{GS1} - V_{GS2} = V_C$ (as in Figure 9c).

$$I_{OUT} = I_{D1} - I_{D2} = KV_C V_{IN} - \frac{K}{2}(V_C + 2V_T)V_C \quad (12)$$

Again a linear relation results from V_{IN} to I_{OUT}. This structure occurs as a subcircuit in variable-G_m cells described by Wang and Guggenbühl [84,85], Czarnul et al. [88,89], Adams and Ramirez-Angulo [95] (preceded by a voltage copier), Wu et al. [96] and Szczepanski et al. [106].

d) Force the sum and the difference of the two gate-source voltages (as in Figure 9d): $V_{GS1} + V_{GS2} = V_C$, $V_{GS1} - V_{GS2} = V_{IN}$.

$$I_{OUT} = I_{D1} - I_{D2} = \frac{K}{2}(V_C - 2V_T) \cdot V_{IN} \tag{13}$$

Although a linear relation exists from both independent variables V_{IN} and V_C to I_{OUT}, the V_{GS} difference is commonly used as input variable, since no offset exists then. This principle was used in various forms by Nedungadi and Viswanathan [65], Viswanathan [70], Park and Schaumann [71], Seevinck and Wassenaar [75], Nauta and Seevinck [79], Noceti Filho et al. [83], Wilson and Chan [86], Czarnul and Takagi [88], Szczepanski et al. [94] and Sevenhans and van Paemel [99].

Note that in case c the V_{GS} voltages of M1 and M2 have a common component containing the signal information, while the differential component is used for transconductance control. In cases b and d it is the other way around (for case a it depends on the details). As a result, the signal components of I_{D1} and I_{D2} are added in-phase to form I_{OUT} in cases b and d, but subtracted for case c. This signal subtraction yields a large transconductance control range in principle, but as with other gain-subtraction schemes the noise and mismatches do not go away when the two gain paths cancel.

Besides these voltage-controlled G_m circuits, source-coupled pairs are also useful in large-signal variable-G_m circuits [65,67]. Such a pair, each obeying (8), with a constant tail current I_C, yields the drain-current difference

$$I_{OUT} = I_{D1} - I_{D2} = V_{IN} \sqrt{K} \sqrt{I_C - \tfrac{1}{4} K V_{IN}^2} \ . \tag{14}$$

I_{OUT} now is an odd-order nonlinear function of V_{IN}. Linearization follows by adding a term proportional to V_{IN}^2 to I_C (Figure 10):

$$I_C = I_{C0} + \tfrac{1}{4} K \cdot V_{IN}^2 \;\Rightarrow\; I_{OUT} = I_{D1} - I_{D2} = V_{IN} \cdot \sqrt{K \cdot I_{C0}} \tag{15}$$

In 1985 Nedungadi and Viswanathan proposed a linear transconductor using mismatched pairs to generate the quadratic tail current [65]. Bult and Wallinga used the circuit of Figure 9b with M3 diode-connected as current squarer [74,78].

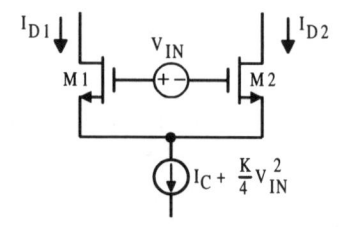

Figure 10: Source-coupled pair as variable G_m circuit. By adding a term proportional to the square of V_{IN}, signal linearization is possible.

Another linearization method, well known from bipolar circuits, is the use of parallel source-coupled pairs with successive input offsets, thus extending the "linear" G_m range beyond that of a single pair [91].

The forgoing circuits are variable transconductors. An electronically variable resistance can be realized using the same relations, but exchanging input and output variables [74,93,102,107]. A variable G_m together with a passive resistor will make a variable voltage gain. Alternatively, a MOSFET resistance and MOSFET variable G_m can implement a MOS *current gain cell* [80,82,93,102,107].

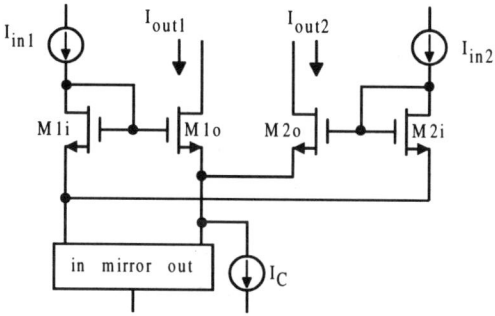

Figure 11: MOS current-gain cell consisting of a linear I-V converter (M1i and M2i, driven by appropriate input currents) and linear V-I converter (M1o and M2o).

The circuit of Figure 11, due to Klumperink and Seevinck [80,82], operates as a large-signal current-difference amplifier with variable gain, like Gilbert's bipolar gain cells. In contrast to the bipolar cells, where signal voltages and currents are nonlinearly related, the linearity in Figure 11 comes from a linear current-voltage conversion followed by a linear voltage-current conversion. The underlying principle is the same as in Figure 10 and requires the input currents to have the algebraic form

$$I_{in1,2} = I_{IN0}\left(1 \pm \frac{(I_{in1} - I_{in2})}{4I_{IN0}}\right)^2 \qquad (16)$$

where I_{IN0} is the input quiescent current. This results in

$$A_i = \frac{I_{out1} - I_{out2}}{I_{in1} - I_{in2}} = \frac{K_o}{K_i}\sqrt{1 + \frac{I_C}{2I_{IN0}}} \qquad (17)$$

-- that is, the gain depends only on a ratio of K parameters. Other related MOS current-gain cells have been reported [93,102,107].

A number of saturation-mode monolithic MOSFET circuits have been developed specifically as multipliers, with varying degrees of large-signal port linearity. The 1982 circuit of Soo and Meyer [64] used two cascaded source-coupled pairs to realize a linear control law. Another class of multiplier circuits employs cross-coupled MOS source-coupled pairs as G_m circuits [62,68,72,77,108], often labeled a "MOS version of Gilbert's six-transistor multiplier" (a misnomer that, like a few more famous ones in electrical engineering, is steadily becoming entrenched.)[4] This configuration was the starting point for multipliers due to Babanezhad and Temes in 1985 [68], Wong et al. in 1986 [72], and Qin and Geiger in 1987 [77]. This approach can deliver variable gain with gain-independent phase shift, and was used to good effect recently in a communications application with that requirement [108]. Other circuits exploit saturation-mode squarer implementations in various ways [66,69,75,76,87,90,97,105], some of them overlapping the principles of the G_m cells described above.

---

[4] The six-transistor multiplier cell that Gilbert introduced and is known for (the "six-pack" of Section 3) is a current-input current-output translinear core [13,18]. It refers specifically to a different set of six transistors than the MOS authors do, some of whom even cite Gilbert when they use this terminology.

7. FETs in other or mixed modes of operation

Some circuits have exploited a combination of saturated and nonsaturated MOSFETs, adjusting geometries to realize a nonlinearity reduction [109,110,111]. Weakly-inverted MOSFET channels afford an exponential transconductance nonlinearity useful in translinear circuits resembling the classic BJT versions of Section 4. This idea has been explored for many years and has found its chief practicality in the lower-bandwidth applications, since weak inversion entails reduced current densities in the FETs without concomitantly reduced capacitances. Some recent examples of interest in weak-inversion circuits for variable G_m have appeared [112,113]. Special monolithic FET structures have also been proposed for variable-gain service, such as the ultralinear nonsaturation-mode FET structure of Tsividis and Vavelidis [114] and the eight-gate GaAs MESFET attenuator of Miyatsuji and Ueda, with a bandwidth greater than 3 GHz [115].

8. Variable-gain amplifiers for high-frequency applications

Some of the variable-gain circuit techniques described earlier in this chapter are capable of signal frequencies well into the 100s or 1000s of MHz. High-frequency communications systems however impose special requirements on their variable-gain circuits, not only in bandwidth but sometimes in constant port impedances, when these circuits operate in a context of matched impedances such as 50 Ω. Linearity and noise are considerations in common with lower-frequency applications, though differently specified. Perhaps the most profound difference between RF and low-frequency or "baseband" gain control is that there are simply fewer ways to build gain control well at 1000 MHz or 3000 MHz than at 1 MHz. What follows is a set of case studies of solutions to variable-gain requirements for high-frequency, communications-driven tasks.

Abidi in 1984 demonstrated all-MOS transresistance amplifiers at bandwidths of about 1 GHz using nonsaturated MOSFETs as variable shunt-feedback resistors in multistage AGC, as well as to define the transresistance value [116]. Reimann and Rein in 1989 [117] used a four-quadrant form of current steering for gain control in a 2.5 GHz bipolar AGC amplifier.

The 1991 bipolar DC - 1 GHz variable-gain amplifier of Meyer and Mack [118] was designed for 50-Ω source and load in communication systems. This circuit built on the extensive history of current-steering variable-gain amplifiers based on

Figure 2 [14-17]. As with the circuit of Reimann and Rein, its differential current-steering stage fed a shunt-feedback transresistance stage that loaded the current-steering BJTs with a low impedance, reducing the influence of device parasitic capacitances. The input transconductance stage (shown as current sources in Figure 2) operated at an elevated current, for the noise advantage noted earlier in Section 3. This circuit delivered a noise figure at full gain of 9.3 dB at 50 MHz.

A later 50-Ω amplifier in BiCMOS technology by Meyer and Mack [120] used MOSFETs as switches, rather than for variable gain, but with a bare BJT input stage that is the lowest-noise configuration possible in these technologies. The noise figure was 2.2 dB at 900 MHz. Distortion was mitigated by the fact that at such frequencies the BJT input acts as a current-mode amplifier and is relatively linear. Also, this input stage realized a constant 50-Ω input impedance over a wide frequency range by including a 1.5-nH bonding-wire inductance (which is noiseless) in the emitter of the single input BJT; this combined with the falloff of current gain above f_T/β yielded a net AC input impedance both resistive and constant.

MOSFETs employed as switches (for two gain states plus shutdown) in the previous example were later used more actively by Piriyapoksombut for variable gain in the same technology, while the high-frequency amplifier element continued to be a BJT [122]. In this 1995 1-GHz amplifier, two BJT wideband gain stages, again with bonding-wire emitter inductance, each incorporated MOSFET shunt feedback (Figure 12), which permits regulating the RF gain while maintaining constant input impedance.

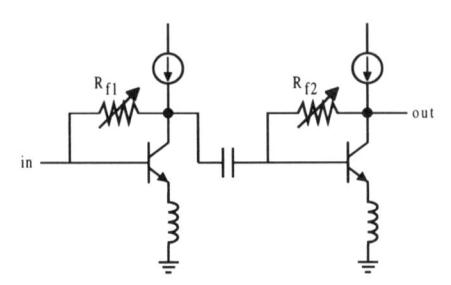

Figure 12: Basis of Piriyapoksombut's wideband AC amplifier using MOS variable resistances to control gain with constant input and output impedances.

Baltus and Tombeur made different use of a MOSFET in a BiCMOS variable-gain amplifier for a 1.9-GHz DECT receiver front end [119]. This MOSFET adjusted the DC collector current of a wideband BJT input stage. The MOSFET was capacitively bypassed so that it did not enter into the RF signal path.

A BiCMOS wideband AGC-transimpedance amplifier due to Meyer and Mack for use as a current-input preamplifier employed multiple nonsaturated MOSFETs in an "automatic-transmission" arrangement. This varied the transresistance of the amplifier continuously with input signal current level, over a range as wide as 70 dB, while varying the resistance of the individual MOSFETs no more than 10:1 [121]. Also, regulating the transresistance simply through a variable feedback resistance around an amplifier tends to cause excessive variation in the loop bandwidth; this was overcome by using additional MOSFETs to shape the loop gain and bandwidth simultaneously with the transresistance.

These high-frequency examples reveal little use of translinear or even generalized-translinear principles (in the usual sense of forming linear gain out of nonlinear devices) other than in low-frequency ancillary circuits in the gain-control paths. The reasons for this are straightforward. The high frequency requirements place a premium on raw amplifier bandwidth, where the translinear circuits, with their inherent multiple active devices and internally nonlinear intermediate signals, tend to deliver less than the maximum usable gain bandwidth of simpler amplifier circuits. Moreover, variable gain in high-frequency communication systems typically occurs in a front-end section needing AGC. Such sections are extremely sensitive to the noise that is added to the signal path by the multiple active devices associated with translinear gain control. These bandwidth and noise constraints subordinate the inherent process-parameter independence and large-signal linearity offered by classic translinear techniques such as gain cells [12].

9. Acknowledgment

The authors wish to thank Yannis P. Tsividis, Mojtaba Atarodi, and Mark N. Seidel for constructive suggestions during the writing of this chapter, and James L. Wallace for production assistance.

10. References

Automatic gain control

[1] Hauser [60], pp. 17-33.

[2] B. M. Oliver. Automatic volume control as a feedback problem. *Proceedings of the IRE* vol. 36 no. 4 pp. 466-473, April 1948.

[3] W. K. Victor and M. H. Brockman. The application of linear servo theory to the design of AGC loops. *Proceedings of the IRE* vol. 48 no. 2 pp. 234-238, February 1960.

[4] R. McFee and S. Dick. Reducing nonlinear distortion due to automatic gain-control circuits. *Electronics Letters* vol. 1 no. 4 pp. 102-104, June 1965.

[5] V. G. Vol'pyan. Synthesis of the characteristics of an automatic control amplifier. *Telecommunications and Radio Engineering* (part 2) vol. 1968 no. 9 pp. 148-150, September 1968. This is the English translation of *Radiotekhnika* vol. 23 no. 9 pp. 103-105, September 1968.

Circuits based on miscellaneous solid-state devices

[6] Allen P. Edwards. Precise, convenient analysis of modulated signals. *Hewlett-Packard Journal* vol. 30 no. 11 pp. 3-18, November 1979. Page 14 in particular.

[7] V. Székely. New type of thermal-function I.C.: the four-quadrant multiplier. *Electronics Letters* vol. 12 no. 15 pp. 372-373, 22 July 1976.

[8] V. N. Kornev. Automatic gain control by thermistors in transistor amplifiers. *Soviet Journal of Instrumentation and Control* vol. 1967 no. 2 pp. 64-65, February 1967. This is the English translation of *Pribori i Systemy Upravleniya* vol. 1967 no. 2 p. 55, February 1967.

[9] Hewlett-Packard Company. The PIN diode. Application note 904, 15 February 1966.

[10] Hewlett-Packard Company. Applications of PIN diodes. Application note 922, 1971.

[11] B. M. Potts. PIN diode weight circuits. Technical note #1979-26, MIT Lincoln Laboratory, Lexington, Massachusetts, 28 December 1979.

Bipolar-transistor circuits

[12] Barrie Gilbert. A new wide-band amplifier technique. *IEEE Journal of Solid-State Circuits* vol. SC-3 no. 4 pp. 353-365, December 1968.

[13] Barrie Gilbert. A precise four-quadrant multiplier with subnanosecond response. *IEEE Journal of Solid-State Circuits* vol. SC-3 no. 4 pp. 365-373, December 1968.

[14] W. Richard Davis and James E. Solomon. A high-performance monolithic IF amplifier incorporating electronic gain control. *IEEE Journal of Solid-State Circuits* vol. SC-3 no. 4 pp. 408-416, December 1968.

[15] Willy M. C. Sansen and Robert G. Meyer. Distortion in bipolar transistor variable-gain amplifiers. *IEEE Journal of Solid-State Circuits* vol. SC-8 no. 4 pp. 275-282, August 1973.

[16] William J. McCalla. An integrated IF amplifier. *IEEE Journal of Solid-State Circuits* vol. SC-8 no.6 pp. 440-447, December 1973.

[17] Willy M. C. Sansen and Robert G. Meyer. An integrated wide band variable-gain amplifier with maximum dynamic range. *IEEE Journal of Solid-State Circuits* vol. SC-9 no. 4 pp. 159-166, August 1974.

[18] Barrie Gilbert. A high-performance monolithic multiplier using active feedback. *IEEE Journal of Solid-State Circuits* vol. SC-9 no. 6 pp. 364-373, December 1974.

[19] Chu-Sun Yen. Distortion in emitter-driven variable-gain pairs. *IEEE Journal of Solid-State Circuits* vol. SC-14 no. 4 pp. 771-773, August 1979.

[20] Barrie Gilbert and Peter Holloway. A wideband two-quadrant multiplier. *1980 IEEE International Solid-State Circuits Conference Digest of Technical Papers* pp.200-201, February 1980.

[21] Chu-Sun Yen. High-frequency distortion in emitter-driven variable-gain pairs. *IEEE Journal of Solid-State Circuits* vol. SC-15 no. 3 pp. 375-377, June 1980.

[22] Un-Ku Moon. Noise in bipolar-transistor variable-gain circuits. Master of Engineering report, Cornell University, Ithaca, New York, May 1989.

[23] B. Gilbert. Translinear circuits: a proposed classification. *Electronics Letters* vol. 11 no. 1 pp. 14-16, 9 January 1975.

[24] A. Grasselli and R. Stefanelli. New analogue multiplier-divider. *Electronics Letters* vol.2 no. 1 pp. 2-3, January 1966. Corrections appear vol. 2 no. 5 p. 188.

[25] D. J. Hamilton and K. B. Finch. A single-ended current gain cell with AGC, low offset voltage, and large dynamic range. *IEEE Journal of Solid-State Circuits* vol. SC-12 no. 3 pp. 322-323, June 1977.

Generalizations of translinearity

[26] Hauser [60], pp. 84-94.

[27] Evert Seevinck and Remco J. Wiegerink. Generalized translinear circuit principle. *IEEE Journal of Solid-State Circuits* vol. 26 no. 8 pp. 1098-1102, August 1991.

[28] R. J. Wiegerink. *Analysis and synthesis of MOS translinear circuits.* Kluwer Academic Publishers, 1993.

Nonsaturation-mode FET circuits

[29] Yannis P. Tsividis. *Operation and Modeling of the MOS Transistor*, p. 122. McGraw-Hill, 1977.

[30] I. C. Hutcheon and D. J. Puddefoot. New solid-state electronic multiplier-divider. *Proceedings of the IEE* (UK) vol. 112 no. 8 pp. 1523-1531, August 1965.

[31] M. M. Abu-Zeid, H. Groendijk, and A. Willemse. Temperature-compensated F.E.T. multiplier. *Electronics Letters* vol. 4 no. 16 pp. 324-325, 9 August 1968.

[32] M. M. Abu-Zeid and H. Groendijk. Field-effect-transistor-bridge multiplier-divider. *Electronics Letters* vol. 8 no. 24 pp. 591-592, 30 November 1972.

[33] T. B. Martin. Circuit applications of the field-effect transistor. *Semiconductor Products* vol. 5 pp. 30-38, March 1962. Cited in Bilotti [36].

[34] W. Y. Elliott, Jr. Field effect transistor as a linear variable resistance. *IBM Technical Disclosure Bulletin* vol. 7 no. 1 p. 111, June 1964.

[35] Carl David Todd. FETs as voltage-variable resistors. *Electronic Design* vol. 13 no. 19 pp. 66-69, 13 September 1965.

[36] Alberto Bilotti. Operation of a MOS transistor as a variable resistor. *Proceedings of the IEEE* vol. 54 no. 8 pp. 1093-1094, August 1966.

[37] H. P. von Ow. Reducing distortion in controlled attenuators using FET. *Proceedings of the IEEE* vol. 56 no. 10 pp. 1718-1719, October 1968.

[38] Howard N. Leighton. An optimized gain-control configuration using the field-effect transistor. *IEEE Journal of Solid-State Circuits* vol. SC-3 no. 4 pp. 441-447, December 1968.

[39] H. P. von Ow. A simple circuit using FETs for fading audio signals in and out exponentially in time. *Proceedings of the IEEE* vol. 57 no. 2 pp. 222-224, February 1969.

[40] Mihai Banu and Yannis Tsividis. Fully integrated active RC filters in MOS technology. *IEEE Journal of Solid-State Circuits* vol. SC-18 no. 6 pp. 644-651, December 1983.

[41] Yannis Tsividis, Mihai Banu, and John Khoury. Continuous-Time MOSFET-C filters in VLSI. *IEEE Transactions on Circuits and Systems* vol. 33 no. 2 pp. 125-140, February 1986.

[42] Arthur W. Crooke and Horst A. R. Wegener. Programmable general-purpose analog filter. US Patent no. 3,987,293, 19 October 1976.

[43] Jay P. Sage and Arthur M. Cappon. CCD analog-analog correlator with four-FET bridge multipliers. *Proceedings of the 1979 International Conference on Solid-State Devices*, Tokyo, 1979; *Japanese Journal of Applied Physics* vol. 19 Supplement 19-1 pp. 265-268, 1980.

[44] Zdzislaw Czarnul. Modification of Banu-Tsividis continuous-time integrator structure. *IEEE Transactions on Circuits and Systems* vol. 33 no. 7 pp. 714-716, July 1986.

[45] Bang-Sup Song. CMOS RF circuits for data communications applications. *IEEE Journal of Solid-State Circuits* vol. SC-21 no. 2 pp. 310-317, April 1986.

[46] Gert Groenewold. Optimal dynamic range integrators. *IEEE Transactions on Circuits and Systems-1: Fundamental Theory and Applications* vol. 39 no. 8 pp. 614-627, August 1992.

[47] Gert Groenewold, Bert Monna, and Bram Nauta. Micro-power analog-filter design. In Rudy J. van de Plassche, Willy M. C. Sansen, and Johan H. Huijsing, editors, *Analog Circuit Design: Low-Power Low-Voltage, Integrated Filters and Smart Power*, pp. 73-88. Kluwer Academic Publishers, 1995.

[48] Gert Groenewold. *Optimal Dynamic Range Integrated Continuous-Time Filters*. PhD thesis, Delft University of Technology, 1992.

[49] Gert Groenewold and W. J. Lubbers. Systematic distortion analysis for MOSFET integrators with use of a new MOSFET model. *IEEE Transactions on Circuits and Systems-II* vol. 41 no. 9 pp. 569-580, September 1994.

[50] W. H. Highleyman and E. S. Jacob. An analog multiplier using two field effect transistors. *IRE Transactions on Communications Systems* vol. CS-10 no. 3 pp. 311-317, September 1962.

[51] S. Østerfjells. Analog multiplier with field effect transistors. *Proceedings of the IEEE* vol. 53 no. 5 p. 521, May 1965.

[52] Patrick Bosshart. An integrated analog correlator using charge-coupled devices. *1976 IEEE International Solid-State Circuits Conference Digest of Technical Papers* pp. 198-199, 20 February 1976.

[53] Patrick William Bosshart. A monolithic analog correlator using charge-coupled devices, Chapter 4. Master's thesis, Massachusetts Institute of Technology, February 1976.

[54] J. Mavor, J. W. Arthur, and P. B. Denyer. Analogue C.C.D. correlator using monolithic M.O.S.T. multipliers. *Electronics Letters* vol. 13 no. 12 pp. 373-374, 9 June 1977. Note figures 3a, 3b interchanged.

[55] D. V. McCaughan, J. C. White, and J. R. Hill. A novel C.C.D. multiplying input technique. *Electronics Letters* vol. 14 no. 6 pp. 165-166, 16 March 1978.

[56] Tadayoshi Enomoto, Masa-Aki Yasumoto, Tsutomu Ishihara, and Kohjiro Watanabe. Monolithic analog adaptive equalizer integrated circuit for wide-band communication networks. *IEEE Journal of Solid-State Circuits* vol. SC-17 no. 6 pp. 1045-1054, December 1982.

[57] Tadayoshi Enomoto and Masa-Aki Yasumoto. Integrated MOS four-quadrant analog multiplier using switched capacitor technology for analog signal processor ICs. *IEEE Journal of Solid-State Circuits* vol. SC-20 no. 4 pp. 852-859, August 1985.

[58] J. L. Pennock. CMOS triode transconductor for continuous-time active integrated filters. *Electronics Letters* vol. 21 no. 18 pp. 817-818, 29 August 1985.

[59] Roberto Alini, Andrea Baschirotto, and Rinaldo Castello. Tunable BiCMOS continuous-time filter for high-frequency applications. *IEEE Journal of Solid-State Circuits* vol. 27 no. 12 pp. 1905-1915, December 1992.

[60] Max W. Hauser. Large-signal electronically variable gain techniques. Master's thesis, Massachusetts Institute of Technology, December 1981.

Saturation-mode FET circuits

[61] Hiroaki Ikeda. MOST variable resistor and its application to an AGC amplifier. *IEEE Journal of Solid-State Circuits* vol. SC-5 no. 1 pp. 43-45, February 1970.

[62] I. Házman. Four-quadrant multiplier using M.O.S.F.E.T. differential amplifiers. *Electronics Letters* vol. 8 no. 3 pp. 63-65, 10 February 1972.

[63] O. A. Seriki and R. W. Newcomb. Direct coupled MOS squaring circuit. *IEEE Journal of Solid-State Circuits* vol. SC-14 no. 4 pp. 766-768, August 1979.

[64] D. C. Soo and R. G. Meyer. A four-quadrant NMOS analog multiplier. *IEEE Journal of Solid-State Circuits* vol. SC-17 no. 6 pp. 1174-1178, December 1982.

[65] A. Nedungadi and T. R. Viswanathan. Design of linear CMOS transconductance elements. *IEEE Transactions on Circuits and Systems* vol. CAS-31 no. 10 pp. 891-894, October 1984.

[66] Z. Hong and H. Melchior. Analogue four-quadrant CMOS multiplier with resistors. *Electronics Letters* vol. 21 no. 12 pp. 531-532, June 1985.

[67] R. R. Torrance, T. R. Viswanathan, and J. V. Hanson. CMOS voltage to current transducers. *IEEE Transactions on Circuits and Systems* vol. CAS-32 no.11 pp. 1097-1104, November 1985.

[68] J. N. Babanezhad and G. C. Temes. A 20-V four-quadrant CMOS analog multiplier. *IEEE Journal of Solid-State Circuits* vol. SC-20 no. 6 pp. 1158-1168, December 1985.

[69] K. Bult and H. Wallinga. A CMOS four-quadrant analog multiplier. *IEEE Journal of Solid-State Circuits* vol. SC-21 no. 3 pp. 430-435, June 1986.

[70] T. R. Viswanathan. CMOS transconductance element. *Proceedings of the IEEE* vol. 74 no.1 pp. 222-224, January 1986.

[71] C. S. Park and R. Schaumann. A high-frequency CMOS linear transconductance element. *IEEE Transactions on Circuits and Systems* vol. CAS-33 no. 11 pp. 1132-1138, November 1986.

[72] S. L. Wong, N. Kalyanasundaram, and C. A. T. Salama. Wide dynamic range four-quadrant CMOS analog multiplier using linearized transconductance stages. *IEEE Journal of Solid-State Circuits* vol. SC-21 no. 6 pp. 1120-1122, December 1986.

[73] S. L. Garverick and C. G. Sodini. Large-signal linearity of scaled MOS transistors. *IEEE Journal of Solid-State Circuits* vol. SC-22 no. 2 pp. 282-286, April 1987.

[74] K. Bult and H. Wallinga. A class of analog CMOS circuits based on the square-law characteristic of a MOS transistor in saturation. *IEEE Journal of Solid-State Circuits* vol. SC-22 no. 3 pp. 357-365, June 1987.

[75] E. Seevinck and R. F. Wassenaar. A versatile CMOS linear transconductor/square-law function circuit. *IEEE Journal of Solid-State Circuits* vol. SC-22 no. 3 pp. 366-377, June 1987.

[76] J. S. Peña-Finol and J. A. Connelly. A MOS four-quadrant analog multiplier using the quarter-square technique. *IEEE Journal of Solid-State Circuits* vol. SC-22 no. 6 pp. 1064-1073, December 1987.

[77] S. Qin and R. L. Geiger. A ±5-V CMOS analog multiplier. *IEEE Journal of Solid-State Circuits* vol. SC-22 no. 6 pp. 1143-1146, December 1987.

[78] K. Bult. *Analog CMOS square-law circuits*. PhD thesis, University of Twente, Enschede, The Netherlands, 1988.

[79] B. Nauta and E. Seevinck. Linear CMOS transconductance element for VHF filters. *Electronics Letters* vol. 25 no. 7 pp. 448-450, March 1989.

[80] E. A. M. Klumperink and E. Seevinck. Linear-gain amplifier arrangement. Patent application no. 78102418, Taiwan, 28 March 1989.

[81] E. A. M. Klumperink, E. v. d. Zwan, and E. Seevinck. CMOS linear transconductor circuit with constant bandwidth. *Electronics Letters* vol. 25 no. 10 pp. 675-676, May 1989.

[82] E. A. M. Klumperink and E. Seevinck. MOS current gain cells with electronically variable gain and constant bandwidth. *IEEE Journal of Solid-State Circuits* vol. SC-24 no. 5 pp. 1465-1467, October 1989.

[83] S. Noceti Filho, M. C. Schneider, and R. N. G. Robert. New CMOS OTA for fully integrated continuous-time circuit applications. *Electronics Letters* vol. 25 no. 24 pp. 24-26, November 1989.

[84] Z. Wang. Novel linearisation technique for implementing large-signal MOS tunable transconductors. *Electronics Letters* vol. 26 no. 2 pp. 138-139, January 1990.

[85] Z. Wang and W. Guggenbühl. A voltage-controllable linear MOS transconductor using bias offset technique. *IEEE Journal of Solid-State Circuits* vol. SC-25 no.1 pp. 315-317, February 1990.

[86] G. Wilson and P. K. Chan. Low-distortion CMOS transconductor. *Electronics Letters* vol. 26 no.11 pp. 720-722, May 1990.

[87] H. Song and C. Kim. An MOS four-quadrant analog multiplier using simple two-input squaring circuits with source followers. *IEEE Journal of Solid-State Circuits* vol. 25 no. 3 pp. 841-848, June 1990.

[88] Z. Czarnul and S. Takagi. Design of linear tuneable CMOS differential transconductor cells. *Electronics Letters* vol. 26 no. 21 pp. 1809-1811, October 1990.

[89] Z. Czarnul and N. Fujii. Highly-linear transconductor cell realised by double MOS transistor differential pairs. *Electronics Letters* vol. 26 no. 21 pp. 1819-1821, October 1990.

[90] P. J. Langlois. Comment on "A CMOS four-quadrant multiplier": effects of threshold voltage. *IEEE Journal of Solid-State Circuits* vol. SC-25 no. 6 pp. 1595-1597, December 1990.

[91] S. T. Dupuie and M. Ismail. High frequency CMOS transconductors. Chapter 5 of C. Toumazou et al., editors, *Analogue IC Design: The Current-Mode Approach*. Peter Peregrinus Lts., London, 1990.

[92] G. Wilson and P.K. Chan. Saturation-mode CMOS transconductor with enhanced tunability and distortion. *Electronics Letters* vol. 27 no. 1 pp. 27-29, January 1991.

[93] E. A. M. Klumperink and H. J. Janssen. Complementary CMOS current gain cell. *Electronics Letters* vol. 27 no. 1 pp. 38-39, January 1991.

[94] S. Szczepanski, R. Schaumann, and P. Wu. Linear transconductor based on crosscoupled CMOS pairs. *Electronics Letters* vol. 27 no. 9 pp. 783-785, April 1991.

[95] W. J. Adams and J. Ramirez-Angulo. Extended transconductance adjustment/linearization technique. *Electronics Letters* vol. 27 no. 10 pp. 842-844, May 1991.

[96] P. Wu and R. Schaumann. Tunable operational transconductance amplifier with extremely high linearity over very large input range. *Electronics Letters* vol. 27 no. 14 pp. 1254-1255, July 1991.

[97] Z. Wang. A CMOS four-quadrant analog multiplier with single-ended voltage output and improved temperature performance. *IEEE Journal of Solid-State Circuits* vol. SC-26 no. 9 pp. 1293-1301, September 1991.

[98] M. C. H. Cheng and C. Toumazou. Linear composite MOSFETs (COMFETs). *Electronics Letters* vol. 27 no. 20 pp. 1802-1804, September 1991.

[99] J. Sevenhans and M. Van Paemel. Novel CMOS linear OTA using feedback control on common source node. *Electronics Letters* vol. 27 no. 20 pp. 1873-1875, September 1991.

[100] E. J. van der Zwan, E. A. M. Klumperink, and E. Seevinck. A CMOS OTA for HF filters with programmable transfer function. *IEEE Journal of Solid-State Circuits* vol. 26 no. 11 pp. 1720-1724, November 1991.

[101] S. Szczepanski, J. Jakusz, and A. Czarniak. Differential pair transconductor linearisation via electronically controlled current-mode cells. *Electronics Letters* vol. 28 no. 12 pp.1093-1095, June 1992.

[102] Z. Wang. Two CMOS large current-gain cells with linearly variable gain and constant bandwidth. *IEEE Transaction on CAS I: Fundamental Theory and Applications* vol. CAS-39 no. 12 pp. 1021-1024, December 1992.

[103] C. S. Kim, Y. H. Kim, and S. B. Park. New CMOS linear transconductor. *Electronics Letters* vol. 28 no. 21 pp. 1962-1964, 8 October 1992.

[104] R. Raut. A novel VCT for analog IC applications. *IEEE Transactions on Circuits and Systems-II: Analog and Digital Signal Processing* vol. 39 no. 12 pp. 882-883, December 1992.

[105] C. W. Kim and S. B. Park. Design and implementation of a new four-quadrant MOS analog multiplier. *Analog Integrated Circuits and Signal Processing* no. 2 pp. 195-103, 1992.

[106] S. Szczepanski, A. Wyszynski, and R. Schaumann. Highly linear voltage-controlled CMOS transconductors. *IEEE Transactions on Circuits and Systems-I: Fundamental Theory and Applications* vol. CAS-40 no. 4, April 1993.

[107] E. A. M. Klumperink. Cascadable CMOS current gain cell with gain insensitive phase shit. *Electronics Letters* vol. 29 no. 23 pp. 2027-2028, November 1993.

[108] E. A. M. Klumperink, C. T. Klein, B. Rüggeberg, and A. J. M. van Tuijl. AM suppression with low AM-PM conversion with the aid of a variable-gain amplifier. *IEEE Journal of Solid-State Circuits* vol. 31 no. 5 pp. 625-633, May 1996.

FET circuits in other or mixed modes of operation

[109] F. Krummenachter and N. Joehl. A 4-MHz CMOS continuous-time filter with on-chip automatic tuning. *IEEE Journal of Solid-State Circuits* vol. 23 no. 3, June 1988.

[110] P. M. Van Peteghem, F. G. L. Rice, and S. Lee. Design of a very linear CMOS transconductance input stage for continuous-time filters. *IEEE Journal of Solid State Circuits* pp. 497-501, April 1990.

[111] J. Silva-Martinez, M. S. J. Steyaert, and W. M. C. Sansen. A large-signal very low-distortion transconductor for high-frequency continuous-time filters. *IEEE Journal of Solid-State Circuits* vol. SC-26 no. 7, July 1991.

[112] I. E. Opris and G. T. A. Kovacs. Large-signal subthreshold CMOS transconductance amplifier. *Electronics Letters* vol. 31 no. 9 pp. 718-720, April 1995.

[113] P. M. Furth and A. G. Andreou. Linearised differential transconductors in subthreshold CMOS. *Electronics Letters* vol. 31 no. 7 pp. 545-546, March 1995.

[114] Y. Tsividis and K. Vavelidis. Linear, electronically tunable resistor. *Electronics Letters* vol. 28 no. 25 pp. 2303-2305, 3 December 1992. Annotation appears vol. 29 no. 6, 18 March 1993.

[115] Kazuo Miyatsuji and Daisuke Ueda. A low-distortion GaAs variable attenuator IC for digital mobile communication system. *1995 IEEE International Solid-State Circuits Conference Digest of Technical Papers* pp.42-43, February 1995.

High-frequency variable-gain circuits

[116] Asad A. Abidi. Gigahertz transresistance amplifiers in fine line NMOS. *IEEE Journal of Solid-State Circuits* vol. SC-19 no. 6 pp. 986-994, December 1984.

[117] Reinhard Reimann and Hans-Martin Rein. A single-chip bipolar AGC amplifier with large dynamic range for optical-fiber receivers operating up to 3 Gbit/s. *IEEE Journal of Solid-State Circuits* vol. 24 no. 6 pp. 1744-1748, December 1989.

[118] Robert G. Meyer and William D. Mack. A DC to 1-GHz differential monolithic variable-gain amplifier. *IEEE Journal of Solid-State Circuits* vol. 26 no. 11 pp. 1673-1680, November 1991.

[119] Peter Baltus and Anton Tombeur. DECT zero IF receiver front end. In Willy Sansen, Johan H. Huijsing, and Rudy J. van de Plassche, editors, *Analog Circuit Design: Mixed A/D Circuit Design, Sensor Interface Circuits, and Communication Circuits* pp. 295-318. Kluwer Academic Publishers, 1994.

[120] Robert G. Meyer and William D. Mack. A 1-GHz BiCMOS RF front-end IC. *IEEE Journal of Solid-State Circuits* vol. 29 no. 3 pp. 350-355, March 1994.

[121] Robert G. Meyer and William D. Mack. A wideband low-noise variable-gain BiCMOS transimpedance amplifier. *IEEE Journal of Solid-State Circuits* vol. 29 no. 6 pp. 701-706, June 1994.

[122] Pramote Piriyapoksombut. A radio frequency variable-gain amplifier. Master of Science report, University of California at Berkeley, May 1995.

CMOS Translinear Circuits

Evert Seevinck

*Philips Research Laboratories
Eindhoven, The Netherlands*

Abstract*

This paper addresses the design of translinear circuits in the context of CMOS-technology. First, extension of the translinear circuit principle to implementation by MOS transistors operating in strong inversion is reviewed.

Second, the supply voltage requirements are discussed and MOS-translinear circuits for low supply voltage are considered. A circuit technique suitable for the minimum possible supply-voltage is investigated. The key to this approach is allowing current-source transistors to operate in the linear region and viewing these transistors as consisting of two saturated transistors acting as current sources and connected in anti-parallel.

---

* Part of this work was performed at the Department of Electrical and Electronic Engineering, University of Pretoria, South Africa, while on leave from Philips Research Laboratories during 1995.

1. Introduction

For a long time analog circuits were limited to linear signal processing; in fact, initially they were called linear circuits. The missing operations of multiplication and division were first provided in a practical way by Gilbert when he introduced translinear (TL) techniques [1], [2]. The logarithmic voltage - current characteristic of bipolar devices was the key. Linear operations performed on base-emitter voltages transform to products and quotients of collector currents. The TL principle paved the way to analog circuits of rare elegance, precision and economy. For probably the first time the nonlinear transistor properties were constructively used instead of being treated as undesirable deviations from ohm's law.

The term translinear is a contraction of transconductance linear with current, which is a fundamental property of bipolar transistors, thus

$$g_m = \frac{dI}{dV} = \alpha I \tag{1}$$

Integrating, we find

$$V = \frac{1}{\alpha} \ln\left(\frac{I}{I_s}\right) \tag{2}$$

with I_s the integration constant. If we identify α with q/kT, (2) is recognized as the logarithmic bipolar relationship leading to TL circuits.

The emergence of MOS circuit-technology has prompted the question whether nonlinear and computational analog MOS circuits comparable to bipolar TL circuits can be realized. Obviously, this is possible when the MOS transistors are restricted to weak-inversion operation since then the voltage-current characteristic is also logarithmic [3], [4]. For high-speed applications however, circuits using MOS transistors in strong inversion would be interesting.

2. MOS translinear circuits

If we extend the meaning of the concept "translinear" to include devices having transconductance linear with voltage rather than current, we have

$$g_m = \frac{dI}{dV} = \beta V \tag{3}$$

integrating as before, we obtain

$$V = \sqrt{\frac{2I}{\beta}} \tag{4}$$

where the integration constant was taken equal to zero. This square-root relationship represents a MOS transistor operating in strong inversion and saturation, with I the drain current and V the gate-source drive voltage $V_{gs} - V_{th}$. A broadened interpretation of the concept "translinear" thus leads naturally to MOS circuits in strong inversion [5]. Since operation is based on the quadratic MOS-characteristic, the term "quadratic translinear" (QTL) is proposed for these circuits. It remains to explore the nature of the signal relationships.

Consider a simple example of a MOS-TL circuit as shown in fig. 1. This is a direct analogue of a bipolar-TL circuit.

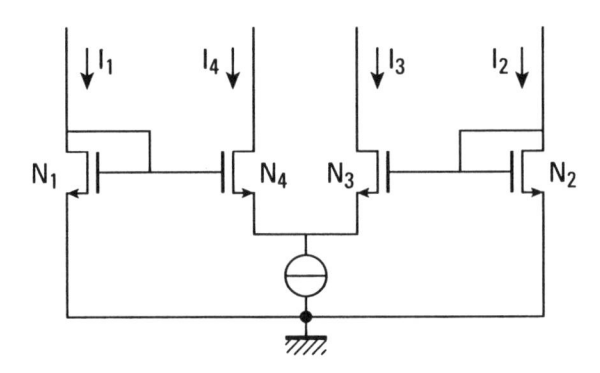

Fig. 1: Example of a MOS-TL circuit.

The circuit consists of four NMOS transistors arranged in a loop of gate-source voltages. Therefore we have

$$V_{gs1} + V_{gs3} = V_{gs2} + V_{gs4} \tag{5}$$

Combining this with (4) and taking $V = V_{gs} - V_{th}$:

$$\sqrt{\frac{2I_1}{\beta_1}} + V_{th} + \sqrt{\frac{2I_3}{\beta_3}} + V_{th} = \sqrt{\frac{2I_2}{\beta_2}} + V_{th} + \sqrt{\frac{2I_4}{\beta_4}} + V_{th} \qquad (6)$$

Noting that $\beta = \frac{W}{L} \mu C_{ox}$ and assuming matched devices, (6) reduces to

$$\sqrt{\frac{I_1}{W_1/L}} + \sqrt{\frac{I_3}{W_3/L}} = \sqrt{\frac{I_2}{W_2/L}} + \sqrt{\frac{I_4}{W_4/L}} \qquad (7)$$

This result can be easily generalized. Consider any loop of MOS transistors operating in strong inversion and saturation where the gate-source voltages are connected in series and with equal numbers of transistors arranged clockwise and counter-clockwise. The last-mentioned requirement is essential since it allows the threshold voltage terms to be cancelled, as in (6). The general statement of the quadratic translinear (QTL) principle then follows as [5]

$$\sum_{cw} \sqrt{\frac{I_d}{W/L}} = \sum_{ccw} \sqrt{\frac{I_d}{W/L}} \qquad (8)$$

where the subscripts cw and ccw indicate the devices connected clockwise and counterclockwise in the loop, respectively. Note that all temperature - and process-dependent quantities have disappeared.

It is interesting to compare (8) with the original TL principle which is expressed by [1], [2]

$$\prod_{cw} \frac{I_c}{A} = \prod_{ccw} \frac{I_c}{A} \qquad (9)$$

with I_c the collector currents and A the emitter areas of the bipolar transistors comprising the TL loop. This product relation lends itself naturally to functions involving multiplication or division. The more peculiar QTL sum-of-roots relation (8) may be more suitable for synthesizing square - law functions or power series.

An important point of comparison is the quality of function implementation. Bipolar-TL circuits will probably be superior owing to the unprecedented law conformance of bipolar transistors over wide ranges of current. For MOS devices the current range for square - law behavior is much smaller. It is bounded at the low end by weak inversion and at the high end by mobility reduction. Also, device matching is better for bipolar devices. On the other hand, one can expect the drive to submicron VLSI to give a spin-off of better large-geometry MOS-device matching. In addition, the zero DC gate current of MOS transistors is a distinct advantage.

3. Nonlinear function synthesis by QTL circuits

Prescribed nonlinear functions can be synthesized by QTL circuits. First, the function has to be expressed in terms of square-roots to be compatible with (8). As an example, consider the geometric-mean function $z = \sqrt{xy}$. We need to manipulate this function into sums of square-roots:

$$\begin{aligned} z &= \sqrt{xy} \\ x+y+2z &= x + y + 2\sqrt{xy} \\ \sqrt{x+y+2z} &= \sqrt{x} + \sqrt{y} \\ 2\sqrt{\frac{x+y+2z}{4}} &= \sqrt{x} + \sqrt{y} \end{aligned} \qquad (10)$$

This is in the required form to correspond with the QTL principle (8). A straightforward circuit implementation is shown in fig. 2 [6]. The figures in brackets indicate relative aspect-ratios.

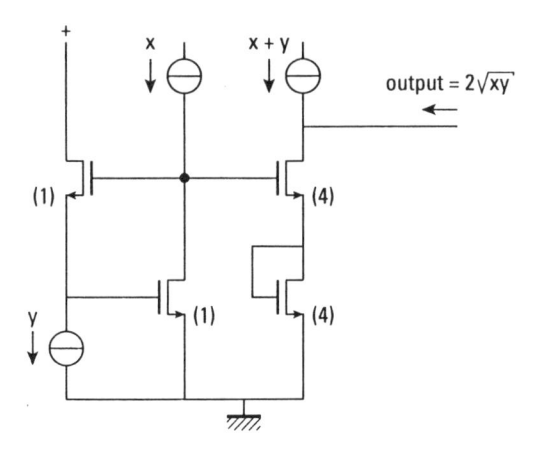

Fig. 2 QTL circuit synthesizing the geometric - mean function $\sqrt{xy}$

This example shows that product forms can also be realized. Although in this case it was easy to find a suitable mathematical form in an analytical way, this is by no means generally true. Wiegerink has introduced CAD-techniques to aid the process of finding sum-of-roots approximations to arbitrary functions [6]. This work is an important step toward fast, interactive design of QTL circuits and should be extended to general TL application.

The circuit of fig. 2 illustrates a second kind of TL loop when compared with fig. 1. Gate-source voltages are stacked instead of being arranged in an up-down sequence. Stacked circuits are adversely affected by the body-effect unless the individual wells are connected to the sources. When this is done however, the large well-capacitances will slow down the circuit. This drawback does not apply to the up-down topology since opposing pairs of transistors have identical source voltage and therefore the body-effect will cancel pair-wise. Another advantage of the up-down topology is the suitability for low supply-voltage. This point will be explored next.

4. Low-voltage MOS-TL circuits

There is a steady trend to lower supply voltages for portable systems. Future integrated circuits will have to operate down to the lowest possible supply voltage. TL circuits included in such ICs will have to conform to this trend. It was mentioned before that the up-down topology shown again in fig. 3 is more suitable for low supply voltage than the stacked topology.

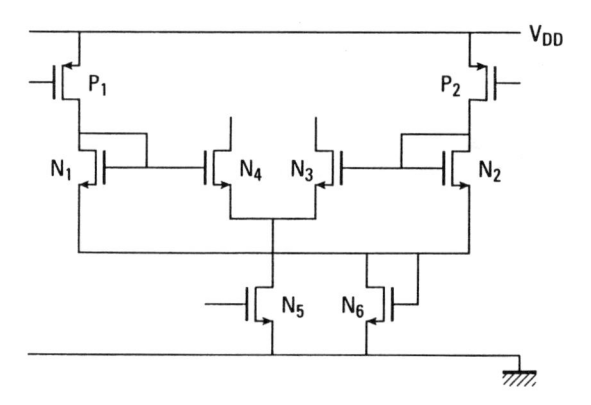

Fig. 3 MOS-TL circuit with up-down topology

Transistors N_5, P_1, P_2 act as current sources and N_6 provides a DC level-shift voltage to ensure that N_5 will always operate in the saturation region. The lowest supply voltage is given by $2V_{gs} + V_{dsat}$. This limit is too high since in general the lowest possible supply voltage for CMOS circuits is $V_{gs} + V_{dsat}$, just sufficient for a saturated transistor to drive a diode-connected transistor. The limit can be lowered somewhat by replacing N_6 by an on-chip reference voltage just high enough to ensure the current-source operation of N_5. However, this will complicate the circuit and the desired minimum of $V_{gs} + V_{dsat}$ will still not be attained.

In addition to the stacked and the up-down topologies there is a third possibility: the simulated TL loop [5], [6]. The simplest version is shown in fig. 4.

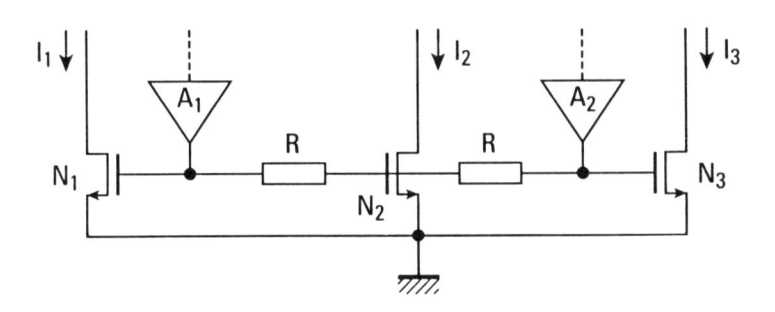

Fig. 4 A simulated - loop MOS-TL circuit.

The buffer amplifiers A_1 and A_2 are suitably connected to force two of the three drain currents. The two resistors are equal, therefore the gate-source voltage of N_2 will be equal to the average of the N_1 and N_3 gate-source voltages. It follows that

$$V_{gs2} = \frac{1}{2}(V_{gs1} + V_{gs3})$$

Therefore, assuming strong inversion

$$\sqrt{\frac{2I_2}{\beta_2}} + V_{th} = \frac{1}{2}\left\{\sqrt{\frac{2I_1}{\beta_1}} + V_{th} + \sqrt{\frac{2I_3}{\beta_3}} + V_{th}\right\}$$

This simplifies to a QTL relation like (7) or (8):

$$2\sqrt{\frac{I_2}{W_2/L}} = \sqrt{\frac{I_1}{W_1/L}} + \sqrt{\frac{I_3}{W_3/L}} \tag{11}$$

When compared to the stacked (fig. 2) and up-down (fig. 3) topologies, the simulated-loop circuit appears the most attractive for low-voltage use. Unfortunately, the need for resistors and buffer amplifiers significantly complicate the circuit and in addition will reduce the bandwidth.

A simple MOS-TL circuit compatible with the minimum possible supply voltage is obtained when the level-shift transistor N_6 of fig. 3 is replaced by a short-circuit. But then N_5 will no longer operate as a current source since it will not be saturated. In fact, it will be biased deeply into the "linear" or unsaturated region and will act as a nonlinear resistor instead of a current source as is required. The consequences of this will be considered next.

5. Current-source behaviour of MOS-transistors in linear region

Changing the point of view will lead to a solution [7]. A transistor operating in the linear region is in fact two saturated transistors acting as current-sources and connected in anti-parallel, as is shown next with the aid of fig. 5

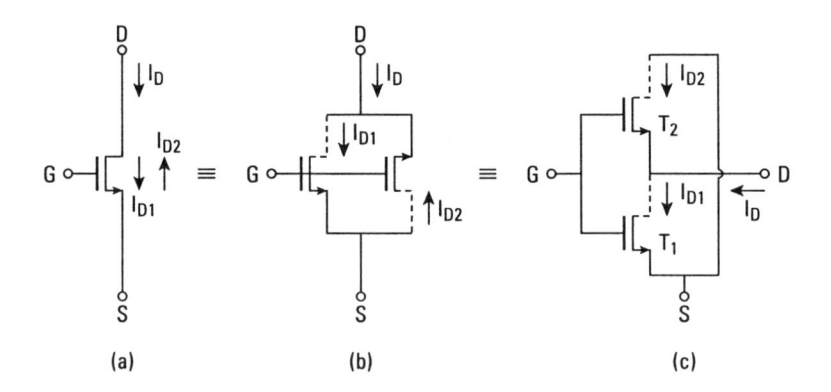

Fig. 5 A linear - region MOS transistor is equivalent to two saturated transistors in anti-parallel

In fig. 5 (a) the NMOS-transistor is biased in the linear region and therefore both V_{gs} and V_{gd} are larger than the threshold voltage. The net drain current can then be viewed as consisting of two anti-parallel components:

$$I_d = I_{d1} - I_{d2} = f(V_{gs}) - f(V_{gd}) \tag{12}$$

Each of these current components relate to a transistor operating in saturation and with controlling (gate-source) voltage V_{gs} and V_{gd} respectively. This is symbolically shown in fig. 5 (b) where the linear-region transistor is decomposed into two identical saturated transistors connected in anti-parallel. The dashed drain symbols indicate that these are fictitious devices in the sense that they operate in saturation notwithstanding the wrong polarity of the drain-gate voltages. The function f in (12) is identical for the two components owing to MOS-transistor symmetry. The form of f is the normal MOS-relation for saturated operation. For strong inversion it is the well-known quadratic relation and for operation in weak inversion it is an exponential function. A symmetrical expression like (12) has recently been used to implement a linear current division technique [8].

This view of the MOS-transistor is similar to the Ebers - Moll model for bipolar transistors which represents a saturated (different meaning) transistor as two anti-parallel component - transistors.

Combining (12) with the MOS quadratic relation for strong inversion, and using $V_{gd} = V_{gs} - V_{ds}$, it is easy to verify the decomposition:

$$\begin{aligned} I_d &= \frac{\beta}{2}(V_{gs} - V_{th})^2 - \frac{\beta}{2}(V_{gd} - V_{th})^2 \\ &= \frac{\beta}{2}[V_{gs} - V_{th} - (V_{gd} - V_{th})](V_{gs} - V_{th} + V_{gd} - V_{th}) \\ &= \beta V_{ds}(V_{gs} - V_{th} - \frac{1}{2}V_{ds}) \end{aligned} \tag{13}$$

This is the relationship for the linear region. A graphical representation of this decomposition is given in [9].

Finally, fig. 5 (c) is the same as (b) but drawn slightly different to be more suitable for use in circuits.

The equivalence of fig. 5 also holds for bipolar transistors via the Ebers - Moll model, as mentioned before. However, practical application will be severely hampered by the fact that while MOS - transistors are perfectly symmetrical structures, this is not true for bipolar transistors. The low inverse beta and speed are serious limitations. Therefore, only MOS - circuits will be considered.

6. MOS-TL circuits for minimum supply-voltage.

Fig. 5 (c) is the key to MOS-TL circuits for minimum supply-voltage. The linear-region transistor appears as a saturated transistor T_2 biased by current source T_1. Referring to fig. 3 with N_6 replaced by a short-circuit, N_4 can be replaced by T_2 and N_5 by T_1. The resulting circuit is shown in fig. 6.

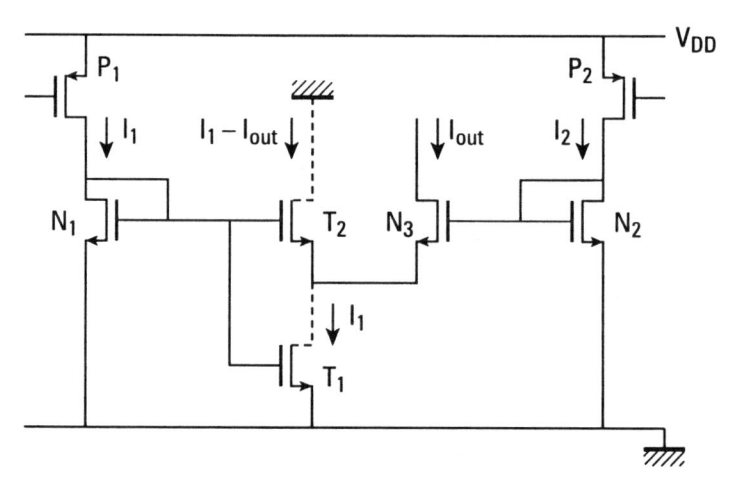

Fig. 6 Circuit of fig. 3 combined with fig. 5 (c)

Next, T_1 and T_2 are replaced by the constituent single transistor of fig. 5 (a). This sequence of replacements results in the generic minimum - voltage MOS-TL circuit shown in fig. 7.

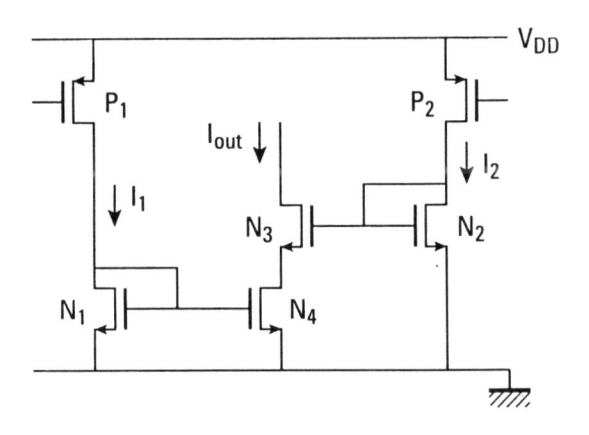

Fig. 7: Generic minimum-voltage MOS-TL circuit equivalent to fig. 6

The circuit was previously proposed for computing signal correlation [7] and it is very economical: N_4 performs the functions of two transistors T_1 and T_2, and N_6 has been eliminated. Note also that minimum supply-voltage operation down to $V_{dd} = V_{gs} + V_{dsat}$ is made possible by the fact that N_4 operates in the linear region. Rather than a drawback, the linear region is used constructively; in fact, it is the key to successful operation.

To analyse the circuit function of fig. 7, the equivalent version of fig. 6 is used. Assume for simplicity equal-sized NMOS transistors operating in weak inversion. Recall that the fictitious transistors T_1 and T_2 are saturated, therefore N_1 and T_1 form a current mirror and T_1 passes current I_1. It follows that the current of T_2 is $I_1 - I_{out}$. The original TL principle (9) is next applied to the loop N_1, T_2, N_3, N_2 because weak inversion is assumed.

$$I_{N1} \cdot I_{N3} = I_{N2} \cdot I_{T2}$$

$$I_1 \cdot I_{out} = I_2 \cdot (I_1 - I_{out})$$

Solving for I_{out}

$$I_{out} = \frac{I_1 \cdot I_2}{I_1 + I_2} \qquad (14)$$

This result shows that the generic minimum-voltage building block of fig. 7 implements the harmonic - mean function when the transistors operate in weak inversion [7]. This function is useful in high-performance class AB circuits

[10]. CMOS class AB output stages based on fig. 7 and suitable for minimum supply - voltage were recently investigated. Very promising results were obtained. For strong-inversion operation, fig. 7 can be analyzed using the QTL principle (8). This results in a more complicated transfer function which is difficult to interpret. The behavior is similar to (14), however. This can be seen by inspection of fig. 7. For I_1 and I_2 equal, I_{out} will be equal to one-half the input current. When one of the input currents is much larger than the other, I_{out} becomes equal to the smaller of the two, thus performing an approximate minimum - function. These conclusions are in agreement with (14) and are independent of the mode of operation (strong or weak inversion).

Class AB techniques thus become possible and can help to maintain acceptable dynamic range in future low-voltage circuits [11]. The generic circuit of fig. 7 could also be useful in implementing other nonlinear and computational functions.

7. Conclusions

Translinear techniques, which were originally formulated for bipolar technology, can also be realized by CMOS circuits. This follows from a broadened interpretation of the concept "translinear". In weak inversion the original translinear product-of-currents functions are generated. In strong inversion however, this is modified into sums of square-roots of currents for which the term quadratic translinear (QTL) is proposed.

Presently-used TL circuits are not very suitable for minimum supply-voltage. A design technique for MOS-TL circuits suitable for minimum supply-voltage has been investigated. The key to this approach is allowing current-source transistors to operate in the linear region and viewing these transistors as consisting of two saturated transistors acting as current sources and connected in anti-parallel. This results in a simple generic minimum-voltage MOS-TL building block which implements the harmonic-mean function, useful in class AB circuits. Additional nonlinear functions and applications are presently being studied.

Acknowledgements

Thanks are due to M. du Plessis, L. Snijman, T-H. Joubert and A.E. Theron of the Department of Electrical and Electronic Engineering, University of Pretoria, South Africa, for their collaboration in the work on CMOS circuits for minimum supply-voltage. The author also wishes to acknowledge discussions during and after the Workshop with B. Gilbert, J. Mulder and B. Nauta, leading to a clearer and more complete paper.

References

[1] B. Gilbert, "A new wide-band amplifier technique", IEEE J. Solid-State Circuits, Vol. SC-3, pp 353-365, 1968

[2] B. Gilbert, "Translinear circuits: A proposed classification", Electron. Lett., Vol. 11, pp. 14-16, 1975; also "Errata", ibid., p. 136

[3] E. Vittoz and J. Fellrath, "CMOS analog integrated circuits based on weak inversion operation", IEEE J. Solid-State Circuits, vol. SC-12, pp. 224-231, 1977

[4] J.A. de Lima, "Design of micro power CMOS four-quadrant multiplier based on the translinear principle", in Proc. ESSCIRC, 1989, pp. 260-263

[5] E. Seevinck and R.J. Wiegerink, "Generalized translinear circuit principle", IEEE J. Solid-State Circuits, vol. SC-26, pp. 1098 - 1102, 1991

[6] R.J. Wiegerink, Analysis and synthesis of MOS translinear circuits, Boston: Kluwer, 1993

[7] A.G. Andreou and K.A. Boahen, "Neural information processing II", Chapter 8 of Analog VLSI: Signal & Information Processing, M. Ismail and T. Fiez, Eds., McGraw-Hill, 1994

[8] K. Bult and G.J.G.M. Geelen, "An inherently linear and compact MOST-only current division technique", IEEE J. Solid-State Circuits, Vol. SC-27, pp. 1730-1735, 1992

[9] H. Wallinga and K. Bult, "Design and analysis of CMOS analog signal processing circuits by means of a graphical MOST model", IEEE J. Solid-State Circuits, Vol. SC-24, pp. 672-679, 1989

[10] E. Seevinck, W. De Jager and P. Buitendijk, "A low-distortion output stage with improved stability for monolithic power amplifiers", IEEE J. Solid-State Circuits, vol. SC-23, pp. 794 - 801, 1988

[11] E. Seevinck, "Companding current-mode integrator: a new circuit principle for continuous-time monolithic filters", Electron.Lett., vol. 26, pp. 2046 - 2047, 1990

Design of MOS Translinear Circuits Operating in Strong Inversion

Remco J. Wiegerink

MESA Research Institute
University of Twente
P.O. Box 217, 7500 AE Enschede, The Netherlands
phone: x-31-53-4894373, fax: x-31-53-4309547

Abstract - Recently, it was proposed to generalize the well-known translinear (TL) circuit principle in such a way that it also applies to MOS transistors operated in strong inversion. The 'sum-of-square-roots' relation which is typical for this kind of TL circuits is much more difficult to handle mathematically than the product relation of traditional TL circuits. Therefore, in this paper a graphical analysis method is presented. Although this method does not result in an exact solution of the sum-of-square-roots relation, it does provide some insight into the behaviour of MOS strong-inversion TL circuits. The graphical method was implemented in a computer program, which is now used as an interactive design tool to implement nonlinear signal processing functions. Two design examples are presented in this paper: a class AB output stage for CMOS operational amplifiers and a variable-gamma circuit for colour television.

I. Introduction

The translinear (TL) principle was originally formulated as a practical means of implementing (non-)linear signal processing functions by bipolar analog circuits [1]. The concept translinear was based on the fundamental property of bipolar transistors, namely *trans*conductance *linear* with collector current. This property, when applied in circuits consisting of loops of junction voltages and having inputs and outputs in the form of currents, allows the implementation of exact, temperature- and process-insensitive signal processing functions [2-5].

For MOS transistors operating in strong inversion, a similar principle can be derived based on the property that the *trans*conductance varies *linear* with the gate-source voltage [6,7]. This MOS translinear principle is valid for a loop of MOS transistors as indicated in fig. 1. In the loop, the gate-source voltages are connected in series,

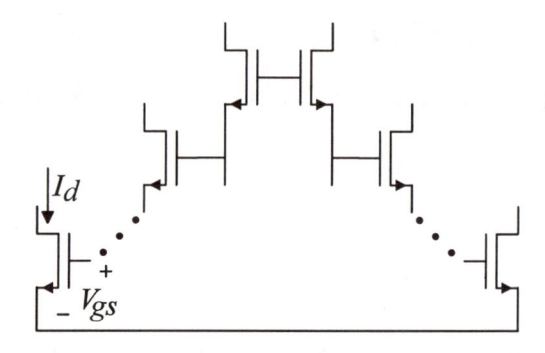

fig. 1 Conceptual MOS translinear loop

with equal numbers of transistors arranged clockwise and counterclockwise in the loop. It follows from Kirchhoff's voltage law that:

$$\sum_{cw} V_{gs} = \sum_{ccw} V_{gs} \qquad (1)$$

where the subscripts *cw* and *ccw* indicate the devices clockwise and counterclockwise in the loop, respectively. Applying the square-law model of an ideal saturated MOS transistor operating in strong inversion,

$$I_d = k(V_{gs} - V_t)^2 \qquad (2)$$

we obtain the following expression for V_{gs}:

$$V_{gs} = V_t + \sqrt{I_d/k} \qquad (3)$$

Substituting (3) into (1) results in:

$$\sum_{cw}(V_t + \sqrt{I_d/k}) = \sum_{ccw}(V_t + \sqrt{I_d/k}) \qquad (4)$$

Since equal numbers of transistors are present in the clockwise and counterclockwise directions, the number of threshold voltage terms on both sides of (4) are equal. Assuming well-matched threshold voltages and neglecting the body effect allows the threshold voltages to be dropped. Also, the parameters μ and C_{ox} will then be common and thus cancel. Now (4) reduces to:

$$\sum_{cw} (\sqrt{I_d/(W/L)}) = \sum_{ccw} (\sqrt{I_d/(W/L)}) \qquad (5)$$

with W/L the temperature- and process-independent aspect ratio determined by the designer. Relation (5) is a statement of the MOS translinear principle. It is a simple algebraic relation between the transistor currents and it is insensitive to temperature and processing.

II. Analysis of MOS translinear circuits

Although a systematic analysis method has been developed [7], for most practical MOS translinear circuits a "sum-of-square-roots" equation like (5) can be obtained by simple inspection of the circuit. For example, for the circuit in fig. 2 it is clear that M1, M2, M3 and M4 are connected in a translinear loop. Therefore, (5) is valid and we have the following relation between the drain currents in the circuit:

$$\sqrt{I_1/4} + \sqrt{I_2/4} = \sqrt{I_3} + \sqrt{I_4} \qquad (6)$$

We can express the drain currents in terms of the input and output currents using Kirchhoff's current law:

$$I_1 = I_2 = I_x + I_y,$$

$$I_3 = I_x - \tfrac{1}{2} I_z,$$

$$I_4 = I_y - \tfrac{1}{2} I_z \qquad (7)$$

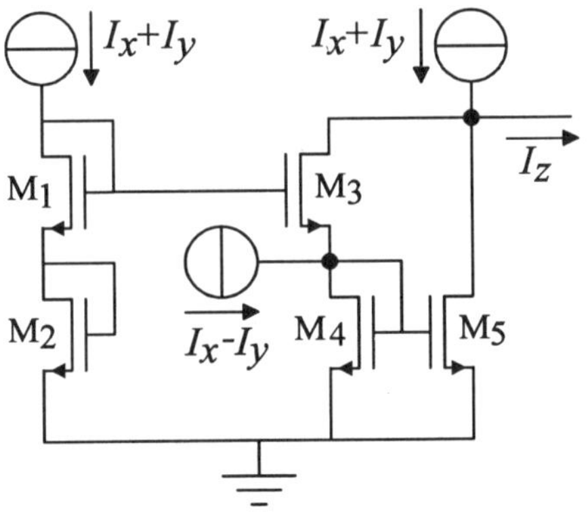

fig. 2 *MOS strong-inversion translinear circuit realizing the harmonic mean function.*

Substituting (7) into (6) results in the following temperature- and process-independent relation between the output current I_z and the input currents I_x and I_y:

$$\sqrt{I_x + I_y} = \sqrt{I_x - \tfrac{1}{2} I_z} + \sqrt{I_y - \tfrac{1}{2} I_z} \qquad (8)$$

Solving (8) for the output current I_z gives:

$$I_z = \frac{2 I_x I_y}{I_x + I_y}, \qquad (9)$$

which is the harmonic-mean function.

Another example is shown in fig. 3. This is a well-known current-squaring circuit [8]. In this circuit, again the translinear loop is formed by M_1, M_2, M_3 and M_4, resulting in the following relation between their drain currents (all transistors have equal aspect ratio's):

$$\sqrt{I_1} + \sqrt{I_2} = \sqrt{I_3} + \sqrt{I_4} \qquad (10)$$

Applying Kirchhoff's current law we can express the drain currents in terms of the input, bias and output currents:

$$I_1 = I_2 = I_{bias}$$

$$I_3 = \frac{I_z - I_x}{2}$$

$$I_4 = \frac{I_z + I_x}{2} \qquad (11)$$

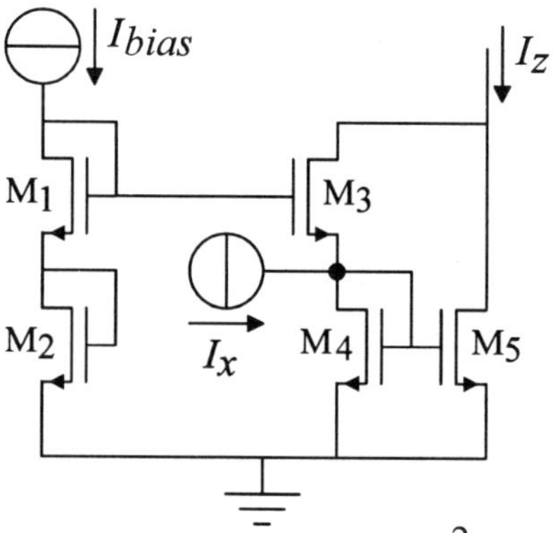

fig. 3 Current-squaring circuit: $I_z = 2 I_{bias} + \dfrac{I_x^2}{8 I_{bias}}$.

Substituting (11) into (10) results in:

$$2\sqrt{I_{bias}} = \sqrt{(I_z - I_x)/2} + \sqrt{(I_z + I_x)/2} \tag{12}$$

Solving (12) for I_z gives:

$$I_z = 2\,I_{bias} + \frac{I_x^2}{8\,I_{bias}} \tag{13}$$

In these examples, solving the sum-of-square-roots equations (8) and (12) is relatively easy. However, in general it is not easy to deal with these equations mathematically and numerical methods have to be used to obtain a solution for the output current. Therefore, a graphical representation was developed. This graphical representation provides some insight into the bahaviour of MOS translinear circuits. Furthermore, it significantly speeds up the computation of a numerical solution of the output current.

III. A simple graphical representation

A graphical representation was developed for MOS translinear circuits having one input variable (x) and one output variable (z) [7]. In that case the drain currents of the transistors in the translinear loop can be expressed as linear combinations of a constant (bias) current, a current proportional to x and a current proportional to z. Thus, for each transistor in the translinear loop:

$$I_{d_i} = a_i + b_i \cdot x + c_i \cdot z \tag{14}$$

where a_i represents the constant current and b_i and c_i are dimensionless multiplication factors of the input and output currents x and z. The translinear loop equation (5) now becomes:

$$\sum_{cw} \sqrt{\frac{a + b \cdot x + c \cdot z}{W/L}} = \sum_{ccw} \sqrt{\frac{a + b \cdot x + c \cdot z}{W/L}} \tag{15}$$

This equation is only valid if all drain currents are larger than or equal to zero. For the transistors having $b_i = c_i = 0$ this means that a_i must be positive. If either b_i or c_i is not equal to zero, the condition

$$I_{d_i} = a_i + b_i \cdot x + c_i \cdot z = 0 \tag{16}$$

represents a line in the x-z plane. This line is a boundary to the region where the loop-equation (15) is valid. The solution of (15) must be in this region.

For example, consider the circuit in fig. 3. The drain currents in this circuit are given by (11). From (11), the coefficients in (16) are easily obtained: $a_1 = a_2 = I_{bias}$, $b_1 = b_2 = c_1 = c_2 = 0$, $a_3 = a_4 = 0$ and $-b_3 = b_4 = c_3 = c_4 = 1/2$. The lines given by $I_3 = 0$ and $I_4 = 0$ are the boundaries to the region where the loop equation (15) is valid. Fig. 4 shows a plot of these boundary lines and the solution of the loop-equation (15). The solution must be in the area where both I_3 and I_4 are larger than zero.

With the help of the graphical representation, many properties of the solution can be derived [7]. For example, if we have two intersecting boundary lines corresponding to two transistors connected in opposite directions in the loop, as indicated in fig. 5, a solution curve will either start in the intersection point or at one of the boundary lines. If the boundary lines correspond to transistors connected in the same direction in the loop, as indicated in fig. 6, the solution will start at one line and end at the other.

Closed solution curves as indicated in fig. 7 only occur in a few special situations. This property is used by the computer program presented in the next section to find initial points of the solution. Once that the starting points are known, the entire solution can easily be calculated.

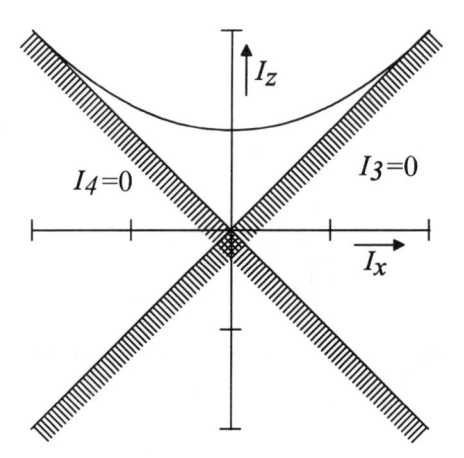

fig. 4 Graphical representation for the current-squaring circuit of fig. 3. The dashed line is the solution to the loop equation (15) and is the response of the circuit.

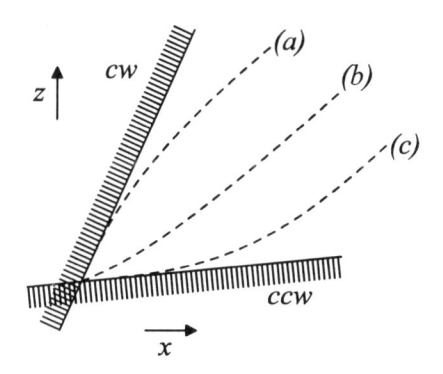

fig. 5 Typical forms of a solution in the case of two intersecting boundaries representing the drain currents of transistors connected in opposite directions in the loop. The intersection point is either part of the solution (b) or the solution ends at one of the boundary lines with a tangent equal to the line (a), (c).

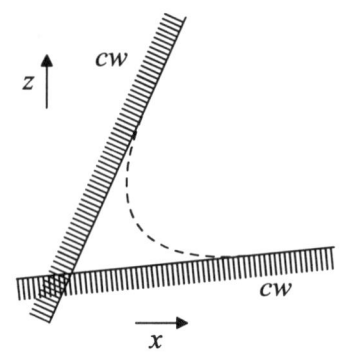

fig. 6 Typical form of the solution curve at the intersection of two boundary lines corresponding to transistors connected in the same direction in the loop.

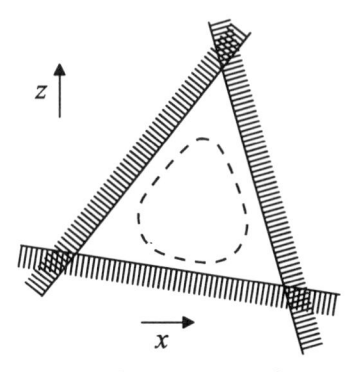

fig. 7 Closed solution curves only occur in a few special situations. Here, the area where the loop equation (15) is valid is completely surrounded by boundary lines originating from transistors connected in the same direction in the translinear loop.

IV. Computer aided analysis and synthesis of MOS translinear circuits.

The graphical representation discussed in the previous section was implemented in a computer program. The program shows the lines (16) on the video display, as indicated in fig. 8. The lines marked *cw* correspond to clockwise connected transistors; the line marked *ccw* corresponds to a counterclockwise connected transistor.

As mentioned before, the computer program exploits the fact that closed solution curves rarely occur and that most solution curves either end at a boundary-line or continue into infinity. The latter simply means that the solution ends at the border of the video screen. The program simply searches for the start of solution curves at a trajectory along the border of the validity area of the loop equation. In fig. 8 this means that the program would start in point A and then continue through points $B, C, .., H$ until it reaches A again. In points C, E, G and H it detects the start or end of a solution-curve (indicated by the dashed lines). Once a single point on a solution-curve has been found, the entire curve can easily be calculated.

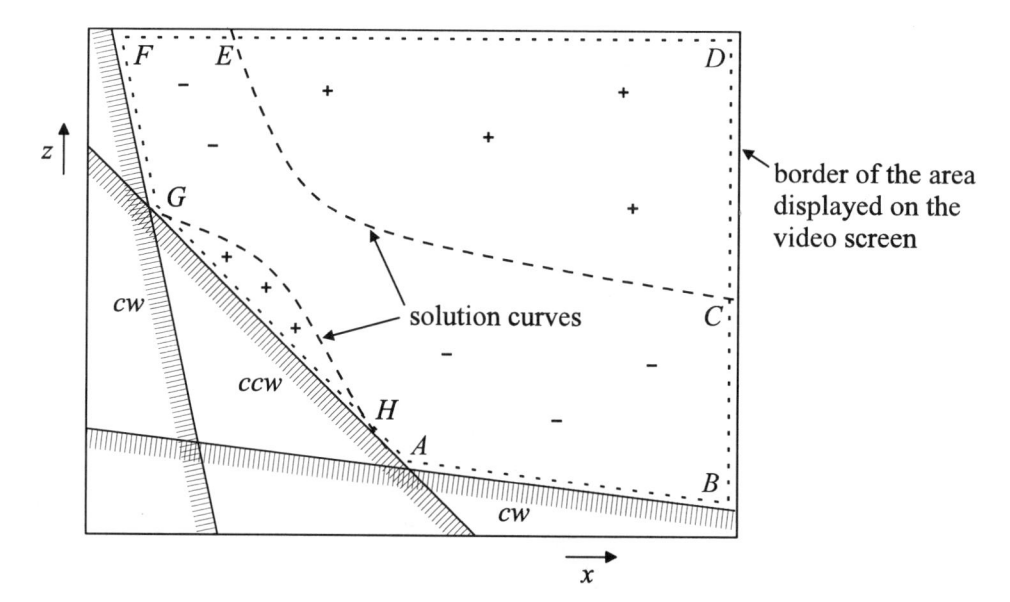

fig. 8 The computer program evaluates the MOS translinear loop equation at a trajectory (dotted line) along the border of the area where the equation is valid. The program starts in point A and continues via points B, C, ... , H untill it reaches A again. At points C, E, G and H it detects the start of a solution curve. Once that initial points of the solution curves are known, the entire curves (dashed lines) can be easily calculated.

The way in which the computer program finds the solutions to a loop equation results in very short calculation times. Therefore, the program was extended with features that make it a powerful interactive design tool for MOS translinear circuits. The user can change the position of the lines (16) by simply pointing at them with the mouse pointer and dragging them to the desired position. In this way, the user can change the coefficients a, b and c and the aspect ratio's W/L in the loop equation (15) very easily. The resulting solution, which is in fact the transfer function of the resulting translinear circuit, is plotted immediately after each change. Furthermore, it is possible to plot the desired transfer function at the background and the user can then position the boundary-lines in a way such that the solution best approximates this function. Other useful features of program are an automatic optimization routine and a sensitivity analysis.

V. Design example: a class-AB control circuit for rail-to-rail CMOS output stages

An output stage wich combines a rail-to-rail output voltage range and a low quiescent power consumption requires class-AB controlled output transistors in a common source configuration. Such an output stage is indicated in fig. 9. Several class-AB output stages based on this principle have been published [9-11]. In this section, an MOS translinear realization of the nonlinear function block that is needed to obtain the class-AB control will be presented [12]. The computer program presented in the previous section was used to find a suitable loop equation (15).

First, the operation of the circuit principle of fig. 9 is explained. The currents through the output transistors M_1 and M_2 are measured by transistors M_3 and M_4 and fed to the class-AB control circuit. This circuit regulates the currents by means of I_c. The objective is that neither of the output transistors ever cuts off completely [13]. Increasing I_c leads to an increase in the drain currents of the output transistors. Decreasing I_c gives a decrease in the currents through the output transistors.

The desired relation between the drain currents of the output transistors is indicated in fig. 10(a). If one of the drain currents becomes very large, the other transistor still conducts a minimum current. The exact relation between the currents is not essential. Fig. 10(b) shows the same relation but now plotted as a function of the output current I_{out} (= I_P - I_N).

With the help of the computer program the following sum-of-square-roots equation is easily found:

$$\sqrt{I_n - \tfrac{1}{2}I_{min}} + \sqrt{I_p - \tfrac{1}{2}I_{min}} = \sqrt{\tfrac{1}{2}I_{min}} + \sqrt{I_n + I_p - I_{min}} \qquad (17)$$

The graphical representation and the solution of this equation are shown in fig. 11. Note that the solution (solid line) does not reach the boundaries of the validity area of the loop equation given by (17) (dashed lines). Therefore, all transistors in the translinear loop keep conducting for all possible values of I_N and I_P.

Equation (17) can now be implemented by a translinear circuit. A possible implementation is shown in fig. 12. In this circuit, transistor and M_9 and M_{10} sense the value of I_P and M_{11} senses the value of I_N. The translinear loop is formed by M_1, M_2, M_3 and M_4. Transistors M_1 and M_2 correspond to the left-hand side of (17). The drain current of M_1 is forced equal to $I_N - \frac{1}{2}I_{min}$ and the drain current of M_2 is forced equal to $I_P - \frac{1}{2}I_{min}$. The drain currents of the transistors connected in the opposite direction in the loop (M_3 and M_4) are forced equal to $I_N + I_P - I_{min}$ and $\frac{1}{2} \cdot I_{min}$.

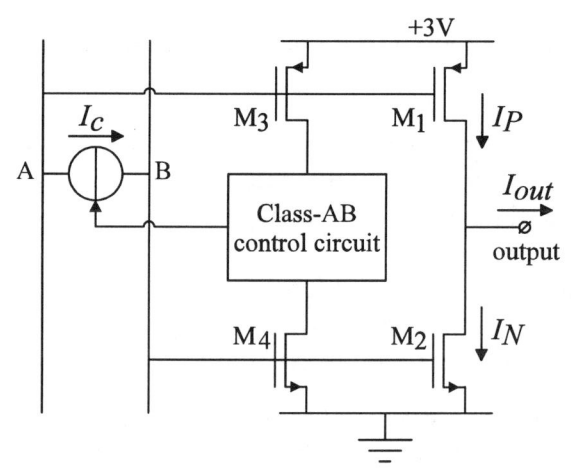

fig. 9 Principle of a class AB rail-to-rail output stage with feedback control circuit.

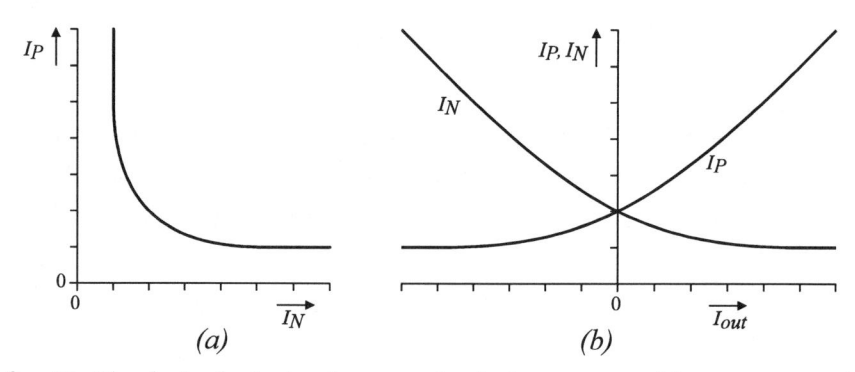

fig. 10 The desired relation between the drain currents of the output transistors (a) and the drain currents plotted as a function of the output current (b).

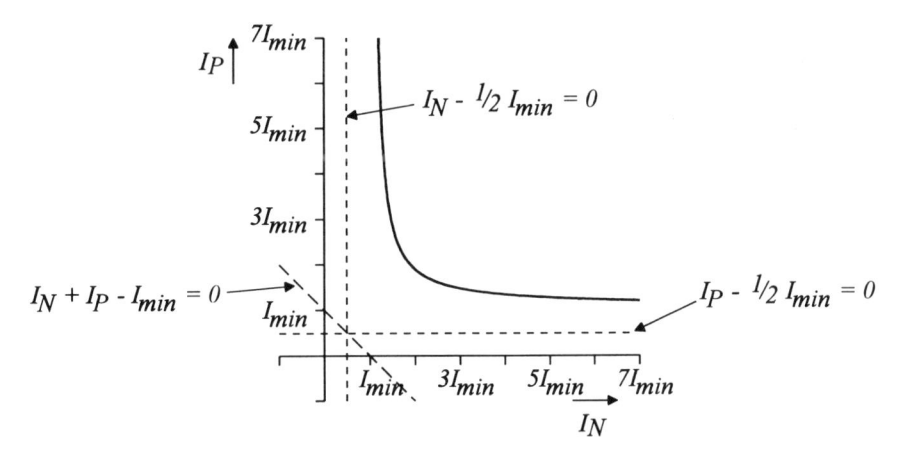

fig. 11 Graphical representation and solution (solid line) of the loop equation (17).

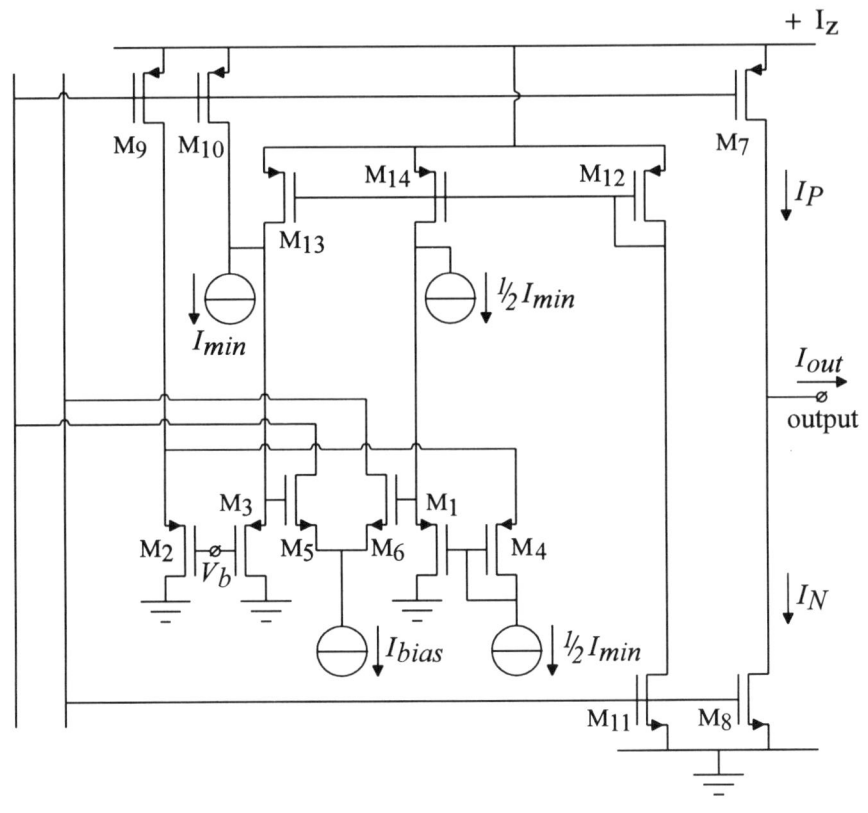

fig. 12 Rail-to-rail output stage with a class AB control circuit based on (17).

If (17) is satisfied, the translinear loop M_1, M_2, M_3, M_4 will be in equilibrium and there will be no voltage difference between the inputs of differential pair M_5, M_6. If the equilibrium is disturbed this will result in a differential input voltage across this differential pair. This in turn results in a differential current between the gates of the output transistors M_7 and M_8 and the drain currents of these transistors will be adjusted until the equilibrium state defined by (17) is reached. Due to the high loop gain the differential input voltage of differential pair M_5, M_6 will always be approximately zero. The value of the bias voltage V_b should be approximately 0.5 V to ensure that all current sources operate correctly.

The circuit of fig. 12 was realized on our semi-custom CMOS array (ACMA, [14]) as a part of a complete operational amplifier [12]. Fig. 13 shows the measured relation between I_N and I_P. The measured curve matches the response calculated by the computer program almost exactly. All transistors in the translinear loop were 100 μm wide and 10 μm long. The value of I_{min} was 10 μA.

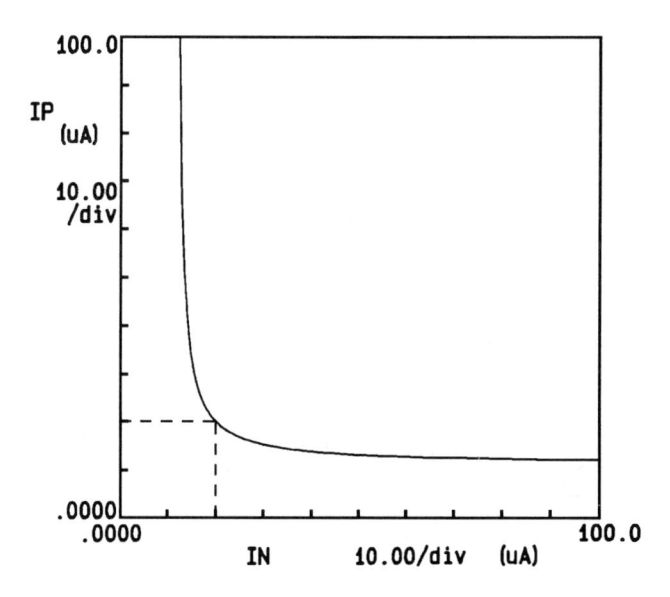

fig. 13 Measured relation between I_N and I_P.

VI. Design example: a variable-gamma circuit for colour television

The light output L_{out} of a television picture tube is a power function of the applied voltage:

$$L_{out} = V_{in}^{\gamma_p} \qquad (18)$$

This nonlinearity is compensated by a gamma-correction circuit in the television camera. The relation between the light output of the television picture tube and de light input L_{in} of the television camera can be expressed as [15]:

$$L_{out} = L_{in}^{\gamma_o \cdot \gamma_c \cdot \gamma_p} \qquad (19)$$

with: γ_o = exponent resulting from camera,
γ_c = exponent realized by the gamma correction circuit, and
γ_p = exponent resulting from the television picture tube.

A linear transfer would be obtained if $\gamma_c \gamma_o = \gamma_p^{-1}$. However, for modern television receivers this is generally not the case. The total exponent at the camera side $\gamma_o \gamma_c$ has a typical value of $1/2.2$. The value of γ_p depends on the type of display that is used. For modern colour television tubes a typical value is 2.8. For liquid crystal displays this value is larger: between 3 and 4. Fortunately, colour distortions due to a non-optimal gamma correction can be eliminated by adjusting the colour-saturation level in the television receiver. The variable-gamma circuit presented here is intended to be inserted not in the camera but in the television receiver. In the television receiver the circuit can replace the traditional intensity and contrast adjustments.

A variable-gamma circuit realizes the following function:

$$z = x^{\gamma} \qquad (20)$$

where the value of γ is adjustable by the television viewer between approximately 0.25 and 1.0. Fig. 14 shows a plot of the function for different values of γ. A signal value of 0.0 corresponds to a black video screen. A value of 1.0 corresponds to the maximum light output of the screen. From the figure it can be seen that if the exponent γ is changed the extreme values 0.0 and 1.0 remain fixed. This is an important property of a variable-gamma circuit. Between the extreme levels, changing gamma results in more or less expansion of the signal near the black level and compression near the maximum intensity level.

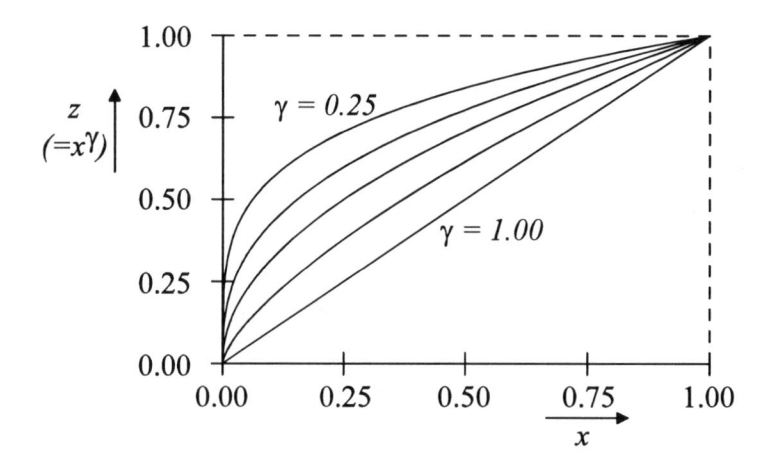

fig. 14 Variable-gamma curves. A signal value of 0.0 corresponds to the black-level on the the video screen; a value of 1.0 corresponds to the maximum light output of the screen.

For colour television three well-matched circuits are required, one for each colour channel. Of course, without further precautions changing gamma would result in a distorted colour reproduction. However, it can be shown that this distortion is independent of the video signal amplitude and it can, therefore, easily be corrected by simultaneously adapting the colour-saturation level.

A commonly used method to realize a gamma-correction circuit is a piece-wise linear approximation. However, circuits based on this approach are generally inaccurate and temperature dependent. Furthermore, adjustment of the value of gamma is hard to obtain. Therefore, another method was proposed consisting of a series connection of a logarithmic, a linear and an anti-logarithmic amplifier [15, 16]. The value of gamma can then be adjusted by changing the gain of the linear amplifier. In practical implementations, however, a problem appears to be the large gain for small input signals (see fig. 14), which deteriorates the signal-to-noise ratio. Furthermore, a series connection of circuits is undesirable because this reduces the feasible bandwidth and, even more important, results in a poor matching between the three colour channels.

An alternative approach is indicated in fig. 15. A linear and nonlinear function block are connected in parallel and the output signal is a weighted sum of the two functions. A suitable choice for the nonlinear function appears to be the inverse hyperbolic sine function:

$$z = 0.25 \cdot \sinh^{-1}(27.32 \cdot x)$$

$$= 0.25 \cdot \ln(27.32 \cdot x + \sqrt{1 + (27.32 \cdot x)^2}) \quad (21)$$

Simulations show that the resulting deviation from an ideal gamma function (fig. 14) causes no visible errors in the colour reproduction. Fig. 16 shows a plot of the resulting transfer function.

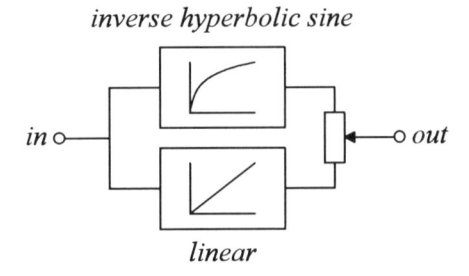

fig. 15 Transfer curves resulting from a variable-gamma implementation based on the weighted sum of a linear transfer and a transfer proportional to the inverse hyperbolic sine function.

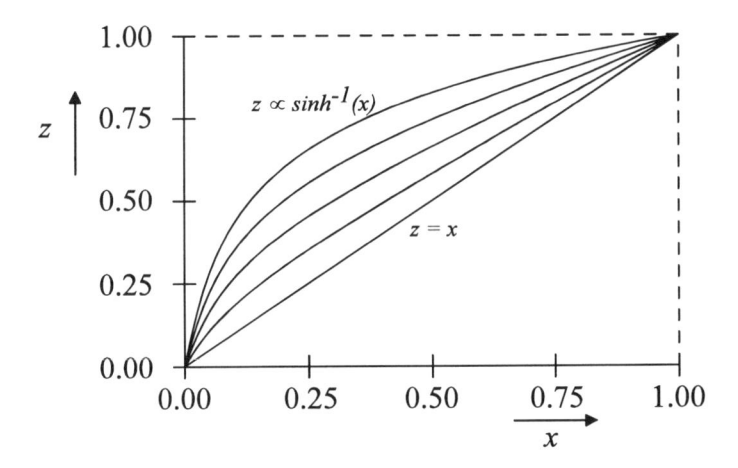

fig. 16 Transfer curves resulting from a variable-gamma implementation based on the weighted sum of a linear transfer and a transfer proportional to the inverse hyperbolic sine function.

We now use the computer program presented in section IV to obtain a suitable sum-of-square-roots equation, which can be implemented by an MOS translinear circuit. From the properties derived in section III, we know that the solution of a sum-of-square-roots equation never starts at the intersection of two boundary lines resulting from transistors connected in the same direction. Instead, the solution will start at one line and end at the other (see fig. 6). Therefore, the required funcion (21) can be approximated by using only two transistors in a loop of four transistors. The other two transistors are biased at a constant current. Fig. 17 shows the graphical representation of a typical solution that can be obtained in this way. The corresponding sum-of-square-roots equation is:

$$\sqrt{\frac{7x-z}{9}} + \sqrt{\frac{4.72-0.25x-3.75z}{2}} = \sqrt{\tfrac{1}{2}} + \sqrt{\tfrac{1}{2}} \qquad (22)$$

In principle (22) can be implemented by a translinear circuit. However, this is not easy because of the non-integer coefficients in the second square-root term and because both the input current x and the output current z flow through two transistors in the translinear loop. It is desirable to have drain currents proportional to either x or z, corresponding to horizontal or vertical lines in the graphical representation.

Therefore, we continue searching for other solutions allowing an extra drain current to be dependent on x or z. With the computer program several suitable equations

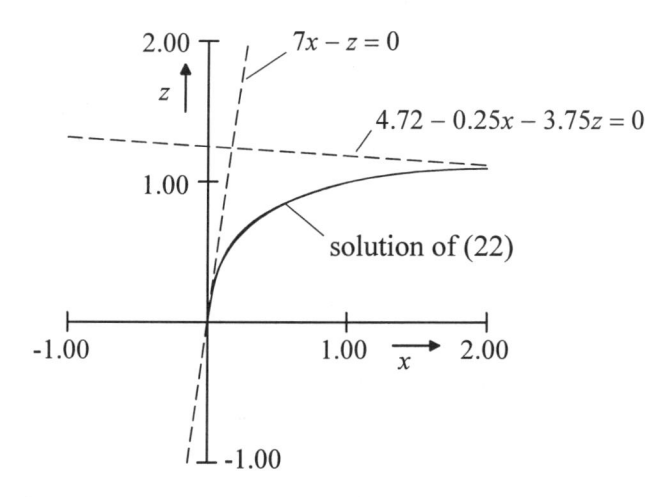

fig. 17 Graphical representation corresponding to equation (22). Although the equation results in a good approximation of the inverse hyperbolic sine function, implementation in a circuit is difficult due to the non-integer coefficients.

can be found of which the following is the simplest with only one drain current dependent on both x and z:

$$\sqrt{\frac{7x-z}{5}} + \sqrt{\frac{3-2z}{1}} = \sqrt{\frac{2+4x}{5}} + \sqrt{1} \tag{23}$$

The corresponding graphical representation is shown in fig. 18(a).

Equation (23) can now be implemented by an MOS translinear circuit. Because of the high desired bandwidth we can not eliminate the body effect by using transistors in individual wells connected to their sources. Instead, the influence of the body effect is minimized by minimizing the differences in source-bulk voltages. This is accomplished by using an up-down topology [7] as indicated in fig. 18(b). In this

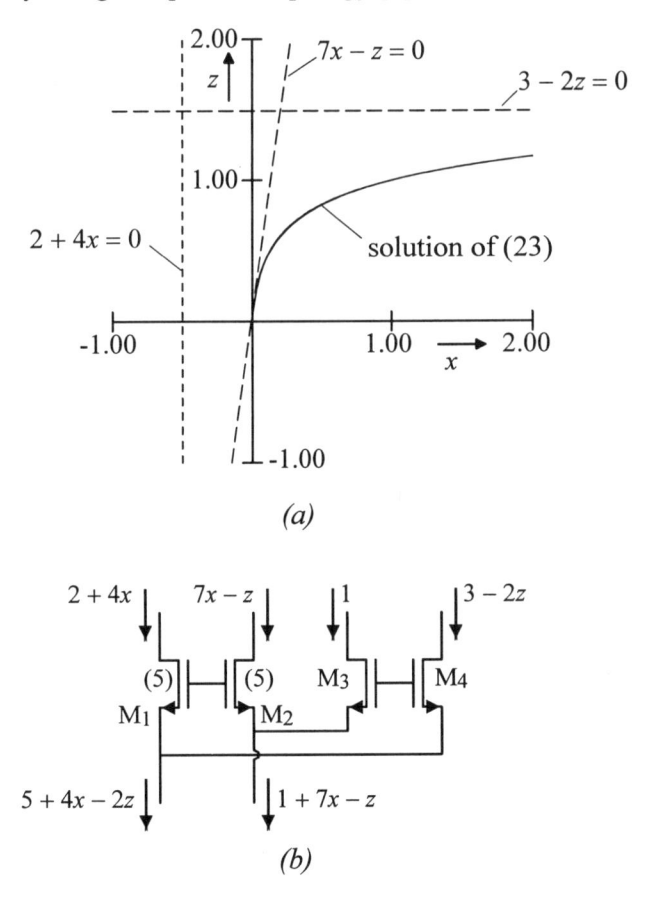

fig. 18 Graphical representation of equation (23) (a). This equation contains only integer coefficients and is therefore easy to implement in a translinear loop (b).

figure transistors M₁ and M3 correspond to the right-hand side of (23) and M2 and M4 correspond to the left-hand side.

Fig. 19 shows how the desired drain currents can be forced into the circuit. Transistors M1 and M3 are simply diode-connected and the drain currents are defined by the current sources at their drains. The current of M4 is used to obtain the value of the output current z by subtracting the constant part of it. With the help of the current mirror below the translinear loop and the current source with value $10x-3$ the difference between the drain currents of M2 and M4 is forced to the correct value. In this way it is not necessary to force a current proportional to the output current z into the circuit.

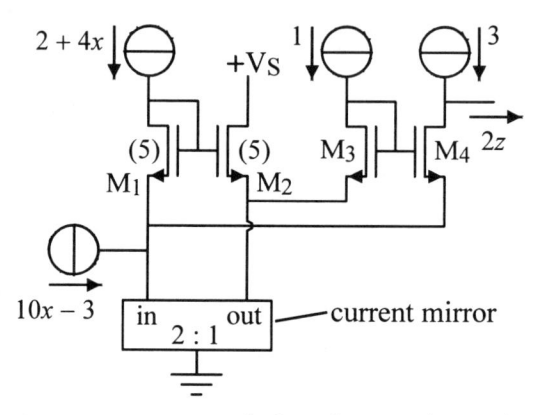

fig. 19 An attenuating current mirror below the translinear loop can be used to balance the contributions proportional to z. In this way it is not necessary to force a current proportional to z into the circuit.

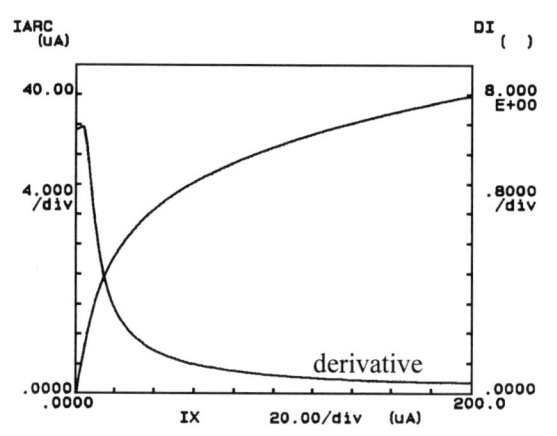

fig. 20 Measured response of the inverse hyperbolic sine circuit and the first derivative (DI).

A complete inverse hyperbolic sine circuit based on fig. 19 was realized in a standard 2.5μm CMOS process. Fig. 20 shows the measured dc response of the circuit and it's derivative. The supply voltage was 5V. Fig. 21 shows the measured response at different temperatures. For small values of the input current x the temperature dependence is negligible. For large values of the input current the output current increases slightly with increasing temperature. The measured -3dB bandwidth of the realized circuit was approximately 20MHz, which is more than sufficient for video applications.

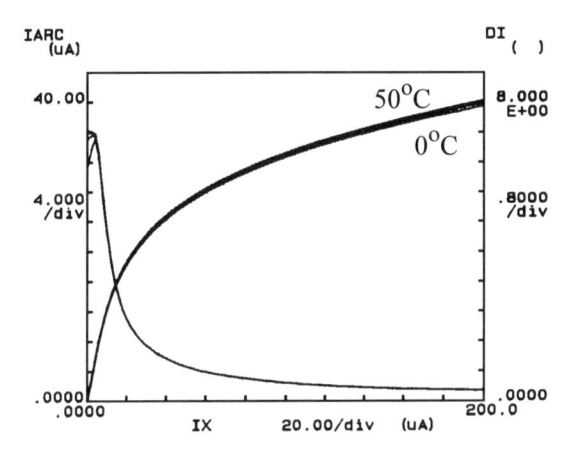

fig. 21 Measured response of the inverse hyperbolic sine circuit at different temperatures: $0\,°C$, $25\,°C$ and $50\,°C$. The output current slightly increases at higher temperatures.

VII. Conclusion

In this paper a graphical analysis method for MOS translinear circuits operating in strong inversion has been presented. The graphical representation was implemented in a computer program. Because of its high calculation speed the program can be used as a powerful interactive design tool for implementing prescribed nonlinear signal processing functions by MOS translinear circuits.

Two design examples have been presented: a class AB output stage for CMOS operational amplifiers and a variable-gamma circuit. In both cases the theoretical circuit response matches the measured response very well.

References

[1] B. Gilbert, "Translinear circuits: a proposed classification," *Electronics Letters*, vol. 11, pp. 14-16, 1975.

[2] E. Seevinck, *Analysis and Synthesis of Translinear Integrated Circuits*, Amsterdam: Elsevier, 1988.

[3] E. Seevinck, "Synthesis of nonlinear circuits based on the translinear principle," *Proceedings ISCAS*, 1983, pp. 370-373.

[4] B. Gilbert, "A new wide-band amplifier technique," *IEEE J. Solid-State Circuits*, vol. 3, pp. 353-365, 1968.

[5] B. Gilbert, "A precise four-quadrant multiplier with sub- nanosecond response," *IEEE J. Solid-State Circuits*, vol. 3, pp. 365-373, 1968.

[6] E. Seevinck, and R.J. Wiegerink, "Generalized translinear circuit principle," *IEEE J. Solid-State Circuits*, vol. 26, pp. 1098-1102, 1991.

[7] R.J. Wiegerink, *Analysis and Synthesis of MOS Translinear Circuits*, Kluwer Academic Publishers, 1993.

[8] K. Bult, and H. Wallinga, "A class of analog CMOS circuits based on the square-law characteristic of an MOS transistor in saturation," *IEEE J. Solid-State Circuits*, vol. 22, pp. 357-365, 1987.

[9] F.N.L. Op't Eynde, P.F.M. Ampe, L. Verdeyen, W.M.C. Sansen, "A CMOS large-swing low-distortion three-stage class AB power amplifier," *IEEE J. Solid-State Circuits*, pp. 265-273, 1990.

[10] M.D. Pardoen, M.G. Degrauwe, "A rail-to-rail input/output CMOS power amplifier," *IEEE J. Solid-State Circuits*, pp. 501-504, 1990.

[11] R. Hogervorst, R.J. Wiegerink, P.A.L. de Jong, J. Fonderie, R.F. Wassenaar, and J.H. Huijsing, "CMOS low-voltage operational amplifier with constant-gm rail-to-rail input stage," *Proceedings ISCAS*, 1992, pp. 2876-2879.

[12] J.H. Botma, R.F. Wassenaar, and R.J. Wiegerink, "A low-voltage CMOS Op Amp with a rail-to-rail constant-gm input stage and a class-AB rail-to-rail output stage," *Proceedings ISCAS*, 1993.

[13] E. Seevinck, W. de Jager, and P. Buitendijk, "A low- distortion output stage with improved stability for monolithic power amplifiers," *IEEE J. Solid-State Circuits*, vol. 23, pp. 802-815, 1988.

[14] E.A.M. Klumperink, *ACMA Design Manual*, University of Twente, Enschede, The Netherlands, 1990.

[15] H.C. Nauta, "An integrated gamma corrector," *IEEE J. Solid-State Circuits*, Vol. 16, pp. 238-241, 1981.

[16] K.G. Freeman, and R.E. Ford, "Variable gamma corrector improves television video signals," *Electron. Eng.*, pp. 90-93, 1970.

TRANSLINEAR CIRCUITS IN LOW-VOLTAGE OPERATIONAL AMPLIFIERS

Klaas-Jan de Langen, Ron Hogervorst, Johan H. Huijsing

Delft University of Technology
DIMES
Laboratory for Electronic
Instrumentation
Mekelweg 4
2628 CD Delft
The Netherlands
Phone: +31 15 278 5747

Abstract—Operational amplifiers are limited in their dynamic voltage range and bandwidth by the supply voltage and power. To obtain the maximum dynamic range, power-efficient rail-to-rail class-AB output stages and voltage-efficient rail-to-rail input stages are needed. In order to implement the desired class-AB behavior in rail-to-rail output stages and the desired constant transconductance in rail-to-rail input stages, translinear circuits are used extensively.

I. Introduction

The continuing down-scaling of dimensions in IC technology results in a lowering of breakdown voltages. This development in technology pushes the supply voltages to lower values. Further, the smaller dimensions lead to higher component densities which results in less allowable power per electronic function. Another reason for the reduction of supply voltage and power dissipation is the increasing popularity of portable battery-powered electronics. Supply voltages will go from the present 5 V to 3.3 V, further to 1.8 V, and ultimately to 1 V.

The lowering of supply voltages and power has an enormous impact on the signal handling capability in analog circuit design[1]. Firstly, the dynamic range decreases because of lower allowable signal voltages and larger noise voltages caused by the lower supply currents. Secondly, the bandwidth of circuits is reduced due to lower supply currents. Therefore, the available supply voltage and the available supply current have to be used as efficient as possible. Thus, class-AB output stages that can handle output voltages as large as the supply-voltage range are essential to make economic use of the available supply power. Further, rail-to-rail input stages are important to efficiently employ the supply-voltage range at the input of the amplifier.

This paper discusses the use of translinear circuits[2] and MOS translinear circuits[3] in low-voltage low-power operational amplifiers. In Section II power-efficient class-AB output stages based on the conventional class-AB output stage are addressed. Low-distortion class-AB output stages are presented in Section III. In Section IV the application of translinear loops to achieve constant transconductance g_m in rail-to-rail input stages is discussed. Finally in Section V conclusions are drawn.

II. Power-efficient class-AB output stages

Output stages for low-voltage low-power applications should have a rail-to-rail output range and should be class-AB biased. The class-AB biasing has two important functions. First, it has to fix a well-defined quiescent current in the output transistors to achieve a stable AC response in and near the quiescent operating point and, to keep the cross-over distortion as small as possible. To avoid high-frequency distortion, it should be prevented that one of the transistors is ever cut off. Preferably, a minimum bias current should be maintained that is close to the quiescent current.

Conventional bipolar output stages, such as the circuit used in the µA741, have output transistors, Q_1 and Q_2, connected in a common-collector configuration as shown in Fig. 1a. A similar output stage in CMOS is shown in Fig. 1b. The biasing of the output transistors is controlled by the translinear loop consisting of the base-emitter voltages of the output transistors Q_1 and Q_2 and the diodes D_3 and D_4. These diodes keep the sum of the base-emitter voltages of Q_1 and Q_2 constant, which results in a constant product of their collector currents, given by

$$I_1 I_2 = I_Q^2 \qquad (1)$$

where I_1 is the collector current of Q_1, I_2 the collector current of Q_2, and the quiescent current I_Q is given by

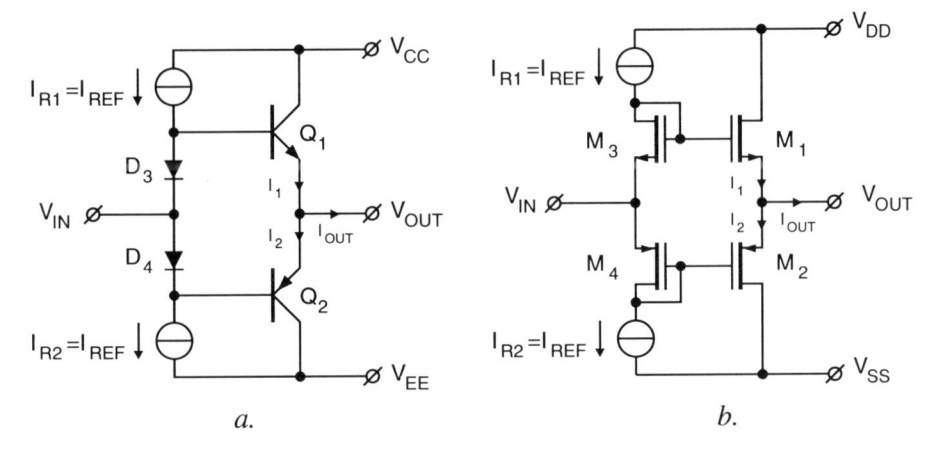

Fig. 1. Conventional class-AB output stage.
 a. Bipolar version.
 b. CMOS version.

$$I_Q = \sqrt{\frac{A_1 A_2}{A_3 A_4}} I_{REF} \tag{2}$$

in which I_{REF} is the reference current that flows through the diodes, and A the emitter area of the transistors and diodes. The class-AB characteristic of the conventional output stage is shown in Fig. 2. For the CMOS output stage operating

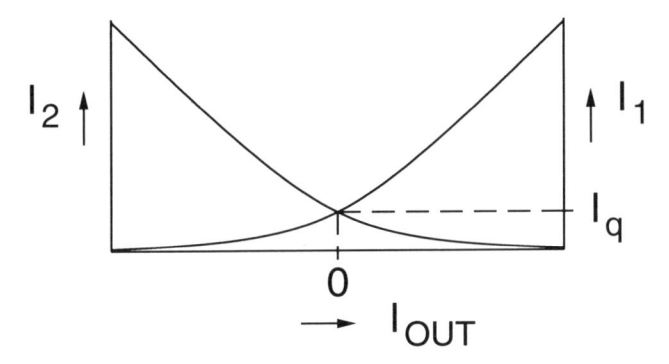

Fig. 2. Conventional bipolar class-AB characteristic.

in strong inversion we find

$$\sqrt{I_1} + \sqrt{I_2} = 2\sqrt{I_Q} \tag{3}$$

where I_1 is the drain current of M_1 and I_2 is the drain current of M_2 and the quiescent current is given by

$$I_Q = \left(\frac{W}{L}\right)_3 \left(\frac{L}{W}\right)_1 I_{REF} \qquad (4)$$

in which $(W/L)_1$ is the width over length ratio of M_1, $(W/L)_3$ is the width over length ratio of M_3. Further, it is assumed that the ratio between the width over length ratios of M_2 and M_4 is equal to the ratio between the width over length ratios of M_1 and M_3, and their transconductance values obey

$$K\frac{W}{L} = \frac{1}{2}\mu_n C_{ox}\left(\frac{W}{L}\right)_n = \frac{1}{2}\mu_p C_{ox}\left(\frac{W}{L}\right)_p \qquad (5)$$

where μ_n is the mobility of electrons, μ_p the mobility of holes, $(W/L)_n$ the width over length ratio of M_1 and M_3, and $(W/L)_p$ the width over length ratio of M_2 and M_4, respectively. In most output stages the latter condition is met, because it reduces the distortion of the amplifier. Eq. 3 is valid until one of the drain currents exceeds the value $4I_Q$. Above that value the drain current increases linearly while the other transistor is cutoff as can be seen in Fig. 3. Cutoff of the output

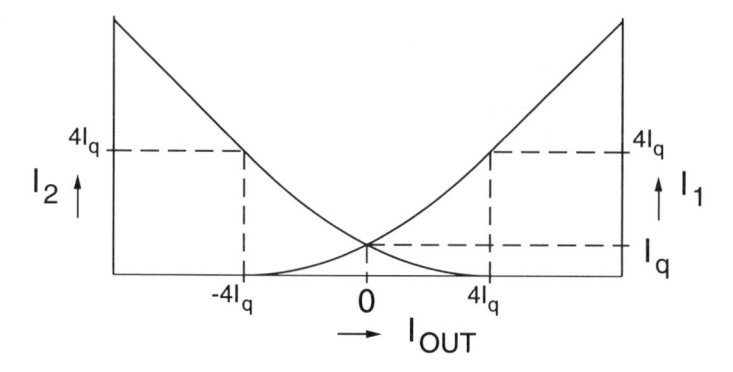

Fig. 3.   Conventional CMOS class-AB characteristic.

transistors generates cross-over distortion due to turn-on delay. Therefore, low-distortion class-AB output stages maintain a minimum current in the output transistors. These circuits are discussed in the next section. We now first discuss class-AB rail-to-rail output stages based on the conventional output stage.

The main disadvantage of the conventional output stage is the limitation of the output-voltage swing, which is limited to the supply voltage minus two base emitter voltages and two saturation voltages in the bipolar circuit and, which is limited in the CMOS case to the supply voltage minus two gate-source voltages

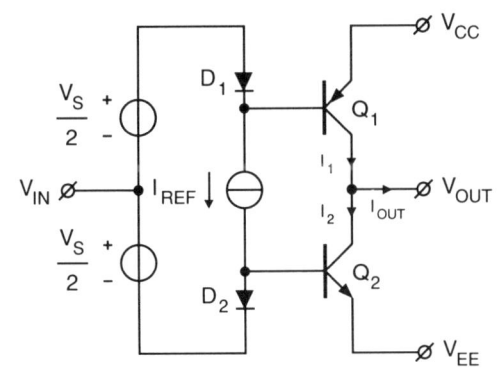

Fig. 4. Rail-to-rail class-AB output stage.

and two saturation voltages. To overcome this limitation, the output transistors should be connected in a common-emitter configuration. Such a circuit can be obtained by moving the lower part of Fig. 1a consisting of Q_2 and D_2 to the positive supply rail and the upper part, Q_1 and D_1 to the negative supply rail. The result is shown in Fig. 4. The translinear loop that controls the current in the output transistors consists of Q_1, D_1, the two voltage sources with value $V_S/2$, D_2, Q_2, and the supply voltage V_{CC}-V_{EE}. If the sum of the two voltage sources is equal to the supply voltage, the sum of the base-emitter voltages of the output transistors is equal to the sum of the diode voltages, D_1 and D_2, and the conventional class-AB behavior is exactly implemented. The principle depicted in Fig. 4 can be simplified to the circuit shown in Fig. 5. In this circuit the reference

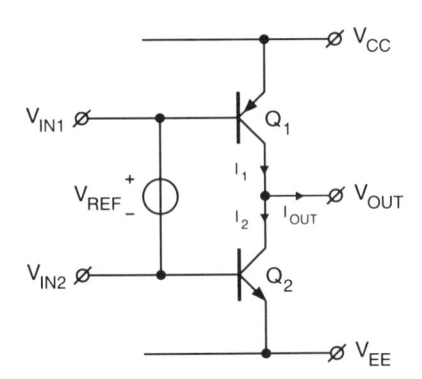

Fig. 5. Simplified principle of rail-to-rail class-AB output stage.

voltage V_{REF} should be equal to the supply voltage minus the base-emitter voltages of the output transistors. Similar circuits can be derived in CMOS.

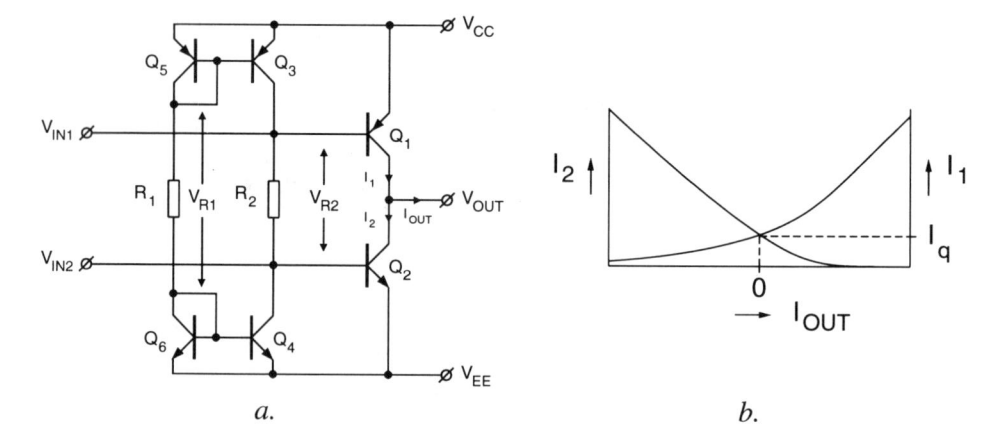

Fig. 6. *Resistive coupled rail-to-rail bipolar class-AB output stage.*
   a. Circuit.
   b. Class-AB characteristic.

An implementation of this principle is shown in Fig. 6a[4, 5]. The reference voltage is realized by resistor R_2. The bias current for this resistor is generated by connecting two diode connected transistors Q_5 and Q_6 and another resistor R_1 between the supply rails. The bias current is passed to R_2 using current mirrors Q_5, Q_3 and Q_6, Q_4. The relation between the base-emitter voltages V_{BE1} and V_{BE2} of the output stage is given by

$$V_{BE1} + V_{BE2} + V_{R2} = V_{BE5} + V_{BE6} + V_{R1} \qquad (6)$$

If V_{R1} is equal to V_{R2}, the conventional class-AB behavior described by Eq. 1 is realized. A disadvantage of this circuit is that the quiescent current depends on variations of the supply voltage, because the current through R_1 and therefore the base-emitter voltages of Q_5 and Q_6 are influenced by the supply voltage. The class-AB characteristic is plotted in Fig. 6b. The asymmetry is caused by the base current of the PNP transistor. The minimum allowable supply voltage is determined by the two base-emitter voltages of the output transistors and is of the order of 1.5 to 1.8 V. The principle can also be applied in CMOS as shown in Fig. 7. The minimum supply voltage in CMOS can be between 1.6 V and 2.8 V depending on the process and on the maximum output current.

If we want to reduce the minimum supply voltage, the direct coupling between the bases of the output transistors is not possible anymore. A first example of a circuit without direct coupling is shown in Fig. 8. The output transistors, Q_1 and Q_2, are biased by the diodes, D_1 and D_2, through the coupling resistors, R_1 and R_2. If the base currents of the output transistors are neglected, the two in-phase

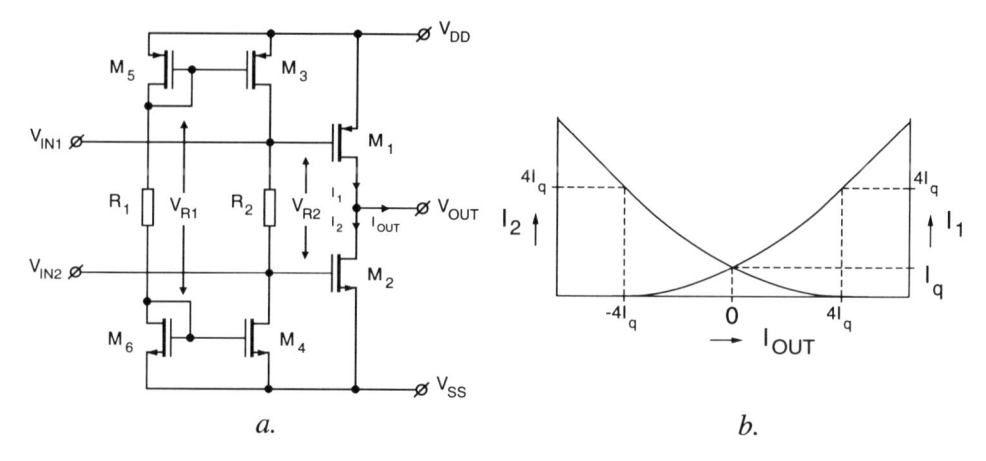

Fig. 7. Resistive coupled rail-to-rail CMOS class-AB output stage.
a. Circuit.
b. Class-AB characteristic.

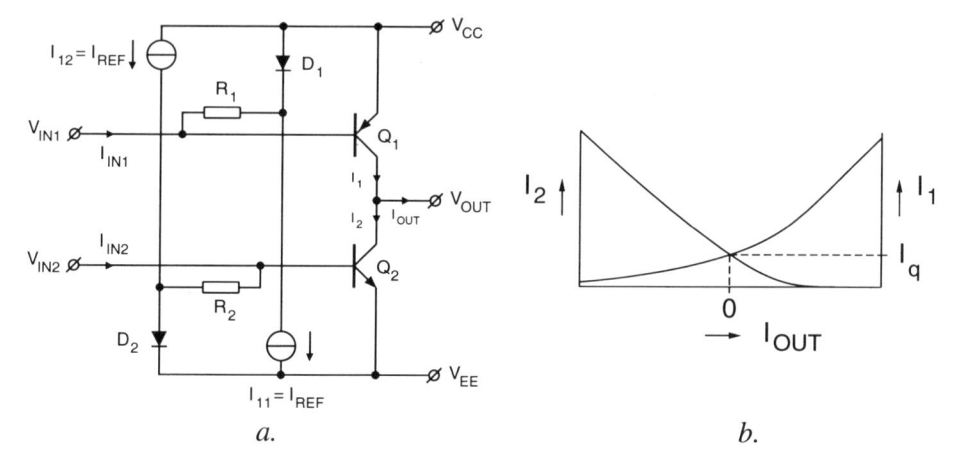

Fig. 8. Bipolar rail-to-rail output stage with separate resistive coupled class-AB control.
a. Circuit.
b. Class-AB characteristic.

input currents I_{in1} and I_{in2} generate equal voltages across R_1 and R_2, respectively. In such a way, a weak coupling is achieved between the two separate halfs of the control circuit so that the sum of the base-emitter voltages of the output transistors is approximately constant. This circuit can operate on supply voltages as low as 0.9 V. Drawbacks of this circuit are the relatively large loss of signal current in the diodes and the inaccurate control of the class-AB behavior which is caused by the separate biasing of the output transistors.

These problems can be solved by preserving the signal current that flows through the resistors R_1 and R_2 and that is lost in diodes D_1 and D_2. This can be done by inserting cascodes, Q_3 and Q_4, to redirect the signal to the other half of the output stage as shown in Fig. 9. Any signal current flowing through resistors R_5, R_6 is now passed by cascodes Q_3, Q_4, respectively, to the other output transistor.

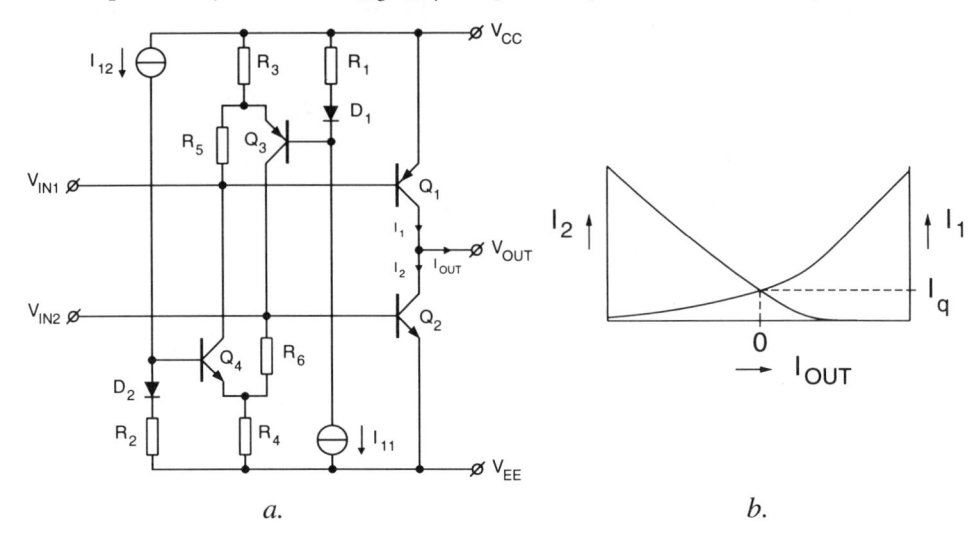

Fig. 9. Bipolar rail-to-rail output stage with resistor coupled feedforward class-AB control.
a. Circuit.
b. Class-AB characteristic.

Thus, the coupling between the two halves of the control circuit is restored and, a higher gain is obtained because the signal loss is prevented. A disadvantage is the inaccurate biasing caused by the large resistors. If we remove the resistors and use transistors and current sources to realize a similar topology, the circuit shown in Fig. 10 is obtained. In this circuit the output transistors are accurately biased, but it is difficult to realize a levelshift that is needed for current sources I_4 and I_3. A possible solution is applying a large scaling between the emitter areas of Q_4 and Q_6, and Q_5 and Q_3. The minimum supply voltage of these circuits is of the order of 1.1 V.

III. Efficient and low-distortion class-AB output stages

To prevent cutoff of the output transistors, a minimum current I_{min} should be maintained in the output transistors. Mathematically, this can be described by applying a translation given by

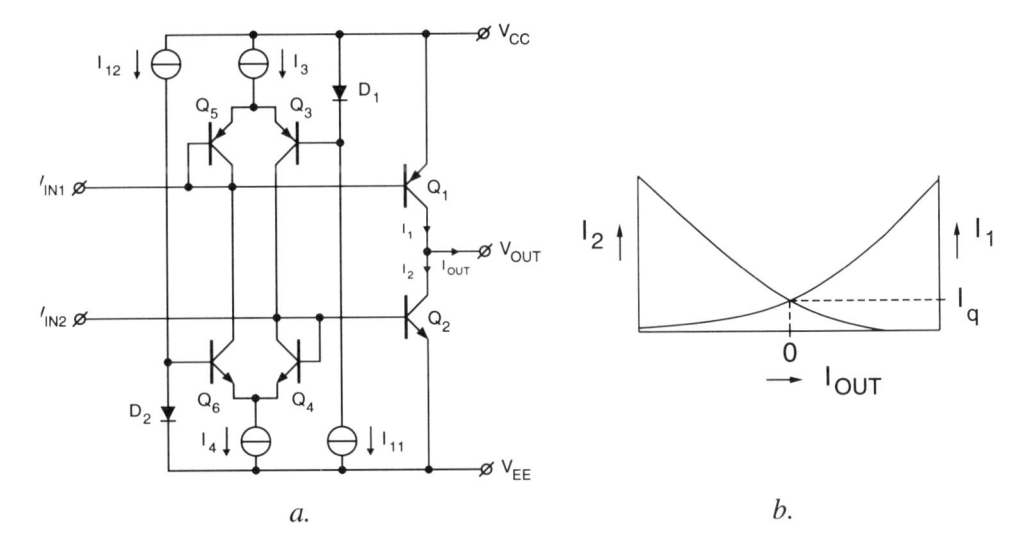

Fig. 10. Bipolar rail-to-rail output stage with diode-coupled feedforward class-AB control.
a. Circuit.
b. Class-AB characteristic.

$$I_1 = I_1' - I_{min} \tag{7}$$

$$I_2 = I_2' - I_{min} \tag{8}$$

to Eq. 1, where the primed currents represent the new variables. This translation also has to be applied to I_Q. If I_{min} is equal to $1/2\ I_Q$ the result is[6]

$$\frac{I_1' I_2'}{I_1' + I_2'} = I_{min} \tag{9}$$

The desired characteristic is plotted in Fig. 11.

A feedforward class-AB control circuit that implements this behavior is shown in Fig. 12. In this circuit the collector current of the output transistors is controlled by two translinear loops Q_2, Q_4, D_2, D_4 and Q_1, Q_3, D_1, D_3[4,5]. Assuming

$$\frac{A_1 A_3}{A_5 A_7} = \frac{A_2 A_4}{A_6 A_8} \tag{10}$$

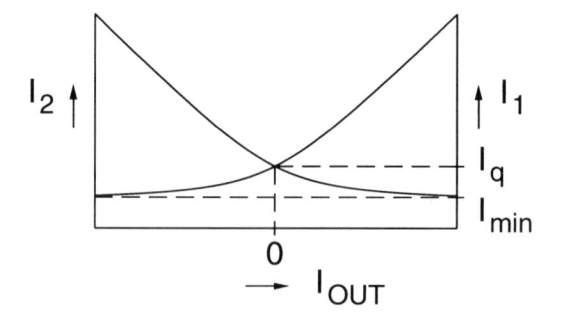

Fig. 11. Bipolar class-AB characteristics without cutoff.

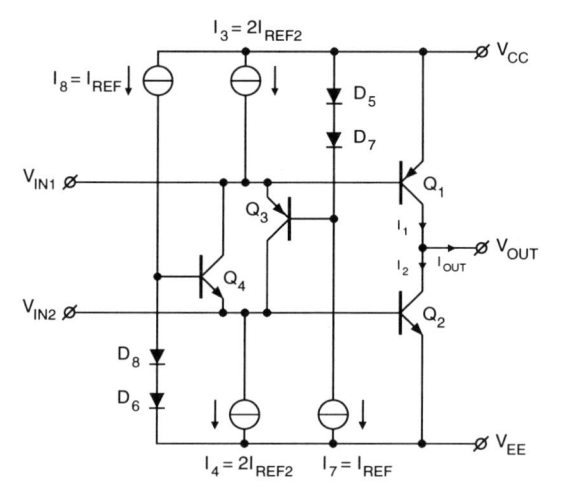

Fig. 12. Bipolar rail-to-rail output stage with transistor coupled feedforward class-AB control.

where A_1-A_4 are the emitter areas of Q_1-Q_4 and A_5-A_8 are the emitter areas of D_5-D_8, the relation between the collector currents of the output transistors is given by

$$\frac{I_1 I_2}{I_1 + I_2} = \frac{A_1 A_3 I_{REF}^2}{A_5 A_7 I_{REF2}} \tag{11}$$

This equation is similar to Eq. 9. The class-AB characteristic is plotted in Fig. 13. Because the class-AB control circuit provides a strong coupling between the two inputs V_{in1} and V_{in2}, the output stage can be driven using only one input. One problem seems to be that the impedance at the inputs V_{in1} and V_{in2} of the output stage is reduced by the emitters of Q_3 and Q_4. However, due to the posi-

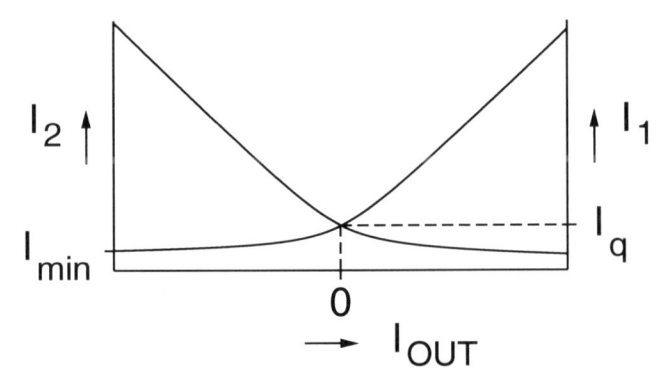

Fig. 13. Class-AB characteristics of bipolar rail-to-rail transistor coupled feedforward class-AB output stage.

tive feedback in the loop created by Q_3 and Q_4, any signal current flowing into the emitters will return. The signal just cannot be lost. Thus, the input impedance is not degraded, so that this configuration also works very well in CMOS as shown in Fig. 14[7]. A disadvantage is that the circuit needs two stacked diodes

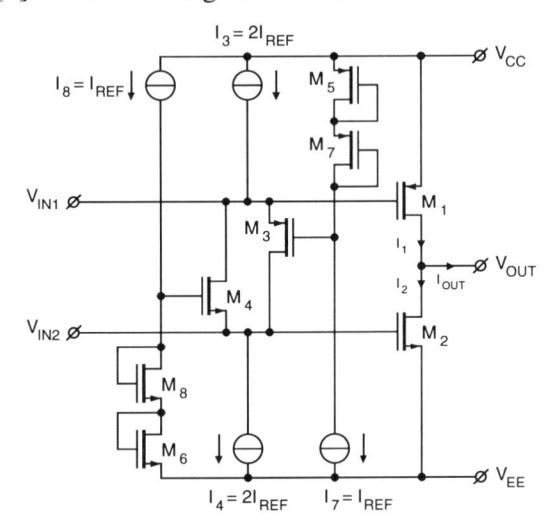

Fig. 14. CMOS rail-to-rail output stage with transistor coupled feedforward class-AB control.

which allows the circuit only to operate on supply voltages down to 1.8 V in bipolar technology. In CMOS the relation between the drain currents I_1 and I_2 is given by

$$\left(\sqrt{I_1} - \alpha\sqrt{I_q}\right)^2 + \left(\sqrt{I_2} - \alpha\sqrt{I_q}\right)^2 = 2\left(\frac{W}{L}\right)_6 \left(\frac{L}{W}\right)_8 I_q \qquad (12)$$

where it is assumed that the transistors operate in strong inversion and that

$$\frac{\left(\frac{W}{L}\right)_1}{\left(\frac{W}{L}\right)_2} = \frac{\left(\frac{W}{L}\right)_3}{\left(\frac{W}{L}\right)_4} = \frac{\left(\frac{W}{L}\right)_5}{\left(\frac{W}{L}\right)_6} = \frac{\left(\frac{W}{L}\right)_7}{\left(\frac{W}{L}\right)_8} \qquad (13)$$

and with

$$\alpha = 1 + \sqrt{\left(\frac{W}{L}\right)_6 \left(\frac{L}{W}\right)_8} \qquad (14)$$

Further, it is assumed that M_1 and M_5 as well as M_2 and M_6 have the same gate-source voltage in order to compensate for the body effect. The drain currents of the output transistors obey Eq. 12 until one of them exceeds a value of $\alpha^2 I_q$. At that moment one of the transistors M_4, M_8 is cut off and the drain current of the other output transistor stays at a minimum current given by

$$I_{min} = (\alpha - \sqrt{2}(\alpha - 1))^2 I_q \qquad (15)$$

If M_6 and M_8 have the same sizes the minimum current is about $0.34 I_q$. The class-AB relation is plotted in Fig. 15. The minimum supply voltage of this cir-

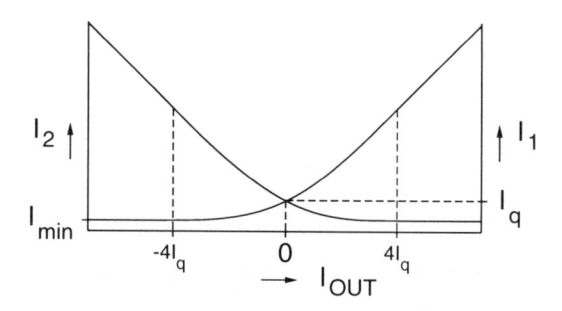

Fig. 15. Class-AB characteristic of CMOS rail-to-rail transistor coupled feedforward class-AB output stage.

cuit is between 1.6 V and 2.8 V depending on the process and the maximum output current.

To obtain accurate class-AB control at lower supply voltages, a feedback control should be used [8, 9, 6]. The first step towards a low-voltage class-AB control circuit is shown in Fig. 16. This circuit uses the same translinear topology as the feedforward class-AB output stage depicted in Fig. 12 to bias a Darlington output stage. Transistor Q_{11} measures the current through the output transistor Q_1.

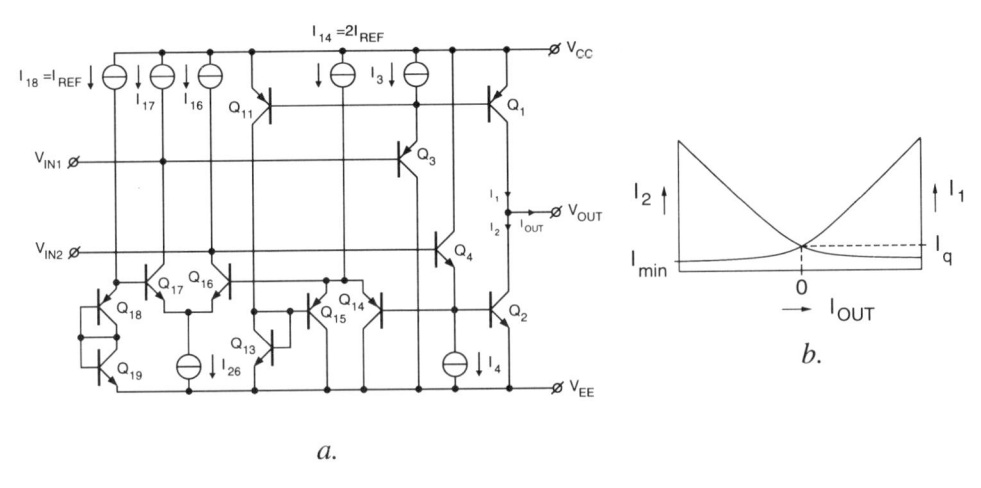

Fig. 16. *Bipolar feedback class-AB control applied in Darlington output stage.*
  a. Circuit.
  b. Class-AB characteristic.

This current is converted into a voltage by transistor Q_{13}. The base-emitter voltage of Q_2 is directly used in the translinear control loop consisting of Q_2, Q_{13}-Q_{14} and Q_{18}, Q_{19}. The control amplifier Q_{16}, Q_{17} together with the output stage Q_1-Q_4 and the translinear loop form a feedback loop that regulates the voltage at the emitter of Q_{14} and Q_{15} equal to the reference voltage created by Q_{18} and Q_{19}. Comparing this circuit with the circuit shown in Fig. 12, the function of Q_{18}, Q_{19} is equal to the diodes D_2, D_4 and D_1, D_3 in the feedforward control circuit and, the function of Q_{14}, Q_{15} is equal to Q_3 and Q_4 in the feedforward control circuit. Differential pair Q_{14}, Q_{15} is often called a decision pair because it is controlled by the smaller of its two input voltages. Thus, it decides which input is the smaller. Assuming a certain scaling is applied between the emitter areas of Q_1 and Q_{11}, and Q_2 and Q_{13}, and assuming all other transistors have equal emitter areas, the quiescent current is given by

$$I_Q = \frac{A_1}{A_{11}} I_{REF} \tag{16}$$

and the minimum current by

$$I_{min} = \frac{1}{2}I_Q \qquad (17)$$

In order to obtain a circuit that can operate on low supply voltages, the voltage at the bases of Q_{14} and Q_{15} must be shifted from one base-emitter voltage above the negative rail potential to a voltage close to the negative rail potential. This

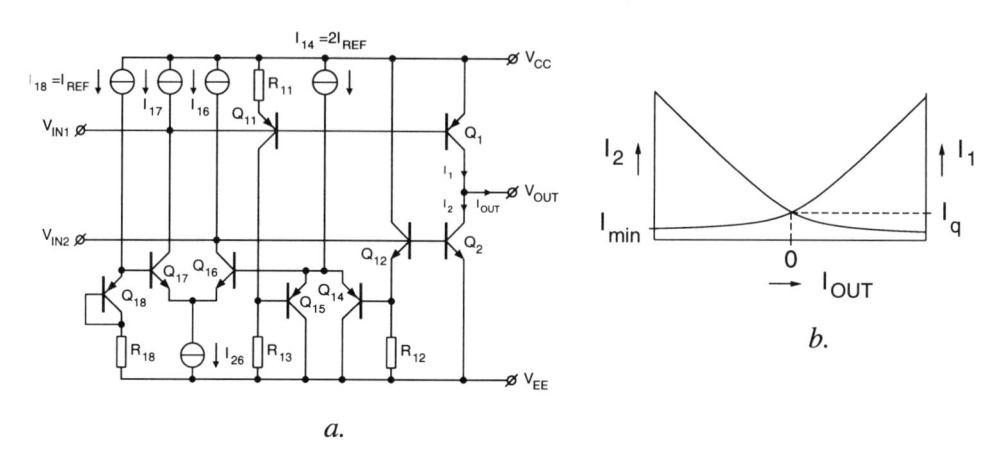

Fig. 17. Low-voltage bipolar feedback-biased class-AB output stage.
  a. Circuit.
  b. Class-AB characteristic.

can be achieved by replacing transistors Q_{19} and Q_{13} by resistors as shown in Fig. 17. A voltage that depends on the collector current of Q_2 is generated across R_{12} using emitter follower Q_{12}. By inserting R_{11} in series with the emitter of Q_{11}, the relation between the collector current of Q_1 and the voltage across R_{13} can be made equal to the relation between the collector current of Q_2 and the voltage across R_{12}. If that is done correctly, the class-AB behavior resembles the class-AB behavior of the previously discussed circuit. The quiescent current is given by

$$I_Q = \frac{A_1}{A_{11}} I_{REF} \exp\left(\frac{V_{R12}}{V_T}\right) \qquad (18)$$

where V_T is the thermal voltage and V_{R12} the voltage across resistors R_{12} and R_{13} in the quiescent state. The minimum current is found as

$$I_{min} = \frac{1}{2} \frac{V_{R12} - V_T \ln 2}{V_{R12}} I_Q \qquad (19)$$

If V_{R12} is 100 mV and with V_T is 26 mV at room temperature I_{min} is $0.4I_Q$. The output stage can operate on supply voltages down to 1 V. The same principle can be applied in CMOS as shown in Fig. 18[10]. However, this circuit has two

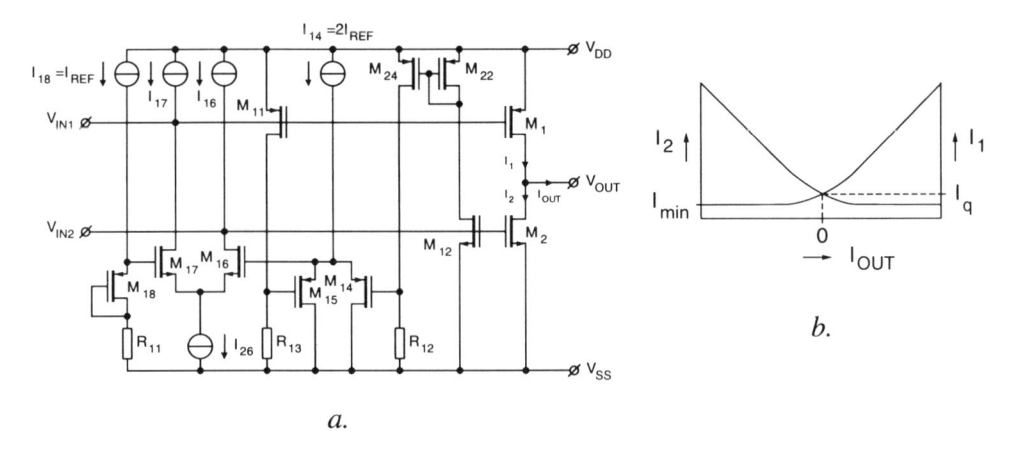

Fig. 18. *CMOS feedback-biased class-AB output stage.*
  a. Circuit.
  b. Class-AB characteristic.

drawbacks. First, due to the bad matching between the gate-source voltage of NMOS and PMOS transistors, the voltage across current source I_{26} is not accurately determined. Further, because of the resistors in the translinear loop, it is not possible to fix both the quiescent current and the minimum current accurately. If all the resistors are equal and the gate-source voltages of M_{14}, M_{15} are equal to M_{18} in the quiescent state, the quiescent current is accurately fixed

$$I_Q = \frac{W_1}{L_1}\frac{L_{11}}{W_{11}}I_{REF} \qquad (20)$$

Then the minimum current is given by

$$I_{min} = \left(1 - \frac{\sqrt{2}-1}{R_{12}}\sqrt{\frac{2L_{14}}{\mu_p C_{ox} W_{14} I_{REF}}}\right)I_Q \qquad (21)$$

Clearly, the minimum value is influenced by R_{12} and the gate-source voltage of M_{14} or M_{15}. By making R_{12} large enough, a reasonable amount of minimum current can be secured. The first drawback can be solved by replacing PMOS transistors M_{14} and M_{15} by two diode coupled NMOS transistors as shown in Fig. 19. Assuming again that the gate-source voltage of M_{14} and M_{15} is equal to the gate-source voltage of M_{18}, the quiescent current is equal to

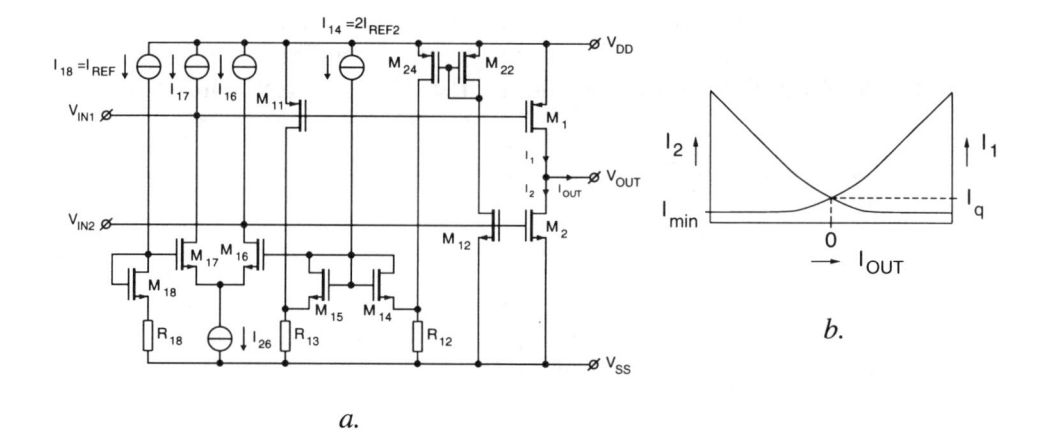

Fig. 19. *CMOS feedback-biased class-AB output stage with diode connected decision pair.*
a. Circuit.
b. Class-AB characteristic.

$$I_Q = \frac{W_1 L_{11}}{L_1 W_{11}} (I_{REF} - I_{REF2}) \qquad (22)$$

and the minimum current

$$I_{min} = \left(1 - \frac{I_{REF2}}{I_{REF} - I_{REF2}} - \frac{\sqrt{2}-1}{R_{12}(I_{REF} - I_{REF2})} \sqrt{\frac{2(I_{REF2} L_{14})}{\mu_n C_{ox} W_{14}}}\right) I_Q \qquad (23)$$

Now the reference current I_{14} also flows through resistors R_{12} and R_{13} counteracting the setting of the minimum current. Thus, the reference current must be small compared to the total current through R_{12} and R_{13} in order to maintain a minimum current.

A CMOS feedback circuit without resistors is shown in Fig. 20[11]. In this class-AB control the decision pair and the feedback amplifier are combined, M_{16}-M_{18}. The output stage functions the same way as the previously discussed output stages. Assuming the gate-source voltages of M_{14}, M_{15} and M_{19} are equal in the quiescent state, the quiescent current is found as

$$I_Q = \frac{W_1 L_{11}}{L_1 W_{11}} (2I_{REF2} - I_{REF}) \qquad (24)$$

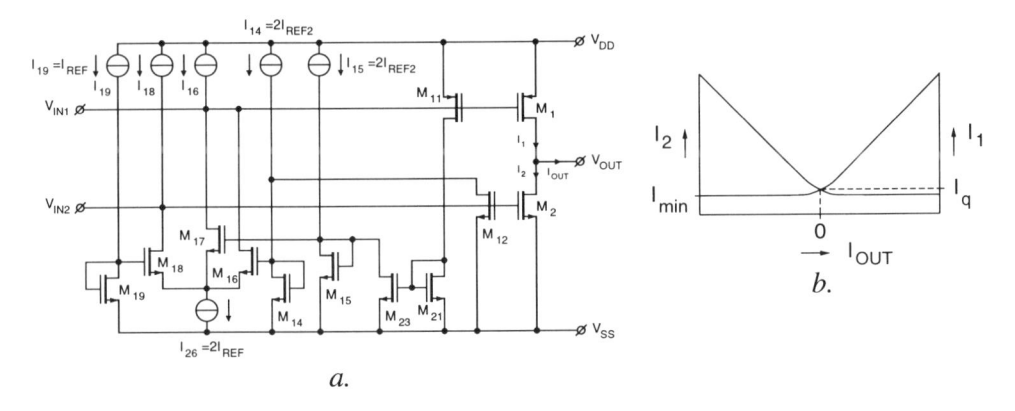

Fig. 20. *CMOS feedback-biased class-AB output stage using folded diode structure.*
a. *Circuit.*
b. *Class-AB characteristic.*

and the minimum current as

$$I_{min} = \left(1 - \frac{I_{REF}\left(2(\sqrt{2}-1)\sqrt{\frac{W_{19}L_{18}}{L_{19}W_{18}}} + (3-2\sqrt{2})\frac{W_{19}L_{18}}{L_{19}W_{18}}\right)}{2I_{REF2} - I_{REF}}\right)I_Q \quad (25)$$

If for example I_{REF} is equal to I_{REF2} and the W/L ratio of M_{18} is ten times the W/L ratio of M_{19}, the minimum current is approximately $0.7I_Q$. So, without tuning sufficient minimum current is easily obtained. The minimum allowable supply voltage of this circuit is 1.3 V, depending on the process.

IV. Voltage Efficient Input Stages

To make efficient use of the supply voltage at the input of the amplifier for applications as input and output buffers, input stages should be able to handle common-mode input voltages from rail to rail. A rail-to-rail common-mode input voltage range can be achieved by placing an N-type input stage Q_2, Q_4 in parallel with a P-type input stage Q_1, Q_3 as shown in Fig. 21. The outputs of the input stages are added in the summing circuit consisting of Q_{11}-Q_{14} and R_{11}-R_{14}. A drawback of this technique is that the transconductance g_m varies a factor two over the common-mode input voltage range, as is shown in Fig 22. This impedes an optimal frequency compensation of the amplifier.

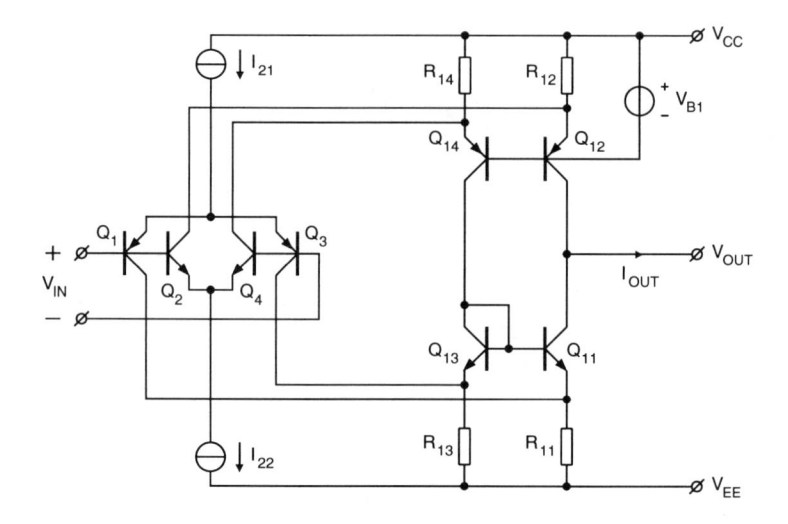

Fig. 21. Rail-to-rail bipolar complementary input stage.

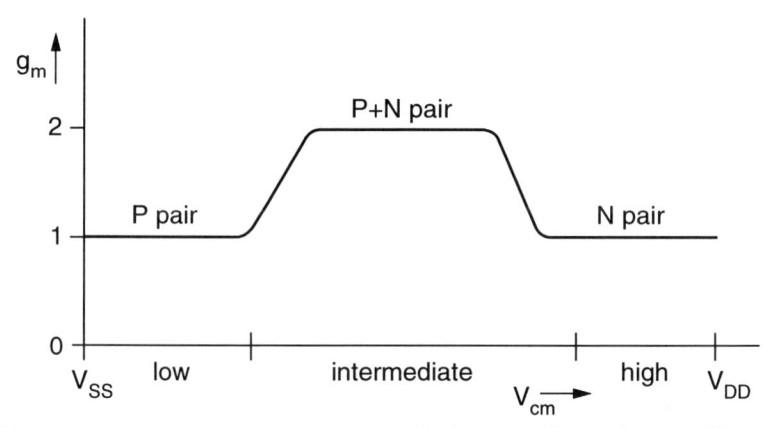

Fig. 22. g_m versus the common-mode input voltage for a rail-to-rail input stage.

In bipolar technology, the transconductance g_m is proportional to the collector current. Therefore, a constant g_m can be obtained by keeping the sum of the tail-currents of the complementary input stages constant as given by

$$I_{Q1,Q3} + I_{Q2,Q4} = I_{21} \tag{26}$$

where $I_{Q1,Q3}$ is the bias current of the PNP input pair and $I_{Q2,Q4}$ the bias current of the NPN input pair and I_{21} is the tail current source. This gives the sum of the transconductance of the input pair as

$$g_{m,Q1,Q3} + g_{m,Q2,Q4} = g_{21} \tag{27}$$

where $g_{m,Q1,Q3}$ is the transconductance of the NPN input stage, $g_{m,Q2,Q4}$ is the transconductance of the PNP input stage and g_{21} is the transconductance corresponding to the current I_{21}. A realization is shown in Fig. 23[12, 8, 13]. Depending on the common-mode input voltage the current switch, Q_5, directs the tail-current, I_{21}, to either one of the input stages. The result is a constant g_m over the common-mode input range, as shown in Fig. 24.

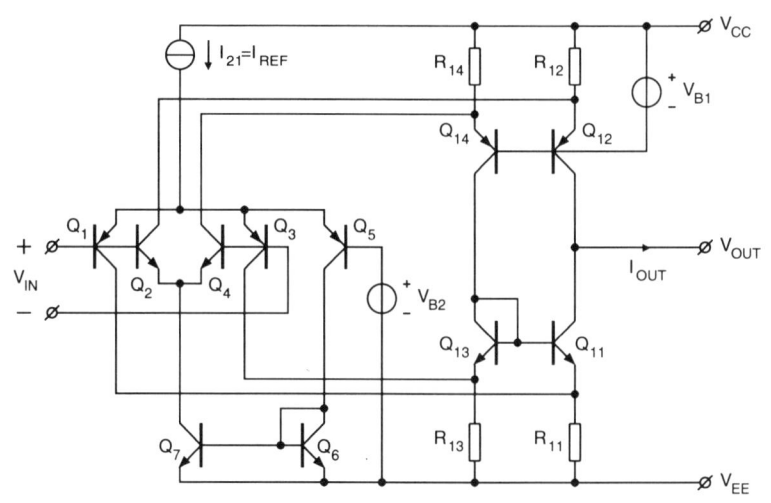

Fig. 23. Rail-to-rail bipolar input stage with g_m control.

In CMOS technology, the g_m of a transistor is also proportional to the drain current when it operates in weak-inversion and therefore, a similar circuit can be used as in bipolar technology. In strong-inversion, however, the bipolar g_m-control technique leads to a g_m which varies approximately 40% over the common-mode input voltage range, as shown in Fig. 24. The reason is that the g_m of an MOS transistor which operates in strong-inversion, is proportional to the square-root of the drain-current instead of proportional to it.

Another possible g_m-control circuit is shown in bipolar technology in Fig. 25. In this circuit the current switches Q_5-Q_8 remove the current that is not needed in such a way that the sum of the tail currents is constant as can be concluded by evaluating the translinear loop consisting of the PNP input transistors Q_1, Q_3 the PNP current switch Q_5 the NPN current switch Q_6 and the NPN input transistors Q_2, Q_4. This gives the relation

$$I_{Q1,Q3}I_{Q2,Q4} = (I_{21} - I_{Q1,Q3})(I_{22} - I_{Q2,Q4}) \tag{28}$$

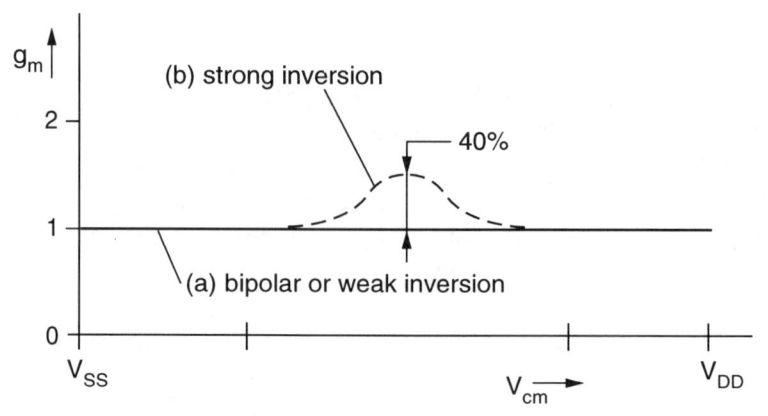

Fig. 24. g_m versus common-mode input voltage for the complementary input stage with one-times current mirror.

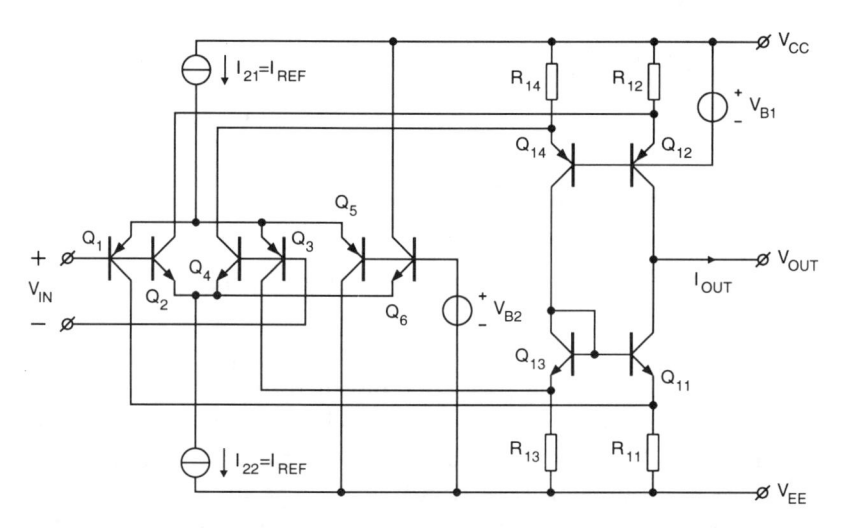

Fig. 25. Rail-to-rail bipolar input stage with g_m control using two current switches.

where it is assumed that the emitter area of the current switches is two times the emitter are of the input transistors. If I_{22} is equals to I_{21} Eq. 26 is obtained. The transconductance as a function of the common-mode input voltage is plotted in Fig. 26. Instead of removing the current to the supply rails, the current can also be passed around the input transistors as shown in Fig. 27. An advantage of this circuit is that the currents in the intermediate stage do not change as a function of the common-mode input voltage. This simplifies the design of the intermediate stage. Further, it allows a simple and compact implementation of an extra output

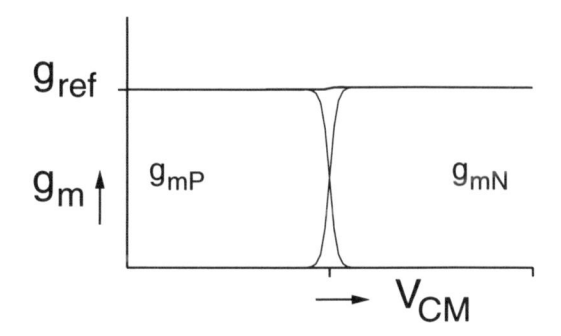

Fig. 26. g_m versus the common-mode voltage for the rail-to-rail input stage with two current switches.

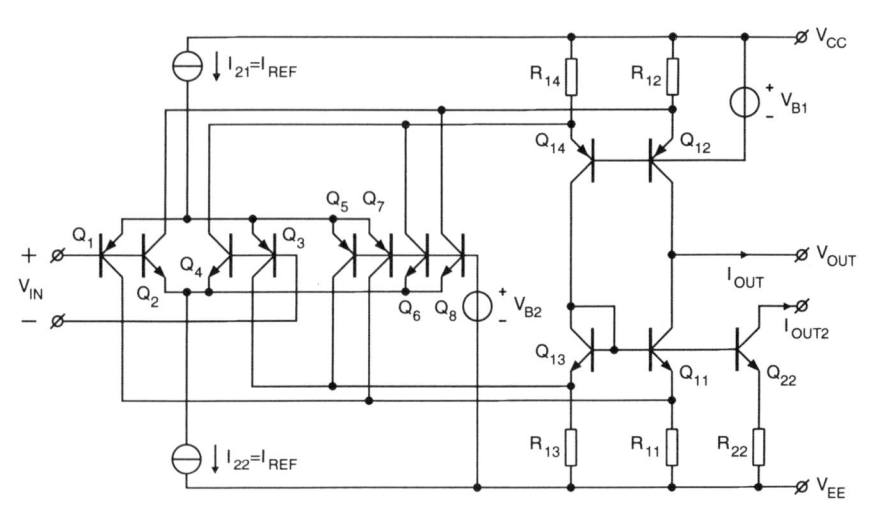

Fig. 27. Rail-to-rail bipolar input stage with g_m control using two current switches yielding constant common-mode currents in the intermediate stage.

signal without duplicating the complete rail-to-rail input stage as indicated by the output at the collector of Q_{22}. Such an extra output is used in multipath compensated amplifiers [14, 15, 16]. A drawback of the circuit is that the current switches add noise that is of the same order as the noise of the input transistors. This can be improved by inserting emitter resistors as shown in Fig. 28. To compensate for the voltage drop across these resistors a levelshift V_{LS} has to be inserted between the bases of the PNP switches and the NPN switches. Another effect is that the cross-over range is extended as shown in Fig. 29. This helps to reduce the distortion in the cross-over range caused by differences between the

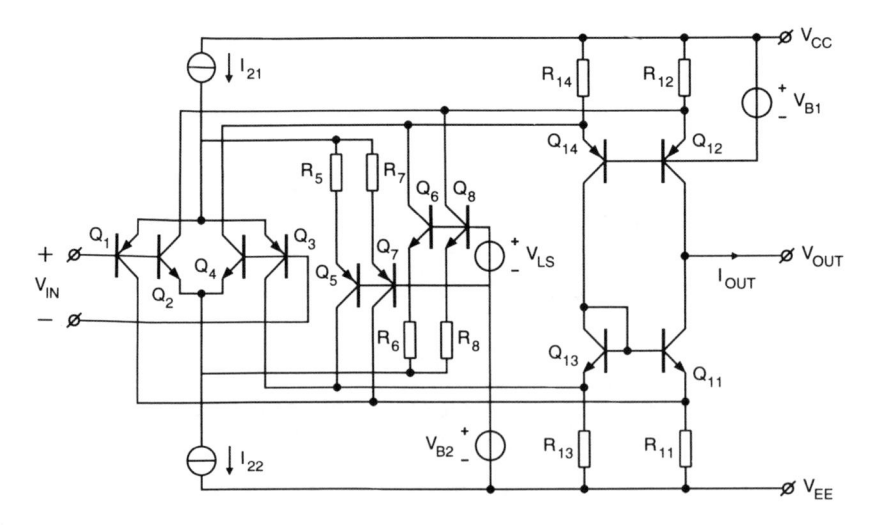

Fig. 28. Rail-to-rail bipolar input stage with g_m control using two current switches and emitter resistors.

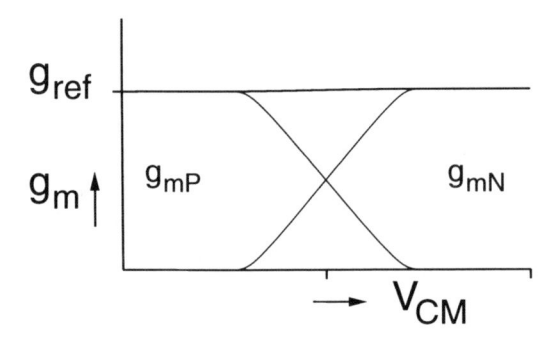

Fig. 29. g_m versus the common-mode voltage for the rail-to-rail input stage with two current switches and emitter resistors.

offset of the two input pairs. The same circuit can also be used in CMOS as shown in Fig. 30. In the CMOS version also a levelshift is needed. One way to implement this levelshift is by making the W/L ratio of the current-switch transistor three times larger than the W/L ratio of the input transistors. In this way, the transconductance in the middle of the common-mode range is equal to the transconductance of a single input stage. However, in the cross-over range the transconductance varies 15% as can be clearly seen in Fig. 31.

A good solution in CMOS is to regulate the sum of the square root of the tail currents to a constant value. A rail-to-rail input stage with square-root control is

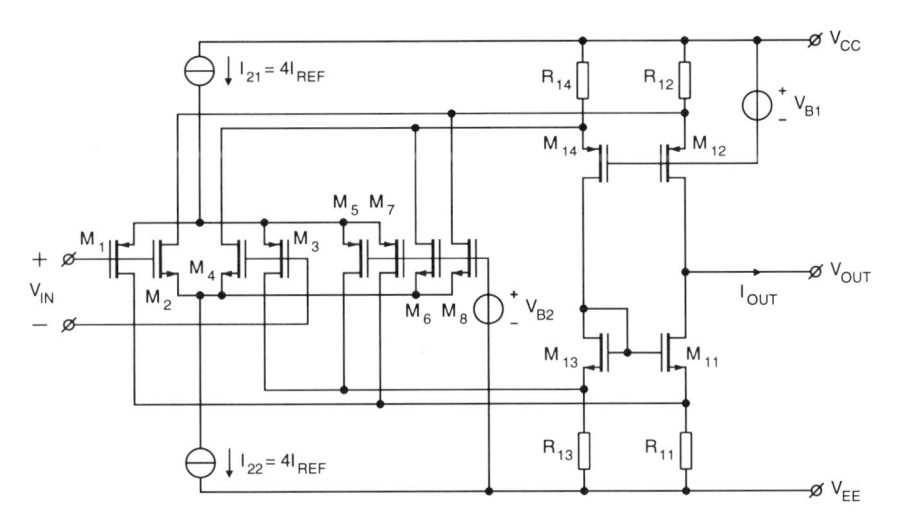

Fig. 30. Rail-to-rail CMOS input stage with g_m control using two current switches.

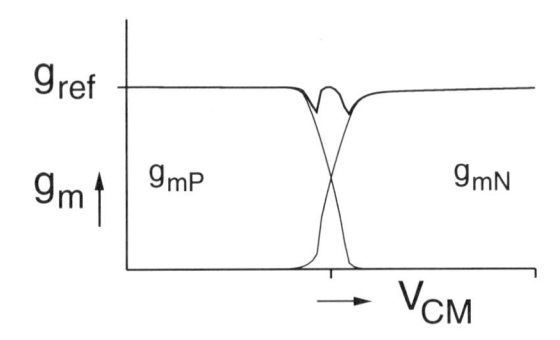

Fig. 31. g_m versus the common-mode voltage for the CMOS rail-to-rail input stage with two current switches.

shown in Fig 32[10]. The heart of the circuit is the translinear loop consisting of transistors M_{21}-M_{24}. The current switch M_{27}, together with current source I_3 measures the tail current of the N-channel input pair. Assuming that transistors M_{21}-M_{24} are matched, a current is generated in M_{24} that obeys

$$\sqrt{I_{M1,M3}} + \sqrt{I_{M2,M4}} = 2\sqrt{I_{REF}} \qquad (29)$$

where $I_{M2,M4}$ is the drain current of the N-channel pair and $I_{M1,M3}$ the tail current of the P-channel input pair generated by M_{24} and assuming their transconductance values obey

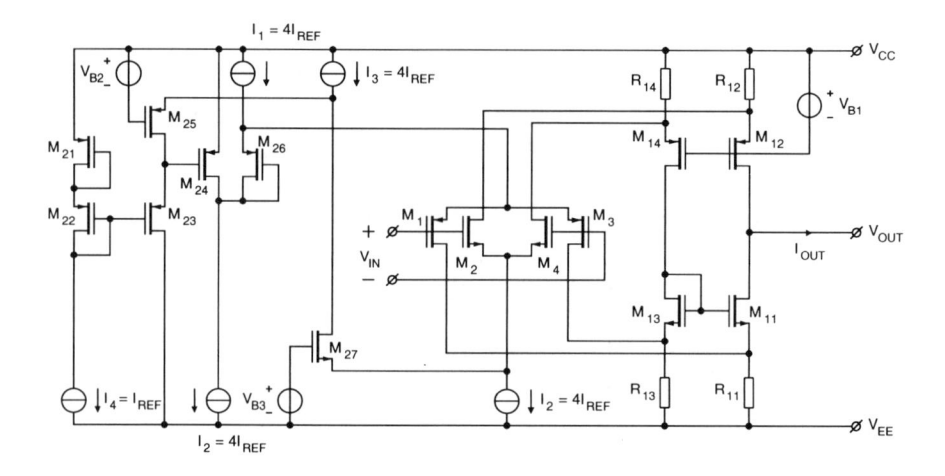

Fig. 32. Rail-to-rail CMOS input stage with square-root g_m control.

$$K\frac{W}{L} = \frac{1}{2}\mu_n C_{ox}\left(\frac{W}{L}\right)_n = \frac{1}{2}\mu_p C_{ox}\left(\frac{W}{L}\right)_p \qquad (30)$$

where $(W/L)_n$ is the width over length ratio of the N-channel transistors and $(W/L)_p$ the width over length ratio of the P-channel transistors. Through diode M_{26} the drain current of M_{24} is used as tail current for the P-channel input pair. M_{26} functions as a current limiter. This transistor prevents that the tail current of the P-channel input pair becomes larger than $4I_{REF}$ when the gate-source voltage of transistor Q_{23} is smaller than its threshold voltage. Fig. 33 shows the transcon-

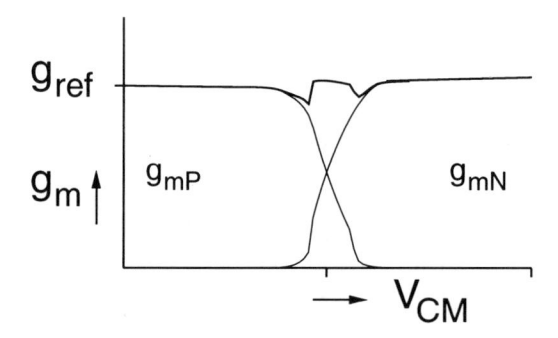

Fig. 33. g_m versus the common-mode voltage for the CMOS rail-to-rail input stage with square-root control.

ductance as a function of the common-mode input voltage. The transconductance varies only 12% over the common-mode input voltage range.

An alternative to controlling the currents of CMOS rail-to-rail input stages is controlling the voltages[17, 18]. Ideally, a constant voltage source should be connected between the sources of the P-channel and the N-channel pair that equals the sum of a reference gate-source voltage of a PMOS transistor and a reference gate-source voltage of an NMOS transistor. A first approach is shown in Fig. 34.

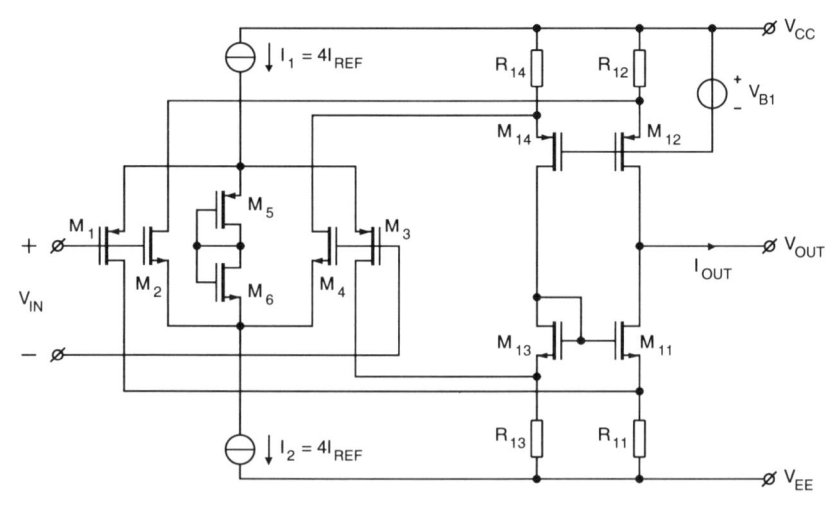

Fig. 34. Rail-to-rail CMOS input stage with constant-voltage g_m control.

In this input stage the reference voltage is formed by a diode connected PMOS transistor in series with a diode connected NMOS transistor. If the W over L ratio of these transistors is six times the W over L ratio of the input transistors, the current through the transistors is $3I_{REF}$ in the intermediate part of the input-voltage range. The remaining current I_{REF} is used for biasing both input pairs. In the other parts of the common-mode input-voltage range no current flows through the reference diodes and one of the input stages is biased by $4I_{REF}$. The result is a constant g_m. In the transition regions the current through the diodes changes. Consequently, the voltage across the diodes changes. This causes a variation of the transconductance of 23% as shown in Fig. 35.

The g_m control can be improved by making the current through the diodes more constant. Fig 36 shows an implementation of an improved g_m-control circuit. Again, reference diodes M_5, M_6 are connected between the sources of the two input stages. Using M_7 a feedback loop is created that forces the current of current source I_3 to flow through M_8. This constant current which equals $1/2\ I_{REF}$, also flows through the reference diodes M_5, M_6. Therefore the voltage across the diodes is constant and the variation of the transconductance as a function of the common-mode input voltage is only 8% as shown in Fig. 37.

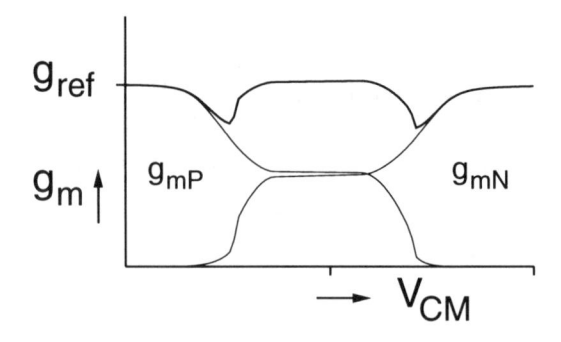

Fig. 35. g_m versus the common-mode voltage for the CMOS rail-to-rail input stage with constant-voltage control.

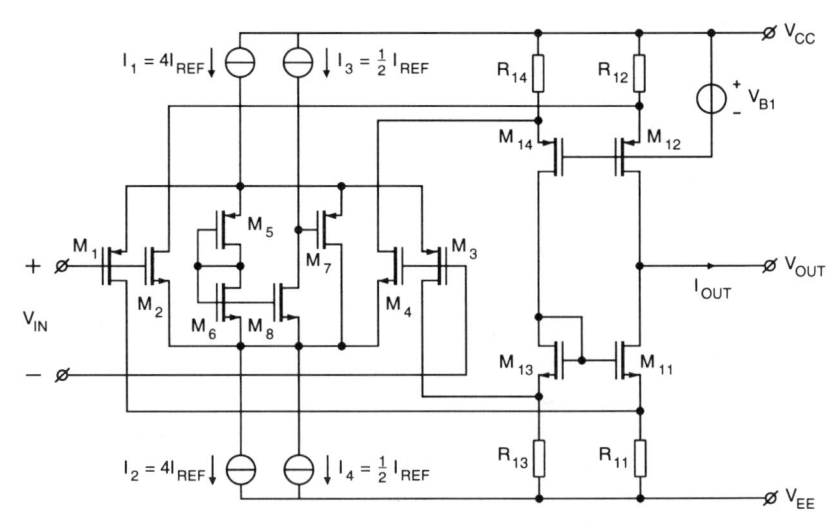

Fig. 36. Improved rail-to-rail CMOS input stage with constant-voltage g_m control.

A final method to control the transconductance is by controlling the size of the input-stage transistors. This idea is used in the circuit depicted in Fig. 38. The transconductance is made constant by switching on an additional input stage in the parts of the common-mode input-voltage range where only the N-type or the P-type input stage is operating. For low common-mode voltages, switch transistor M_{21} is on and M_{22} is off so that both PMOS input pairs M_1, M_3 and M_5, M_7 operate. In the intermediate part of the common-mode range both switches are off and PMOS pair M_5, M_7 and NMOS pair M_6, M_8 function. For high common-mode voltages, switch M_{22} is on and M_{21} is off, so that both NMOS pairs M_2, M_4 and M_6, M_8 operate. The advantage of this technique is that it works properly in weak inversion as well as in strong inversion. Therefore this g_m-control is very

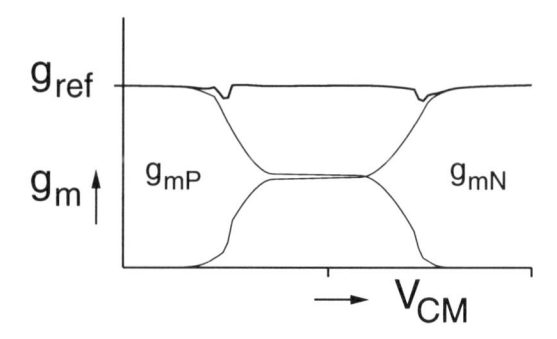

Fig. 37. g_m *versus the common-mode voltage for the CMOS rail-to-rail input stage with improved constant-voltage control.*

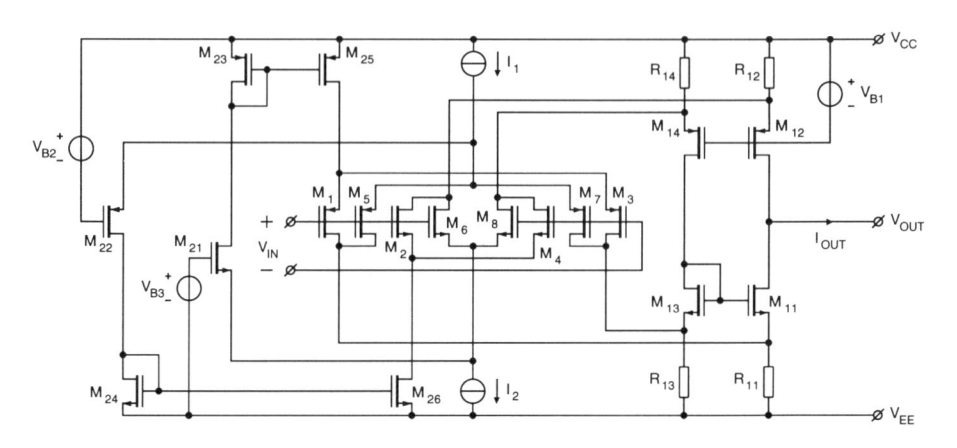

Fig. 38. Rail-to-rail CMOS input stage with constant-voltage g_m control using multiple input pairs.

useful in building blocks with programmable bias current where the operating region of the transistors is not known in advance. The transconductance as a function of the common-mode voltage is plotted in Fig. 39a in weak inversion and in Fig. 39b in strong inversion. In weak inversion the transconductance is nearly constant, while in strong inversion the g_m varies 20%.

Finally, It should be noted that the offset voltage of rail-to-rail input stages changes when the complementary input stage gradually switches from one input pair to the other. This change of the offset voltage degrades the *CMRR* in the transition regions. A way to improve this is by using calibration.

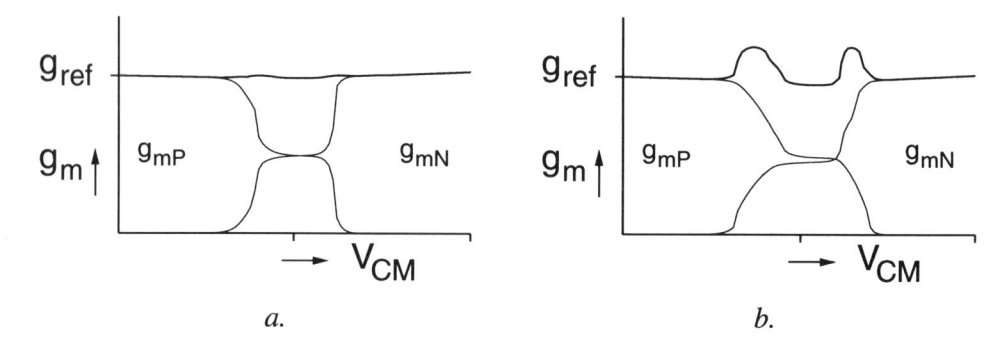

Fig. 39. g_m versus the common-mode voltage for the CMOS rail-to-rail input stage with multiple input stages.
a. Weak inversion.
b. Strong inversion.

V. Conclusions

Due to the limited supply voltage and the limitation on the power consumption, low-voltage operation amplifiers must be power efficient. Therefore, in these amplifiers power efficient class-AB output stages are used. Further, voltage efficient rail-to-rail input stages are applied to use the supply voltage at the input of the amplifier efficiently.

Bipolar and CMOS rail-to-rail class-AB output stages have been shown that can operate on supply voltages down to 1 V. Finally, several bipolar and CMOS implementations of rail-to-rail input stages have been presented.

References

[1] R. Hogervorst, J.H. Huijsing, K.J. de Langen, R.G.H. Eschauzier, "Low-voltage Low-power Amplifiers", in Analog Circuit Design", Kluwer 1995.
[2] B. Gilbert, "Translinear circuits: a proposed classification", Electron. Lett., Vol. 11, pp. 14-16, 1975.
[3] E. Seevinck and R.J. Wiegerink, "Generalized translinear circuit principle", IEEE J. Solid-State Circuits, Vol. SC-26, pp. 1098-1102, 1991.
[4] W.C.M. Renirie, J.H. Huijsing, "Simplified Class-AB Control Circuits for Bipolar Rail-to-Rail Output Stages of Operational Amplifiers", Proc. European Solid-State Circuits Conference, Sept. 21-23, 1992, pp. 183-186.
[5] W.C.M. Renirie, K.J. de Langen, J.H. Huijsing, "Parallel Feedforward Class-AB Control Circuits for Low-Voltage Bipolar Rail-to-Rail Output Stages of Operational Amplifiers", Analog Integrated Circuits and Signal Processing, Vol. 8, 1995, pp. 37-48.
[6] E. Seevinck, W. de Jager, P. Buitendijk, "A Low-Distortion Output Stage with improved

stability for monolithic power amplifiers", IEEE J. Solid-State Circuits, Vol. SC-23, June 1988, pp. 794-801.
[7] D.M. Monticelli, "A quad CMOS single-supply Opamp with rail-to-rail output swing", IEEE J. of Solid-State Circuits, Vol. SC-21, Dec. 1986, pp. 1026-1034.
[8] J.H. Huijsing and D. Linebarger, "Low-Voltage Operational Amplifier with Rail-to-Rail Input and Output Ranges", IEEE J. of Solid-State Circuits, Vol SC-20, No. 6, Dec. 1985, pp. 1144-1150.
[9] J.H. Huijsing and F. Tol, "Monolithic Operational Amplifier Design with improved HF behavior", IEEE J. Solid-State Circuits, Vol. SC-11, No. 2, April 1976, pp. 323-328.
[10] R. Hogervorst, R.J. Wiegerink, P.A.L. de Jong, J. Fonderie, R.F. Wassenaar, J.H. Huijsing, "CMOS Low-Voltage Operational Amplifiers with constant-gm Rail-to-Rail input stage", Proc. IEEE International Symposium on Circuits and Systems, San Diego, May 10-13, 1992, pp. 2876-2879.
[11] R.G.H. Eschauzier, R. Hogervorst, J.H. Huijsing, "A Programmable 1.5 V CMOS Class-AB Operational Amplifier with Hybrid Nested Miller Compensation for 120 dB Gain and 6 MHz UGF", in Digest IEEE International Solid-State Circuits Conference, February 16-18, 1994, pp. 246-247.
[12] J.H. Huijsing and R.J. v.d. Plassche, "Differential Amplifier with Rail-to-Rail Input Capability and Constant Transconductance", U.S. Appl. No. 4,555,673, Nov. 26, 1985.
[13] J. Fonderie, M.M. Maris, E.J. Schnitger, J.H. Huijsing, "1-V Operational Amplifier with Rail-to-Rail input and output Ranges", IEEE J. Solid-State Circuits, vol. SC-24, pp. 1551-1559, Dec. 1989.
[14] J. Fonderie and J.H. Huijsing, "Operational Amplifier with 1-V Rail-to-Rail Multipath-Driven Output Stage", IEEE J. of Solid-State Circuits, vol. 26, No. 12, Dec. 1991, pp. 1817-1824.
[15] J.H. Huijsing and M.J. Fonderie, "Multi-stage amplifier with capacitive nesting and multi-path forward feeding for frequency compensation", U.S. Patent, Appl. No. 5,155,447, Oct. 4, 1992.
[16] R.G.H. Eschauzier, L.P.T. Kerklaan and J.H. Huijsing, "A 100-MHz 100-dB Operational Amplifier with Multipath Nested Miller Compensation Structure", IEEE J. Solid-State Circuits, Vol. 27, No. 12, Dec. 1992, pp. 1709-1717.
[17] J.H. Huijsing, R Hogervorst, J.P. Tero, "Compact CMOS Constant-gm Rail-to-Rail Input Stages by Regulating the Sum of the Gate-Source Voltages Constant", US patent application, Appl. no. 08/430,517, filed April 27, 1995.
[18] R. Hogervorst, J.P. Tero, J.H. Huijsing, "Compact CMOS Constant-gm Rail-to-Rail Input Stages with gm-control by an Electronic Zener Diode", in Proc. ESSCIRC 1995, pp. 78-81.

Low–Voltage Continuous-Time Filters

R. Castello
Dipartimento di Elettronica – Universita' di Pavia
Via Ferrata, 1 – 27100 Pavia – Italia

Abstract

This chapter reviews the design of continuous time analog filters at low supply voltage. In particular it concentrates on gm–C type filters because, at the present time, they have the greatest commercial importance for high frequency medium/low precision applications. Both fundamental and practical limitations to the achievable dynamic range at low supply voltage are explained. The paper reviews well established circuit and architectural techniques as well as some promising new ones which might result in performance improvements in the future.

I. Introduction

Over the last years the interest toward low power low voltage IC has grown due primarily to the increased commercial importance of portable equipment. From the technical point of view, the supply voltage of modern IC should be reduced for two main reasons: technology scaling and power consumption reduction in digital circuits. Regarding to the first point, deep submicron feature sizes force to reduce the supply voltage for reliability reasons. Specifically, the maximum supply is 3.3 V at about 0.5 μm channel and 2 V at about 0.25 μm. Regarding the second point, the widespread use of digital signal processing (DSP) motivates the use of lower supply voltages to reduce power consumption. In fact, it has been shown [1] that there is an optimum supply voltage (about 1.5 V for a 2 μm technology) that gives a minimum power dissipation in digital CMOS circuits.

Even as DSP techniques become more popular many analog blocks are still required. In particular A/D and D/A converters with their associated pre and post-filters will always be needed as interface blocks. In addition, if the required linearity and signal–to–noise ratio are not too high (40 dB or less) analog signal processing may use less power as compared with DSP. To reduce cost, the analog blocks and the DSP should be included on the same die using a single low supply. Therefore new low voltage peripheral analog blocks need to be designed [2].

It should be noted that to reduce the power consumption of analog circuits, there is no reason to reduce the supply voltage. In fact the following fundamental relation exists between the power consumption P and the Dynamic Range DR of analog filters [3,4]:

$$P = \eta k T f_0 DR \qquad (1)$$

where f_0 is the bandwidth and η is a parameter that depends on the filter structure. Eq. 1 is derived assuming that the noise is KT/C limited, that power consumption is proportional to $f_o C V^2$, being dominated by the charging and discharging of the memory capacitors, and that the maximum swing is equal to the supply voltage. In this limit condition for a given dynamic range the power consumption is independent of the supply voltage. In reality, however, to preserve the DR of analog circuits while reducing the supply voltage the power dissipation has to be increased. At low supply voltage is therefore particularly important to optimize power consumption.

This paper will focus on the design of high frequency continuous time filters at low supply voltage and is organised as follows. Section 2 presents some technology considerations for low voltage analog design. In Section 3 the possible alternatives for continuous time filters (g_m–C, g_m-C Opamp, active-R-C, MOSFET–C) are briefly compared from the point of view of low voltage compatibility. On the last part of the section g_m-C and MOSFET-C are discussed in more detail. In Section 4 several g_m-C V/I converters based on both bipolar and MOS (saturated and linear) transistors are compared in term of dynamic range and tunability. In section 5 several techniques to preserve as much as possible Dynamic Range and Tunability at low supply voltage are discussed. Finally Section 6 gives a summary of the paper and draws some general conclusions.

II. Technology considerations

As explained above, the need for low voltage analog blocks is driven by the compatibility with large digital cores. As a consequence, the choice of technology, being dependent on the requirements of the digital part, can only be between pure CMOS and BiCMOS. In a pure digital CMOS technology the threshold voltages are relatively high (close to 1 V at 1 µm minimum channel) to insure a sufficiently small subthreshold current [1]. In addition a relatively high body factor (as high as 1) is generally found. Under these conditions, allowing also for technology spreads and temperature variations, the design of low voltage analog blocks is quite difficult. Things tend to improve with the scaling of the technology because the threshold voltages are also scaled down. Nevertheless the situation remains quite challenging.

One possible solution is a multi–threshold technology where large V_{TH} devices are used for the digital core and low V_{TH} devices are used for the analog part [5].

This, however, is more costly and today is used only when cost reduction is not the key issue [6-8] although in the future it may become more common. An example showing the usefulness of a multi-threshold technology in simplifying the design of analog circuits is shown in Fig. 1.

This is a high swing current mirror that does not require any extra bias voltage and uses only three transistors [6]. The circuit uses a low V_{TH} MOS cascode below a high V_{TH} device and takes advantage of the well controlled difference between the two threshold values.

A less expensive, although less flexible, alternative is to use a technology that provides unimplanted devices. In this case a low threshold nMOS device is obtained by shielding it from the threshold shift implant that is normally applied to the entire wafer. The resulting natural transistor with a typical threshold of 300 mV and a reduced technology spread and body effect can be quite useful for supply voltages below 3V.

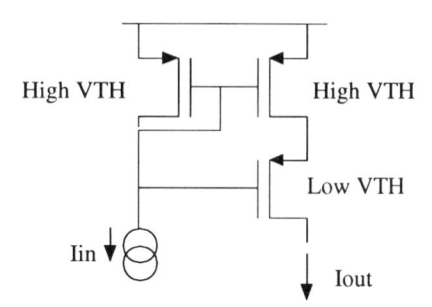

Fig. 1 - Current mirror with two different PMOS threshold voltages

A BiCMOS technology provides several advantages, some of them particularly useful at low voltage. The benefits, however must be weighted against the extra cost. In the end, the commercial success demonstrates the correctness of the choice made. The key benefits that come from the availability of a bipolar transistor are the following: larger gm for a given current, lower offset, larger f_T at low voltage drop, lower voltage noise (both thermal and 1/f). At low supply, the last two properties become particularly relevant. Notice that as the technology is scaled the MOS device reaches the peak of the f_T at lower and lower overdrive voltage. It follows that in the deep submicron region the advantages of BiCMOS will be less significant although some points like better noise (especially 1/f) and less offset may still be important.

III. Comparison of different filtering techniques at low supply voltage

Analog filters can be realized using either sample–domain or continuous–time techniques. Continuous–time (CT) filters have been increasingly used in high–frequency applications [9]. The primary advantage of CT filters is the fact that they do not need a clock at a frequency much higher than the signal bandwidth. A second advantage is the absence of aliasing effects which can degrade the filter dynamic range. On the other hand, CT filters have generally a lower precision as compared with SC (especially at low frequency) even when tuning techniques are used to compensate technology and temperature spreads. In addition CT filters tend to have lower programmability and voltage swing than SC filters. As a consequence, CT filters find their main application in mixed digital/analog systems when high processing speed is required [10-14]. Due to this, the following will only deal with high frequency continuous time filters. Furthermore fully differential circuit will always be assumed due to their better immunity to unwanted signals. This is because at high frequency substrate noise contamination and cross talk from other blocks become more severe.

Most monolithic C T filters are based on integrators. Therefore in the rest of the paper the different CT filters (i.e. g_m–C, g_m-C opamp, MOSFET–C, and Active-R-C) will be compared in terms of integrators. The key performance requirement of a C T integrator are: linearity, noise level, phase response power consumption and tunability. The importance of the last property is often overlooked. In fact just to compensate for technology and temperature variations an almost 3 to 1 tunabiliy is required. Furthermore for programmable filters, as found for instance in disk drives, a10 to 1 tuning range may be required.

The g_m-C integrator is made up of a V/I converter (transconductor) driving a capacitor, as shown in Fig. 2.

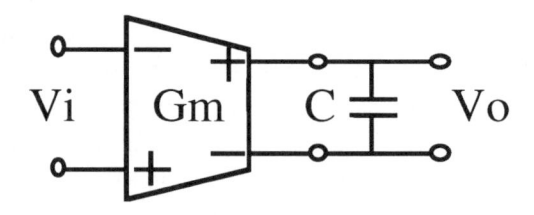

Fig. 2- gm-C integrator

The integrator unity gain frequency is controlled by the transconductance of the V/I converter and by the total capacitance at its output. The transconductor operates in an open loop configuration. As a consequence it must handle a large input and output signal and is sensitive to parasitic capacitance at its output nodes. On the other hand, open loop operation and the fact that the transconductor is driving only high impedance loads, makes the g_m-C approach most suitable for high speed.

To eliminate the effect of the parasitic capacitances on the integrator unity gain frequency a g_m-C opamp topology can be used as shown in Fig. 3 [15].

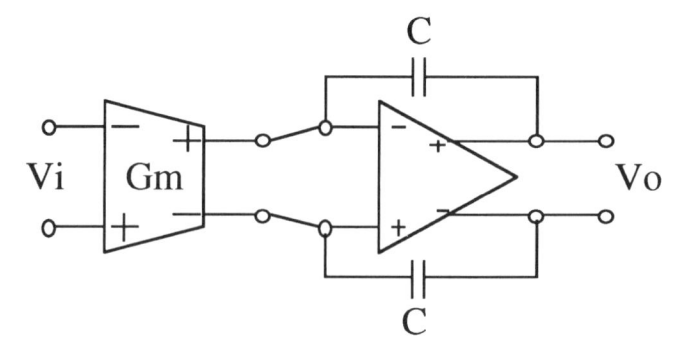

Fig. 3- gm-C Opamp integrator

In this case the load capacitances are connected to the virtual grounds of the additional opamp. This topology effectively shorts out any parasitic at the output of the transcunductor. An additional advantage of this approach is to drastically reduce the output swing of the transconductor thereby simplifying its design. This can be particularly significant at low supply voltage. On the other hand, the additional active element (opamp) increases power consumption and adds extra phase shift thereby reducing the maximum operating frequency.

The third, and possibly most obvious, approach uses as the basic element the active RC integrator of Fig. 4. In this case the V/I convertion is performed in a passive way by the input resistor. In addition the integrating capacitance is connect to virtual grown resulting in small sensitivity to parasitic. The main advantages of this configuration are very good linearity and superior noise performance as compared with the previous solutions. One disadvantage is the need to drive a low impedance load which makes the opamp more difficult to design as compared with the g_m-C opamp case. However the most important limitation is the lack of a tuning mechanism for the resistor. This problem has been addressed in some implementation either using discrete tuning (via a bank of switchable integrating capacitor) [16] or using tunable junction capacitors (varactors) [17]. Discrete capacitor tuning has some important drawbacks. First it complicates the tuning loop. Second it can become quite large and complex if small tuning steps are required. Third it becomes more difficult to do as the frequency is increased due to parsitic effects (capacitors and resistors) associated with the control switches. This last point will be discussed in more detail in section 5 where the problem of extending the tuning range of C T filter at low supply voltage will be addressed.

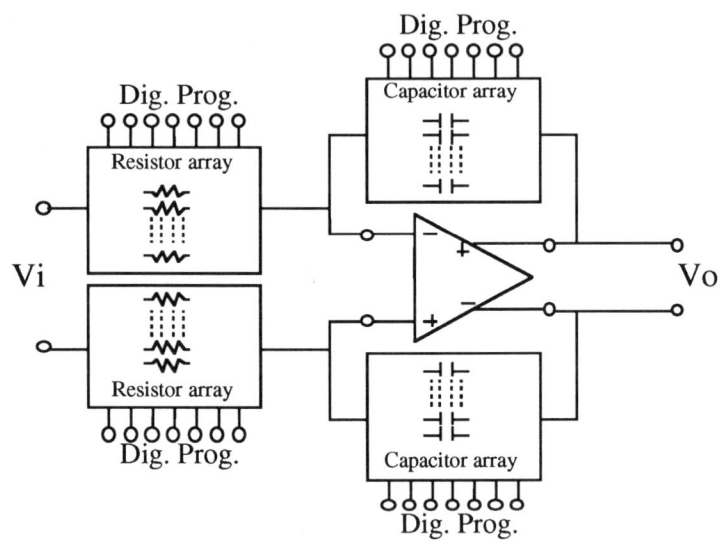

Fig. 4 - Active R-C integrator with discrete programming

A forth possible topology that tries to address the tuning problems of active-RC integrators is the MOSFET-C shown in Fig. 5. In this case the fixed resistors used in active-RC integrators is substituted by MOS transistors operated in the linear region. The value of this equivalent resistor, and therefore of the integrator unity gain frequency, is controlled by the tuning voltage V_c. The control voltage V_c and the quiescent common mode voltage of the signal must be chosen in such a way to maximize the swing while maintaining always triode operation. This becomes more and more difficult at lower supply voltage. MOSFET-C integrators are almost exclusively used in a fully differential implementation. In this case, in fact, the even order non linearity are cancelled out resulting in a potentially very linear V/I characteristic. As for the case of active-RC integrators the presence of the opamp (which must drive a low impedance load) causes additional phase errors.

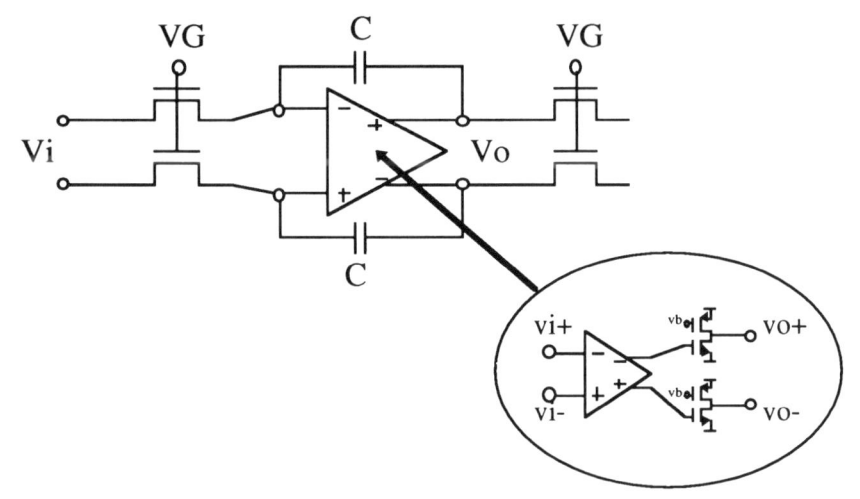

Fig. 5- MOSFET-C integrator with large output swing

The purpose of this paper is to compare the possible CT filter architectures listed above as the supply voltage is reduced. A general comparison is quite difficult to do since its conclusions depend on the specific application and on the performance requirement. In the following, due to space limitations, Active-RC and g_m-C opamp filters will not be discussed any further. This is because, for the moment, they both have had a relatively small practical impact [15,34].

As the supply voltage is reduced the number of devices that can be staked between the rails is drastically reduced. This limits the type of circuit configurations (e.g. cascode) that can be used. In addition, it is no longer possible to turn on MOS transmission gates over the entire voltage swing [12]. However, the most critical effect is the drastic reduction in the linear voltage swing of the transconductors. This reduces the achievable dynamic range since, for a given current consumption, the noise remains approximately constant. Furthermore a large voltage swing is important (especially in mixed mode systems) to increase the immunity of the filter from the disturbances produced by adjacent circuits. Therefore the effect of supply voltage scaling on the *DR* and on the voltage swing is used to compare g_m–C and MOSFET–C filters at low supply.

The maximum peak–to–peak signal for an MOSFET–C integrator (SW_{MC}) is equal to:

$$SW_{MC} = [V_{DD} - V_{TH} - V_{OV} - \Delta V] \quad (2)$$

as shown in Fig.6 where V_{TH} is the threshold voltage of an n–MOS device including body effect.

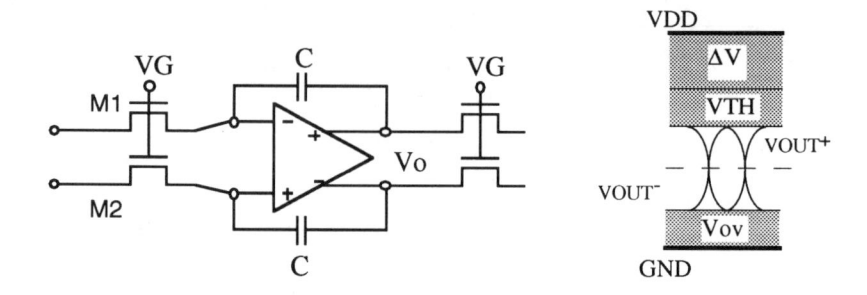

Fig. 6 - MOSFET-C integrator and its possible output swing

This equation assumes that the MOS resistors (M1, M2) reach the limit of the saturation region when the input nodes reach the top of the swing. Furthermore the voltage V_G at the gate of the M1 and M2 is assumed to be ΔV below the positive rail, where ΔV is the minimum voltage drop required by the tuning circuit. Finally, the signal distortion caused by the modulation of the source–to–body voltage of M1 and M2 during operation is assumed to be negligible. In practice the maximum swing is much less than that given in eq. (2). In fact, first the two MOS resistors must remain deep in their linear region during the entire voltage excursion of the input nodes. Second, the gate voltage V_G cannot be kept fixed but must vary to compensate for technology spread and thermal variations. Therefore the minimum possible value for V_G is much lower than V_{DD}. Last, to contain body effect distortion, the source–to–body voltage of M1 and M2 must be always larger than some minimum value. This limits the input voltage swing in the negative direction. Taking into account these non idealities the peak–to–peak swing can vary between 1/3 and 1/5 of $[V_{DD} - V_{TH} - V_{OV} - \Delta V]$. For a 3V supply and assuming a standard technology, this gives a swing between 250 and 400mVpp for a fixed filter cut–off frequency. Furthermore the minimum supply voltage that insures functionality can be estimated to be between 1.5V and 2V. These limitations can be overcome using a special low threshold technology or a double supply, one for the filter and one for the control gate. Both this solutions are more costly and this is not often acceptable for the system design. Another possible solution is the use of on chip voltage multiplication for the control signal. This requires little extra cost but is more susceptible to cross-talk and noise contaminations and has not been proved in practice.

The maximum swing for a g_m–C integrator is very much dependent on the topology used. However, practical circuits can be designed with a swing equal to a large portion of the supply voltage as demonstrated in several MOS–based circuits, reported in literature [14,18,19]. Furthermore using a bipolar structure operation down to a 1.2V supply with a 400mVpp swing was demonstrated.

Regarding noise, MOSFET–C can be made to approach the fundamental kT/C limit quite closely [20]. On the other hand, the noise of g_m–C filters is quite different from case to case and often changes as the circuit is tuned. Gm-C filters with a noise comparable to that of MOSFET–C can be designed although they may not have the largest voltage swing.

From the above considerations there is no clear indication of which is the more suitable approach for low voltage C T filters. However, to limit the length of the paper, the following will concentrate on gm–C filters. This is because g_m–C filter have the highest signal bandwidth [21] (above 100 MHz) and, at the present time, they are the most widely used in commercial circuits. In addition some of the results derived for g_m–C filters can be easily extended to MOSFET–C filters.

IV. Comparison of different gm-C integrators at low supply voltage

The *DR* for a CT filter can be expressed as follows [4,17]:

$$DR = \beta \frac{V_{max}}{\sqrt{F\frac{kT}{C}}} \qquad (3)$$

where V_{max} is the maximum signal amplitude (for a given amount of distortion), C is the total capacitance of the filter, and F is the Noise–Excess–Factor of the transconductor defined as the ratio between the input noise power of the transconductor and the noise power of a resistor equal to $1/g_m$, β depends on the filter structure (e. g. the number of transconductors) and on the shape of the frequency response (e. g. the value of Q). In the following, β will be assumed to be the same for all cases. In addition the maximum voltage swing is assumed to be limited by the input stage. In actual circuits the need to have input–to–output compatibility may further restrict the maximum swing. Typically, for a high order filter, the time constants of the integrators making up the filter have an average value close to 1/BW where BW is the filter bandwidth. i. e. $\frac{1}{n}\sum_i \frac{C_i}{g_{mi}} = \frac{1}{BW}$.

Assuming for simplicity that all transconductors have the same transconductance g_m it follows that $C = \frac{ng_m}{BW}$ since $\sum_i C_i = C$. Substituting in (3) and multiplying and dividing by the DC current consumption (I) of a transconductor we have:

$$DR = \beta\sqrt{\frac{nI}{kTBw}}\sqrt{\frac{g_m}{FI}}V_{max} \qquad (4)$$

The first term in the equation is independent of the transcontuctor topology and depends only on the filter bandwidth and structure (through β) and the total current consumption nI. On the other hand, the second term can be used as a figure of merit for the transconductor.

In addition to its internally generated noise the desired signal can be contaminated by unwanted signal produced by external sources and coupled into the filter. At high frequency, and when very complex mixed analog/digital systems are integrated on the same chip, external noise can become the dominant noise source in a filter. An example where this may be the case is the prefilter of a read channel of an hard disk drive.

Two possible figure of merit can be used to indicate the ability of a filter to properly operate in the presence of disturbances. First assume that the disturbance is a signal source whose output impedance is lower than the impedance level at the internal nodes (voltage source). In this case the relevant figure of merit is the maximum swing in the various nodes V_{max}. V_{max} is set by the maximum allowable distortion which typically is assumed to be 1%. Second assume that the disturbance source impedance is much higher than the node impedance (current source). In this case the appropriate figure of merit is the product of the maximum voltage swing times the admittance level at the various nodes for a given current consumption, i.e. $\frac{g_m}{I}V_{max}$. The second case is the one encountered in the largest majority of cases.

In the following we will compute the three possible figure of merit for different type of differential transconductors using both bipolar and MOS technologies.

a. Bipolar transconductors [17,22]

The voltage signal that can be applied to a bipolar differential pair for a given amount of distortion is proportional to the thermal voltage. V_t i.e. $V_{in} = \alpha_1 V_t$. For example, the maximum differential signal (V_{max}) which keeps the *THD* below 1% is about V_t (α_1 =1). On the other hand in a bipolar implementation the minimum supply voltage V_{DD} is limited by the base–emitter voltage V_{BE} which is much higher than the thermal voltage V_t. This means that the signal level is only slightly affected by the reduction of the supply voltage. For a bipolar differential pair the Noise–Excess–Factor F is equal to 1. As a consequence, the DR does not depend (to first order) on the supply voltage and, using equation (4) can be expressed as follows:

$$DR = \beta \sqrt{\frac{nI}{kTBw}} \sqrt{\frac{V_t}{2}} \qquad (5)$$

Finally, the third figure of merit, i.e. $\frac{g_m}{I} V_{max}$ for 1% maximum allowable distortion can be shown to be equal to 1/2.

A key advantage of bipolar transconductors is that the transconductance can be widely changed with negligible variation in the V_{BE} voltage. As a consequence a low supply voltage does not impose restrictions on the tunability of the transconductor.

These considerations lead to the conclusion that bipolar CT filters will become more and more competitive with respect to CMOS ones as the supply voltage is reduced, especially if wide tunability is required. In fact, at very low supply, the reduced voltage swing of bipolar transconductors ceases to be a limitation since the signal swing has to be reduced anyway. On the other hand, in this situation, their high tunability, high speed and low power consumption can be fully exploited. As an example, a pure bipolar transconductor able to operate from a 1V supply has recently been reported [23]. In this case, using a parallelization technique, a factor of 3 increase in the maximum swing is obtained ($\alpha_1 = 3$) while F is increased by much less. This gives a better value for all the figures of merit.

b. MOS operating in saturation region

An MOS differential transconductor operating in the saturation region is shown in Fig. 7. For a given linearity, the maximum input differential signal is a fraction of the overdrive voltage $V_{ov} = V_{GS} - V_{TH}$ i.e. $V_{in} = \alpha_2 V_{ov}$. For example, the maximum peak signal which keeps the THD below 1% V_{max} is about $0.75 V_{ov}$ ($\alpha_2 = 0.75$). However, to compensate for temperature variations and technology spreads, the overdrive voltage must be varied. This means that V_{max} is limited by the minimum overdrive voltage V_{ovmin} i.e. $V_{max} = \alpha_2 V_{ovmin}$.

For the configuration of Fig. 7 the minimum supply voltage (for a given V_{max}) is given below:

$$V_{DD} = V_S + V_{TH} + V_{ovmax} + \frac{V_{max}}{2} \qquad (6)$$

where V_S is the minimum voltage drop required by the current source I.

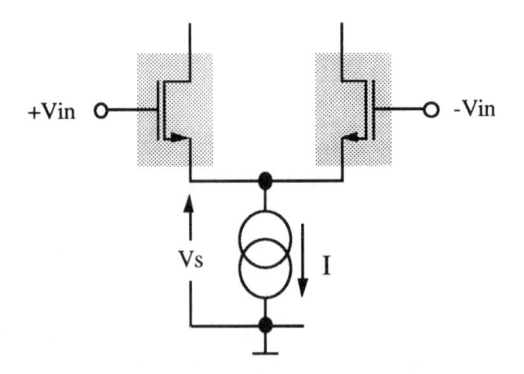

Fig. 7- Example of saturated MOS transconductor

The ratio between maximum and minimum overdrive voltage is the tunability t, that is:

$$t = \frac{V_{ovmax}}{V_{ovmin}} \qquad (7)$$

As a consequence the maximum allowable input voltage can be expressed as follows:

$$V_{max} = 2\frac{V_{DD} - V_S - V_{TH}}{1 + \frac{2t}{\alpha_2}} \qquad (8)$$

The minimum value for g_m/I is given by $1/V_{ovmax}$ giving a minimum value for the second figure of merit, i.e. $\frac{g_m}{I} V_{max} = \frac{\alpha_2}{t}$.

For the case of a saturated MOS differential pair the noise–excess–factor does not depend on tuning and is equal to 4/3. As a consequence the DR can be expressed as follows:

$$DR = \beta \sqrt{\frac{nI}{kTBw}} \sqrt{\frac{V_{DD} - V_S - V_{TH}}{1 + 2t/\alpha_2}} \sqrt{\frac{3\alpha_2}{2t}} \qquad (9)$$

As for the bipolar case, using a more complicated structure instead of a simple differential pair [19] an improvement in the above figures of merit can be obtained.

c. MOS operating in the triode region

The output current of a MOS which operates in the triode region depends both on the gate and drain voltages. These two voltages can be used either as input or tuning signal. Two BiCMOS implementations of this concept are illustrated in Fig. 8a and b.

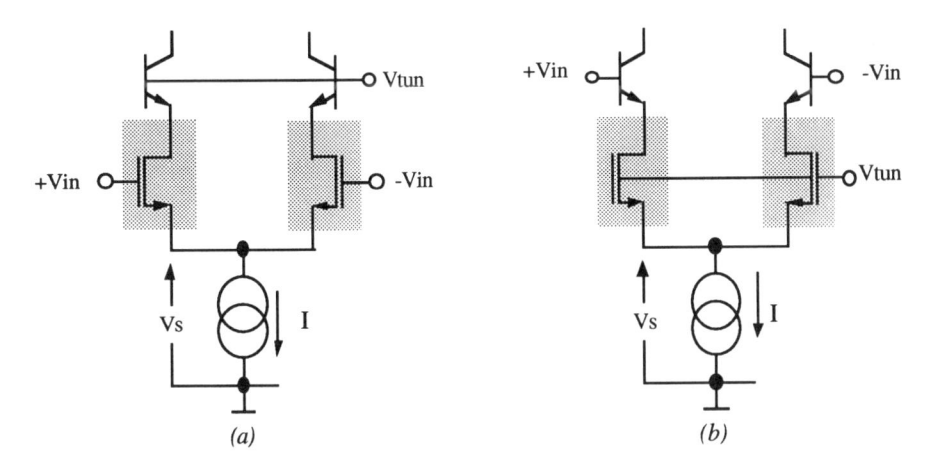

Fig. 8 (a) e (b)- Example of two BiCMOS transconductors with MOS device operating in triode region

In case (a) the input signal is the gate voltage while the transconductance is controlled by V_{DS}. In case (b) the input signal is applied to the drain voltage (through the bipolar transistor) while the transconductance is controlled by the gate voltage. We will consider these two cases separately.

Case (a) [14,18]: Referring to Fig. 8a, the minimum supply voltage (for a given V_{max}) is given by Eq. 6 as in the case of a saturated differential pair. Usually V_{ov} is fixed because the tuning is done varying V_{DS}. The only restriction on V_{max} is to insure that the MOS remains in the triode region during its operation, that is:

$$V_{DSmax} = V_{ov} - \frac{V_{max}}{2} \qquad (10)$$

The tunability range is given by the ratio between the maximum and minimum V_{DS} i. e. $t = V_{DSmax}/V_{DSmin}$. From Eq. 6 and 10 follows that

$$V_{max} = V_{DD} - V_{TH} - V_S - tV_{DSmin} \qquad (11)$$

Neglecting second order effects, V_{DSmin} can be extremely small (tens of millivolts) and a better swing than in the previous cases can be obtained especially for large values of t.

The arbitrary reduction of V_{DS}, to extend both tuning range and input signal, is in contrast with both practical and fundamental limitations. In practice, V_{DSmin} must be larger than some minimum value (50 to 100 mV) for matching reason and to reduce distortion. Extra distorsion, in fact, is caued by the modulation of the drain voltage by the signal current, do to the finite impedance at the bipolar emitter. On the other hand a fundamental limitation comes from the increasing of noise. In fact the Noise–Excess–Factor has its maximum in correspondence to the minimum of V_{DS}. This value, for a differential pair, is given by:

$$F_{MAX} = 2\frac{V_{ov} - V_{DSmin}}{V_{DSmin}} \qquad (12)$$

The Dynamic Range (DR) can be shown to be given by:

$$DR = \beta\sqrt{\frac{nI}{kTBw}}\sqrt{\frac{V_{DSmin}[V_{DD} - V_S - V_{TH} - tV_{DSmin}]^2}{[V_{DD} - V_S - V_{TH} + (t-1)V_{DSmin}][V_{DD} - V_S - V_{TH} + (t-2)V_{DSmin}]}}$$
$$(14)$$

Equation (14) was obtained assuming that $t\,V_{DSmin} = V_{DSmax} \ll V_{DD} - V_{TH} - V_S$. This is equivalent to assume that V_{DSmin} and/or t are sufficiently low.

The value of $\frac{g_m}{I}V_{max}$, can be shown to be given by:

$$\frac{g_m}{I}V_{max} = \frac{V_{DD} - V_{TH} - V_S - tV_{DSmin}}{V_{DD} - V_{TH} - V_S + (t-1)V_{DSmin}} \qquad (13)$$

As opposed to dynamic range $\frac{g_m}{I}V_{max}$ has a maximum in corrispondance to the minimum of V_{DS}. The above results show that there is a trade-off between maximum swing and $\frac{g_m}{I}V_{max}$ on one hand and DR on the other hand as the V_{DSmin} is changed.

If swing and $\frac{g_m}{I}V_{max}$ are the main concern the optimum solution is to minimise V_{DSmin}. An example in which this may be the case is in mixed signal systems where the noise coming from the digital section is dominant with respect to the intrinsic noise of the filter. In the limit situation (V_{DSmin} extremely small) $\frac{g_m}{I}V_{max}=1$, $V_{max} = V_{DD} - V_S - V_{TH}$ and $DR = \beta\sqrt{\frac{nI}{kTBw}}\sqrt{V_{DSmin}}$.

On the other hand, if DR is the main concern an optimum V_{DS} value can be found. The corresponding optimum DR can be shown to be $0.3\sqrt{\dfrac{V_{DD}-V_S-V_{TH}}{t-1}}$.

Case (b) [24]: For this transconductor V_{DS} is fixed since tuning is done changing the gate voltage from V_{ovmin} to V_{ovmax}. In this case, to achieve sufficient linearity the maximum input signal swing is limited in the positive and negative direction (respectively $V^+{}_{max}$ and $V^-{}_{max}$) by the following conditions. In the positive direction the maximum instantaneous drain–to–source voltage should always be sufficiently below the minimum overdrive voltage (i.e. $V_{DS}+V^+{}_{max}/2 = \alpha_4 V_{ovmin}$.). In the negative direction the minimum instantaneous V_{DS} should always be greater than zero to insure class A operation i.e. $V^-{}_{max}/2 = V_{DS}$. In practice speed and linearity considerations limit this value to a fraction of the quiescent V_{DS} i.e. $V^-{}_{max}/2 = \alpha_4' V_{DS}$. However, in most cases, α_4' is close to 1 and can be neglected. There is an optimum value of V_{DS} for which :

$$V^+_{max} = V^-_{max} = V_{max} = \alpha_4 V_{ovmin} \qquad (15)$$

The minimum supply voltage for a given V_{max} is:

$$V_{DD} = V_S + V_{TH} + V_{ovmax} = V_S + V_{TH} + \dfrac{t}{\alpha_4} V_{max} \qquad (16)$$

and therefore:

$$V_{max} = \alpha_4 \dfrac{V_{DD}-V_S-V_{TH}}{t} \qquad (17)$$

In this case, there is a trade-off between the maximum signal swing and the tuning range similar to the case of saturated MOS. $\dfrac{g_m}{I} V_{max}$ can be shown to be:

$$\dfrac{g_m'}{I} V_{max} = 2\dfrac{2-\alpha_4}{4-\alpha_4} \qquad (18)$$

The Noise–Excess–Factor, for the differential transconductor, is given by:

$$F \approx 2\dfrac{V_{ov}}{V_{ov}-V_{DS}} \qquad (19)$$

From the above results the *DR* is:

$$DR = \beta \sqrt{\frac{nI}{kTBw}} \sqrt{\frac{(V_{DD} - V_S - V_{TH})}{t}} \sqrt{\frac{\alpha_4}{2(4-\alpha_4)}} (2-\alpha_4) \quad (20)$$

Many of the considerations which regard this last configuration are valid with minor differences for the case of MOS operating at $V_{DS}=0$ [12]. This situation corresponds to the case of MOSFET–C filters.

V-Low–voltage gm-C integrator design

Independently from the circuit solutions considered, the following conclusions can be drawn. At low supply voltage the *DR* of MOS based transconductors can be extended in three possible ways: (1) by reducing the threshold voltage, (2) by reducing or eliminating the voltage drop V_S, (3) by limiting the analog tuning range *t*.

The reduction of the threshold voltage is a technology dependent option and it will be not further discussed. Notice, however, that this option is not available for bipolar transconductors which are limited by the fundamental Boltzmann constraint of $V_{BE} \geq 0.6V$.

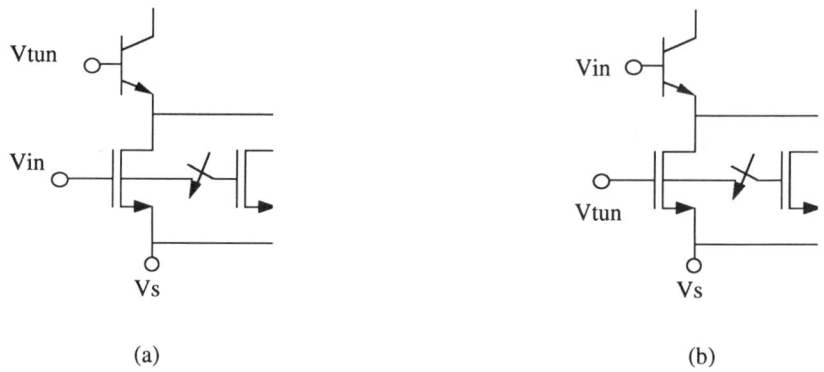

Fig. 9 (a) and (b) - Example of digital tuning of the transconductance by changing the aspect ratio of the MOS devices

The reduction of the analog tuning *t* can be achieved in different ways. A first possibility is to use a programmable capacitor array (as discussed for the active RC case) to split the tuning range in sectors. This solution, however, has the drawback of presenting a MOS switch in series with the load capacitor. The on–resistance of this switch modifies the phase response of the integrators and therefore can not be neglected in the design of high frequency filters. Moreover in the off–state of the switch it presents a parasitic capacitance which adds to the other parasitic at the integrator output worsening the capacitor matching. In the author's experience this option is very difficult to implement at frequencies of several tens of MHz and above for a 1 µm technology. Another way to fraction the tuning range is to use a digitally controlled transconductance. Two ways to implement this concept are illustrated in Fig. 9.a and b. The switches connect together the gates of binary scaled MOS transistors and thus the transconductance can be digitally programmed. However in case (a) the switches are placed in the signal path and therefore they introduce an additional high–frequency pole in the integrator response. This solution should be avoided in high–frequency filters. On the other hand, in the case of Fig. 9.b the MOS transistors can be connected together without affecting the frequency response of the integrator. This solution can be also used in the case of MOSFET–C filters. A way to circumvent the problem associated with the solution of Fig. 9a is shown in Fig 10.

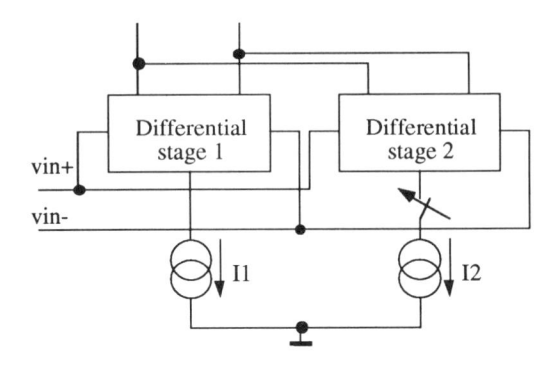

Fig. 10 - Example of digital tuning by adding in parallel two differential stages

In this case two (or more) differential stages with scaled transconductance are connected in parallel and can be turned on and off switching their bias current. The influence of the parasitic when the stages are turned–off should be, however, carefully considered. Since digital tuning is always associate with some amount of analog tuning, controlling the combination of the two to set the cut–off frequency of the filter is an additional problem. This is particularly severe when on–chip automatic tuning must be implemented.

The other way to increase the dynamic range at low supply voltage, is to reduce V_S. The extreme possibility is to set $V_S=0$ connecting to ground the source of the input transconductance elements. This configuration, is known as "pseudo–differential" [25,29]. At low supply voltage it can give a significant increase in the swing since the voltage required by the current generator can be 700mV in worst case. This value depends on the minimum required output impedance of the generator, on the intrinsic noise associated with the tail current and on considerations about matching among different current generators.

The pseudo-differential configuration requires to carefully control the common-mode behaviour of the circuit because it cannot reject the common-mode component of the input signals. The propagation of the common-mode signal can cause distortion and can originate instability in the common-mode positive feedback loops which are present in many filter structures [26]. This problem could be solved using a common-mode local feedback at the integrator outputs. To be effective, however, these loops should provide high-gain also at high frequencies, a difficult task to achieve. This problem can be alleviated by performing a preliminary cancellation of the common-mode input signal using a feed-forward scheme [27], as shown in Fig.11.

In this transconductor MOS M1' and M2' generate a current signal proportional to the common-mode input signal. This current is subtracted at the transconductor output through the current mirror M5-M10 cancelling the common-mode output signal. In this way the suppression of the residual common-mode signals can be easily performed even using low-gain high-bandwidth feedback loops.

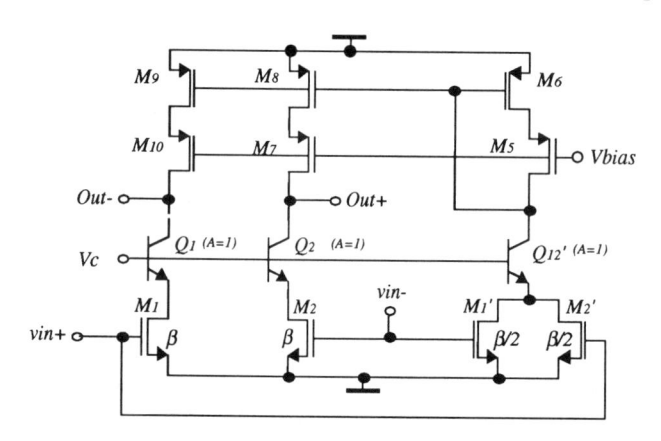

Fig. 12 - Pseudo-differential BiCMOS transconductor with feed-forward common-mode cancellation

When designing high-frequency filters another aspect to be considered is the phase response of the integrators. This aspect is strictly related to the location of the secondary pole of the transconductor as well as to its maximum dc-gain. At low voltage a high dc-gain is difficult to obtain because the use of cascode configurations generally require high voltage drops. On the other hand, a folded structure, has a parasitic pole at much lower frequency which degrades the response of the transconductor. In general, simple transconductor topologies with minimum number of internal nodes (in such a way to approach the f_T of the integrated devices) should be preferred. In principle, the low gain which results in these cases can be compensated at system level if its value can be designed with sufficient precision [28].

Up to now, we have considered voltage–mode CT filter. The dynamic range has been always expressed as the ratio between the maximum signal level and the fundamental noise voltage *kT/C* associated with each state variable (i. e. the voltage across the capacitor). As for the case of sample-domain circuits (SI), the use of current–mode approach has been suggested as a viable alternative to increase the dynamic range [32,33]. However, as long as the transconductors behave linearly the current–mode approach does not enhance the *DR* since both the noise energy and the energy of the current signal scale linearly with Gm^2. More interesting is the case of transconductor with non-linear voltage–to–current characteristic. If the current is used as input signal, in fact, either the saturated MOS or the bipolar transistor perform a signal compression in the current–to–voltage conversion respectively with a square root and a logarithmic law. This non–linearity can be compensated with a complementary voltage–to–current operation preserving the overall linearity of the system. This type of signal processing is known as compounding [30]. Companding is effective in increasing the dynamic range of a filter if class AB operation is implemented. This process can be efficiently implemented, even at low supply voltage, in bipolar technology as reported by Seevick [31]. This makes this approach potentially attractive for the implementation of large *DR* low supply voltage CT filters.

VI. Summary and conclusions

The drive toward low voltage operation for analog circuits and in particular filters is progressing with increasing strength. The two key motivations for this are cost reduction and the desire to use the latest (scaled) technology. On the other hand, for a given dynamic range, power consumption reduction would require not to reduce the supply voltage of analog circuits. This is contrary to the case of digital circuits. At low supply the key problem of analog filters is dynamic range degradation. This is especially severe in mixed mode systems due to the contamination produced by adjacent digital blocks. For this reason often a key target is to keep the largest possible voltage swing even if this increases the intrinsic noise.

In continuous time circuits (g_m–C) faster degradation of the voltage swing as compared with SC tends to occur. This is due to the interaction between the signal and the variation of the control voltage that performs the tuning. Large swing with extended tuning can be achieved using MOS devices in the linear region. This gives extra noise but remain very attractive in mixed mode systems. Bipolar solution have traditionally shown a relatively small swing. They become more competitive below 3V supply down to 1V. This is because they have a very large tuning range and they have a swing almost independent from the supply voltage. Below 1V supply MOS solutions with very small thresholds (multithreshold technologies) or even using depletion devices, are the only possible alternative due to the Boltzmann limitation (0.6V) of bipolar circuits.

An attractive way to cope with dynamic range degradation at low supply voltage is to use companding. This open the way to the use of current mode processing. In the continuous time domain this may further increase the interest for bipolar solution down to 1V supply.

References

1 A.P. Cho, Chandrakasan,S. Sheng and R. W. Brodersen, "Low Power CMOD Digital Design", *IEEE J. of Solid-State Circuits*, Vol. SC-27, pp.473-484, March 1995

2 T. B. Cho, and P. R. Gray, "A 10b, 20Msample/s, 35mW pipeline A/D converter", *IEEE J. of Solid-State Circuits*, Vol. SC-30, no.3, pp.166-172, March 1995

3 R. Castello, and P. R. Gray "Performance limitation in switched-capacitor filters" *IEEE Transactions on Circuits and Systems*, vol. CAS-32, no. 9, pp. 865-876, Sept. 1985

4 Gert Groenewold, "Optimal dynamic range integrators", *IEEE Transactions on Circuits and Systems* - I: Fundamental theory and Applications, Vol. 39, no. 8, August 1992, pp. 614-627

5 Y. Matsuya, and J. Tamada, "1V power supply, low-power consumption A/D conversion technique with swing-suppression noise shaping" *IEEE J. of Solid-State Circuits*, Vol. SC-29, no.12, pp.1524-1530, December 1994.

6 R. Castello, A. G. Grassi, S. Donati, "A 500nA sixth-order bandpass SC filter", *IEEE J. of Solid-State Circuits*, Vol. SC-25, no.3, pp.669-676, June 1990.

7 F. Callias, F. H. Salchli, and D. Girard, "A set of four IC's in CMOS technology for a programmable hearing aid", *IEEE J. of Solid-State Circuits*, Vol. SC-24, no.2, pp.301-312, Apr. 1989

8 T. Adachi, A. Ishikawa, A. Barlow, K. Takasuda, "A 1.4V switched-capacitor filter", *IEEE 1990 Custom Integrated Circuits Conference*, pp. 8.2.1-8.2.4

9 Y. P. Tsividis and J. O. Voorman, "Integrated Continuous Time Filters". New York: IEEE Press, 1993.

10 J. O. Voorman, W. H. A. Bruls, and P. J. Barth, "Integration of analog filters in a bipolar process" *IEEE J. of Solid-State Circuits*, Vol. SC-17, pp.713-722, 1982

11 M. Zuffada, R. Alini, P. Colletti, M. Demicheli, M. Gregori, D. Moloney, S. Portaluri, F. sacchi, S.O. Arf, V. Condito, and R. Castello" A single-chip 9-32MHz Mb/s Read/Write channel for disk drive applications", *IEEE J. of Solid-State Circuits*, Vol. SC-30, no.6, pp.650-659, Feb. 1992

12 J. M. Khoury, "Design of a 15MHz CMOS Continuous-Time Filter with on-chip Tuning", *IEEE J. of Solid-State Circuits*, Vol. 26, No. 12, pp. 1988-1997, Dec. 1991

13 R. G. Yamasaki, T. Pan, M. Palmer and D. Browing, "A 72 Ms/s PRML Disk Drive Channel Chip with an Analog Sampled-Data Signal Processor", *1994 Int. Solid-State Circuits Conf (ISSCC-94)* San Francisco, CA, Feb. 1994, pp. 278-279.

14 R. Alini, A. Baschirotto and R. Castello, "Tunable BiCMOS continuous-time filter for high-frequency applications", *IEEE J. of Solid-State Circuits*, Vol. SC-27, no.12, pp.1905-1915, Dec. 1992

15 C. A. Laber and P. R. Gray, "A 20 MHz 6th Order BiCMOS Programmable Filter Using Parasitic Insensitive Integrators", *International Symposium on VLSI Circuits,* Seattle 1992 pp. 104-105

16 A. M. Durham, W. Redman-White and J. B. Hughes, "High linearity continuous-Time Filters in 5-V VLSI CMOS", *IEEE J. of Solid-State Circuits*, Vol. SC-27, pp.1270-1276, September 1992.

17 J.O. Voorman Continuous-Time Analog Integrated Filters in [9]

18 J. L. Pennock, "CMOS triode transconductor for continuous time active integrated filters", *IEE Electronics Letters*, vol. 21, pp. 817-818, 1985

19 P. Wu, R. Schaumann and S. Szczepansky, "A CMOS OTA with improved linearity based on current addition", *1990 IEEE International Symposium on Circuits and Systems*, Vol. 3, pp. 2296-2300

20 Y. Tsividis, M. Banu, and J. Khoury, "Continuous-time MOSFET-C filters in VLSI", *IEEE J. of Solid-State Circuits*, Vol. SC-21, no.1, pp.15-30, Feb. 1986.

21 B. Nauta, "A CMOS transconductance-C filter technique for very high frequencies", *IEEE J. of Solid-State Circuits*, Vol. SC-27, no.2, pp.142-153, Feb. 1992.

22 T. Aray, M. Koyama, H. Tanimoto and Y. Toshida, "A 2.5V active lowpass filter using all-npn gain cells with 1Vpp linear input range" 1993 ISSCC Digest of Technical Paper , pp. 112-113.

23 H. Tanimoto, M. Koyama and Y. Yoshida, "Realization of 1V active filter using a linearization technique employing plurality of emitter-coupled pairs", *IEEE J. of Solid-State Circuits*, vol. 26, pp 937-945, July 1991.

24 F. Rezzi, V. Pisati, R. Castello and R. Alini: "Novel Linearization Circuit for BiCMOS Transconductors used in High Frequency OTA-C Filters", *IEEE International Symposium on Circuits and Systems* '93, Vol. 2, pp. 1132-1135.

25 F. Rezzi, A. Baschirotto, R. Castello: "A 3V pseudo-differential transconductor with intrinsic rejection of the common-mode input signal", *IEEE Midwest Symposium on Circuits and Systems* '94, Lafayette, Lousiana

26 J. O. Voorman, A. V. Bezooijen and N. Ramalho, "On Balanced Integrator Filters" in [9]

27 F. Rezzi, A. Baschirotto, R. Castello: "Low-voltage balanced transconductor with high input common-mode rejection", *IEE Electronics Letters*, September '94, Vol. 30, No. 20, pp. 1669-1671.

28 A. Baschirotto, R. Castello, F. Rezzi: "Design of hign-frequency BiCMOS continuous-time filters with low-output impedance transconductors", *IEEE International Symposium on Circuits and Systems*'94.

29 G. Nicollini, A. Nagari, P. Confalonieri, C. Crippa "A -80dB THD, 4Vpp switched-capacitor filter for 1.5V battery operated systems" *European Solid State Circuits Conference (ESSCIRC) 1994,* pp. 104-107

30 Y.P. Tsividis, V. Gopiinathan and L. Toth: "Companding in Signal Processing", *IEE Electronics Letters*, August '90, Vol. 26, No. 17.

31 E. Seevinck: "Companding Current-Mode Intagrator: a new Circuit Principle for Continuous Time monolithic Filters", *IEE Electronics Letters*, November '90, Vol. 26, No. 24.

32 S. S. Lee, R. H. Zele, D. J. Allstot, and G. Liang, "A Continuous-time

current-mode integrator" *IEEE Transactions on Circuits and Systems*, Vol. 30, October 1991, pp. 1236-1238

33　J. Ramirez-Angulo, M. Robinson, and E. Sanchez-Sinencio, "Current-mode continuous-time filters: two design approaches" *IEEE Transactions on Circuits and Systems II*, Vol. 39, June 1992, pp. 337-341

34　H. Khorramadi, M.J. Tarsia, N.S. Woo, "Baseband filters for IS-95 COMA Receiver Applications Featuring Digital Automatic Frequency Tuning", *1996 Int. Solid-State Circuits Conf (ISSCC-96* San Francisco, CA, Feb. 1996, pp. 172-173.